DISCRETE SYSTEMS SIMULATION

McGraw-Hill Series in Industrial Engineering and Management Science

Consulting Editors

Barnes: *Statistical Analysis for Engineers and Scientists: A Computer-Based Approach*
Bedworth, Henderson, and Wolfe: *Computer-Integrated Design and Manufacturing*
Black: *The Design of the Factory with a Future*
Blank: *Statistical Procedures for Engineering, Management, and Science*
Denton: *Safety Management: Improving Performance*
Dervitsiotis: *Operations Management*
Hicks: *Industrial Engineering and Management: A New Perspective*
Huchingson: *New Horizons for Human Factors in Design*
Juran and Gryna: *Quality Planning and Analysis: From Product Development through Use*
Khoshnevis: *Discrete Systems Simulation*
Law and Kelton: *Simulation Modeling and Analysis*
Lehrer: *White-Collar Productivity*
Moen, Nolan, and Provost: *Improving Quality through Planned Experimentation*
Niebel, Draper, and Wysk: *Modern Manufacturing Process Engineering*
Polk: *Methods Analysis and Work Measurement*
Riggs and West: *Engineering Economics*
Riggs and West: *Essentials of Engineering Economics*
Taguchi, Elsayed, and Hsiang: *Quality Engineering in Production Systems*
Wu and Coppins: *Linear Programming and Extensions*

DISCRETE SYSTEMS SIMULATION

Behrokh Khoshnevis
University of Southern California

McGraw-Hill, Inc.
New York St. Louis San Francisco Auckland Bogotá Caracas
Lisbon London Madrid Mexico City Milan Montreal
New Delhi San Juan Singapore Sydney Tokyo Toronto

This book was set in Times Roman by Publication Services.
The editors were Eric M. Munson and John M. Morriss;
the production supervisor was Paula Keller.
The cover was designed by Jared Schneidman.
Project supervision was done by Publication Services.
R. R. Donnelley & Sons Company was printer and binder.

DISCRETE SYSTEMS SIMULATION

This book is printed on recycled, acid-free paper containing a minimum of 50% total recycled fiber with 10% postconsumer de-inked fiber.

2 3 4 5 6 7 8 9 0 DOC DOC 9 0 9 8 7 6 5 4

P/N 034533-3
PART OF
ISBN 0-07-833302-4

Library of Congress Cataloging-in-Publication Data

Khoshnevis, Behrokh.
Discrete systems simulation / Behrokh Khoshnevis.
p. cm. — (McGraw-Hill series in industrial engineering and management science)
Includes bibliographical references and index.
ISBN 0-07-833302-4 (set)
1. Discrete-time systems—Computer simulation. I. Title.
II. Series
T57.62K48 1994
003′.83′0113—dc20 93-48562

To Nazanin, My Wife, and to My Beloved Mother and Sister

CONTENTS

PREFACE

This book is intended to serve as an introductory text on the subject of systems analysis and design using simulation. The book concentrates on discrete systems, since the application domains about which most of its intended readers are concerned are expected to relate to this type of system.

The major merits of the book are its broadness, its simplicity, and its coverage of generic and basic concepts that enable the reader to quickly learn and apply the various available simulation tools. To provide a means for hands-on exercises, a new simulation software tool that does not require prior knowledge of computer programming accompanies the book. Based on my experience in teaching simulation during the last 15 years, there is a need for a simulation software that does not demand a substantial portion of instruction time to be devoted to the coverage of its details. Consequently, I designed the EZSIM system during the last eight years, and with the help of several of my fine graduate students I created many implementations of the software on the basis of the original concepts.

The simplicity and the strength of the simulation tool which supplements the text, and the unique approach used for the presentation of the modeling concepts in a generic and tool-independent manner, are the major factors that enhance the applicability of the book to a relatively large and diversified audience. Various engineering disciplines (especially industrial engineering) as well as academic programs in business administration, urban planning, transportation studies, operations research, and systems sciences can use the book. Courses in simulation, operations research, systems engineering, marketing, quantitative methods, operations management, construction management, quality control, reliability analysis, and communications systems can use the textbook, some as the major text and others as a supplement. The prerequisites are junior- or higher-level standing and background in elementary probability and statistics. The book can also be used as a major text for first-year graduate-level standing, provided that the class assignments include a substantial amount of realistic systems simulation projects and reports. One academic quarter or semester should be enough to cover the book in either an undergraduate or a graduate course.

There are currently several very fine simulation textbooks available. These textbooks may be classified into three categories. The books in the first category take a generic approach in their coverage of the related topics, and use either a procedure-oriented programming language (mostly FORTRAN) and/or briefly address one or more of the available high-level simulation languages (mostly GPSS). The advantage of these books is their extensive coverage of important generic concepts. Their disadvantage is the programming sophistication that the reader is expected to have. Another disadvantage of books in this class is the simplicity and triviality of most of the examples that they include. This is because of the cumbersome task of programming complex simulation models in procedure-oriented languages. Moreover, when available high-level simulation languages are briefly presented, the reader needs to refer to the extensive manuals of these languages for the large volume of information which is typically needed but is not covered in these textbooks.

Simulation books in the second class concentrate heavily on an available simulation language and lack sufficient information on general and important topics. These books resemble a software manual more than a textbook. The instructors who use these books as major texts are often compelled to use supplementary books or class notes to cover the generic concepts. Also, students often become bored with the dryness of the subjects related to syntactic details of programming tools in these books. For many students, this becomes a reason to lose interest in the entire subject.

The third category of available simulation books includes those books that attempt to cover the general simulation concepts as well as the details of a simulation language. These books are often extended versions of the related software manual and have a relatively large number of pages, and are consequently expensive. These books are rarely covered entirely in a single course. The instructor usually needs to pick and choose certain sections of the book and to completely ignore the rest. These books, however, are valuable to professional users in the industry.

This book falls within the third category with the exception that it devotes a small portion of its pages to the explanation of the details of its choice of software tool. The friendly environment of EZSIM with its numerous help windows relieves the reader of the task of searching through thick manuals and extensive explanations about the operational details of the software. In fact, most of the software capabilities can be learned in a few short trial sessions. EZSIM is designed to be a pleasant environment, and its easily constructed model network and animation attract the student's interest in exploring the software and in trying to use the simulation concepts and techniques learned in realistic problem scenarios. This has been proven to me over the last five years, during which I effectively used various prototypes of EZSIM in my simulation classes. The valuable feedback received from the students in these classes has helped greatly in shaping the software system to suit student needs.

The distinctive approach of the text is to first present the generic concepts in a format that makes the presentation independent of all software tools, and then present the accompanying software tool for use in the hands-on exercises. The independence of the first part of the book from any available software structure has enforced a rather unique approach to systems realization which is reflected especially in Chapters 2 and 4. The coverage of application examples is divided into two major sections (service and manufacturing) with clear classification of topics within each category.

The simplicity of using the accompanying software allows the instructor to devote more class time to the coverage of the basic concepts and applications case studies instead of spending several class sessions on teaching the details of cumbersome simulation languages. Also, in the design of the software an attempt has been made to make it run even on hardware platforms with minimum configuration (i.e., no hard disk, graphics card, or coprocessor). This enables instruction laboratories with limited resources to effectively use the software. Another important factor of EZSIM as a supplementary software tool is that it operates in both stand-alone and program generator modes. The latter mode supports several popular simulation languages. This should enable those instructors who like to stay with their current choices of simulation languages to use the book and/or software with the simulation software systems for which EZSIM serves as a front-end application. In other words, the program generator modes of EZSIM should be viewed as supplementary tools for their target languages, rather than as competition or as substitutes.

The recommended teaching strategy is to cover the chapters in order and assign the chapter exercises. Whenever possible, the use of realistic systems for studying and modeling is recommended. It is suggested that the instructor encourage the students to run the software and explore its environment during the early stages of the class. Many students will be able to learn the system on their own even before the related chapter is covered. The use of the software to validate the results of the earlier exercises and examples as an optional activity may be recommended to the students. This facilitates the teaching of basic concepts of simulation and common discrete system processes and the teaching of the chapter on EZSIM. This approach is not feasible with most of the other available simulation tools, since from the early stages of using these tools an intensive involvement of the instructor and the teaching assistant is usually needed to get the students started. The model files for all examples in the textbook that use EZSIM are made available on the accompanying diskettes.

Chapter 3 attempts to provide an insight into the structure of simulation programs and what goes into building a simulation language. The representative programs, which are written in BASIC and C, are listed in an appendix. These sections may be considered as optional reading materials which may be skipped for non–engineering students.

Throughout the years my interactions with Professors Joe Mize, Alan Pritsker, and Hamdy Taha have enriched my knowledge of the field of simulation and the related research and educational issues. I thank all of these individuals for their valuable contributions.

I thank Professors Kenneth Case and Philip Wolfe, my former academic advisor; Eric M. Munson, engineering editor of McGraw-Hill; and Chris Cochrane (production coordinator) and Chris DeVito (copyeditor) of Publication Services, for their support and cooperation throughout this publication project.

I thank the following professors who have reviewed various drafts of the manuscript for their invaluable comments: K. N. Balasubramanian, California Polytechnic State University; Andy R. Bazar, California State University, Fullerton; George Bekey, University of Southern California; Jeffery Cochran, Arizona State University; Maged Dessouky, University of Southern California; William S. Duff, Colorado State University; Rasoul Haji, Sharif University of Technology; Wafik Iskander, West Virginia University; Patricia M. Jones, University of Illinois; Andrew Kusiak, The University of Iowa; Hashem Mahlooji, Sharif University of Technology; Mansooreh Mollaghasemi, University of Central Florida; James Morriss, University of Wisconsin; Eberhart Rechtin, University of Southern California; Hamid Seifoddini, University of Wisconsin; Jose A. Sepulveda, University of Central Florida; and Nanua Singh, Wayne State University.

Last but not least my former students Dr. An-Pin Chen (who worked with me on the first program generator mode of EZSIM); Juan Perez (who worked on a version of the simulation engine for EZSIM); Dr. Qingmei Chen (who patiently worked to create several complete versions of the software); Gena Kostelecky (who extensively tested EZSIM and contributed some of the exercise problems); and my current students Sima Parisay (who performed some testing of the software outputs) and Richard Hilschire (who created several examples used in Chapter 8) must be acknowledged for their valuable contributions.

Behrokh Khoshnevis

CHAPTER 1

INTRODUCTION TO SYSTEMS SIMULATION

1.1 INTRODUCTION

In a broad sense, everyone uses simulation to make day-to-day decisions. Prior to taking action, people usually build a mental, or sometimes a physical, model of their conception of the environment to aid in decision making. They then manipulate the model with various configurations and various forms of possible influences in order to generate certain information that may be used to make rational decisions. For example, when preparing to catch a flight, one may envision (build a mental model) the various preparations needed, the route to the airport, and possible delays. To rearrange a living room (an existing system), rather than moving the heavy furniture around to find the most appealing layout, one can sketch scaled-down layouts or cut out a few cardboard figures that mimic the actual furniture and quickly rearrange their positions on a tabletop, choosing the most appealing arrangement for the actual living room. To design a new building (a future, nonexisting system), architects create mock-ups to enhance their mental model of the appearance of the building and to study the layout of the utilities.

These model-building and experimentation processes predict the results of the decisions to be implemented later. As simple as they may look, these examples have a great deal in common with the more technical simulation practices that are the subject of this book. Systems simulation can, therefore, be defined as

> the practice of building models to represent existing real-world systems, or hypothetical future systems, and of experimenting with these models to explain system behavior, improve system performance, or design new systems with desirable performances.

Unlike the alternative analytical methods, which generally use mathematical models, the types of models considered in simulation are those that are best represented by computer programs. The data regarding the system performance is gathered from these computer programs rather than from the real-world system under study.

Despite its lack of mathematical elegance, because of the flexibility that it offers, the diversity of problem domains to which it applies, and the degree of modeling realism that it allows (as compared to other systems analysis and design methods such as mathematical techniques), computer simulation as a powerful systems analysis and design tool has been gaining widespread and increasing popularity since its appearance approximately half a century ago. Simulation is now being used in numerous application domains.

According to a survey of nonacademic members of The Institute of Management Science (TIMS) and the Operations Research Society of America (ORSA), simulation ranks high among the analysis and design tools in frequency of use. Surveys have also indicated that almost 90 percent of service and manufacturing organizations use simulation.

1.2 APPLICATIONS OF SIMULATION

Simulation has been used in a wide range of domains, a few of which are mentioned in this section. Some of the application areas of simulation in service industry and the corresponding possible analysis and design objectives are as follows:

Communications. Simulation applications are becoming increasingly vital in the communications industry. Local and wide area computer networks, telephone systems, national and intercontinental satellite communication systems, television cable networks, and cellular telephone systems are examples of complicated systems that demand the power of computer simulation for efficient design and operation. The emergence of *information superhighways,* which is expected in the mid-1990s, will create an even higher demand for simulation studies. Several special purpose simulation tools are commercially available for communications systems analysis and design.

Education. Studies concerning issues related to the effects of changes in the level of enrollment, registration process, classroom allocations and scheduling, bookstore and cafeteria inventory planning, and library arrangement and operation system design for schools, colleges, and universities may be performed by simulation.

Entertainment. Simulation techniques are being widely used in the design of structure and operations of various components of amusement parks, production studios, and movie theater systems. Ticketing systems, waiting line and vehicle parking design, capacity design and scheduling of rides and shows, and crew, equipment, and film production scheduling are some of the typical purposes of simulation applications in the entertainment industry.

Financial services. There are many reports of applications of simulation in banking, securities, and insurance companies. Transactions analysis, cash-flow analysis, office system design, materials and supplies planning, data processing and computer network design, and automatic teller machine and drive through service system designs are some of the activities that may be performed by simulation.

Food service. Systems such as independent restaurants, take-out restaurants, fast food restaurants, franchised restaurants, and grocery store systems may be subjects of simulation studies for purposes such as materials inventory and procurement planning, distribution planning, site selection, layout arrangement, and manpower planning and scheduling.

Health care. Hospitals, emergency rooms, physician and dentist offices, and paramedics are frequently studied by simulation to determine shift schedules for nurses and doctors, medicine and food inventory policy, and capacity plans for resources such as beds, waiting areas, operating rooms, test and survival equipment, and ambulances. In addition, epidemiological studies such as forecasting rates of disease spread, and analyzing alternative disease-control policies are routinely performed by simulation.

Hotel and hospitality services. Systems such as hotels, motels, and resort areas may be studied by simulation for determination of factors such as suitable capacities, sites, resource inventory management policies, manpower planning and scheduling methods, and reservation and booking systems.

Transportation. These systems may involve one or more types of vehicles (e.g., taxi, shuttle, bus, train, airplane, ship), passengers, cargo, and transportation routes. The simulation study may have objectives such as vehicle capacity design, manpower (operator, maintenance crew, etc.), planning and scheduling, spare-parts planning, maintenance planning, urban planning, vehicle routing, highway design, ground and air traffic control system design, and parking lot and structure design.

Weather, environmental, and ecological forecasting. Weather forecasting routinely and intensively uses computer simulation. A large number of variables are manipulated by simulation programs that usually run on supercomputers to predict the local and global weather situation for various time windows. Studies concerning the control of pollution, the greenhouse effect, insect populations, and other environmental and ecological issues are also performed by computer simulation.

Production and manufacturing systems are another major class of simulation application. Some of the typical industries in this class and the corresponding purposes for simulation studies are as follows:

Natural resource extraction/harvesting. Industries such as mining, timber harvesting, well drilling, and fishing use computer simulation for planning of related activities and creation of policies for timely procurement of expensive resources such as large tractors, loaders, shovels, elevators, derricks, crushers, conveyors, bulldozers, and ships.

Plant and animal farming. Farming systems may be simulated for production forecasting, planning of resources such as land, fertilizer, animal food, drugs, tractors, combines, and transportation vehicles, and for the study and design of operational procedures to determine factors such as production, harvesting, storage, and distribution.

Power generation. Electric power generation systems based on sources such as steam, fossil fuel, hydro, nuclear, solar, or wind are usually simulated for design of capacity, configuration, and distribution systems and for operational systems design and analysis, which may concern issues such as generation rate schedules, distribution schedules, control system design, safety and reliability systems design, maintenance scheduling, and environmental impact control.

Manufacturing. All kinds of manufacturing concerns including chemical processing plants, automotive plants, aerospace vehicle manufacturing, furniture, electronics, machinery, tools, appliances, and so on use simulation extensively in applications such as strategic planning, midrange capacity and production planning, plant layout design, equipment selection, replacement, maintenance policy design, inventory planning and control, production scheduling, assembly line balancing, materials handling and storage, manufacturing information systems design, and numerous other issues related to design, fabrication, assembly, quality control, packaging, storage, and distribution. The popularity of simulation studies in manufacturing systems is rapidly increasing. Consequently, a considerable number of special purpose simulation tools are commercially available for manufacturing systems design and analysis.

1.3 FUTURE POTENTIAL OF SIMULATION

The use of computer simulation is expected to grow at an even faster rate in the near future. The major impetus to this growth may be attributed to the following factors:

1. The degree of complexity of the physical, biological, sociotechnical, and socioeconomic systems studied by today's systems designers and analysts forbids the use of classical, or even modern, mathematical tools in the majority of realistic studies.
2. The ever-increasing capabilities of computer hardware systems, which offer dramatically higher performances (mainly in memory and speed) than preceding generations at a fraction of the cost, allow for large-scale system simulation, and their processing speed provides for timely information generation. Low-cost, powerful microcomputers give small businesses access to the capabilities that simulation offers.
3. The evolution of existing simulation modeling software toward offering newer capabilities and the emergence of new software tools that provide sophisticated and easier user interface capabilities bring users with various technical backgrounds and computer programming skills to the realm of simulation.
4. The increasing awareness of the power of simulation by managers of various organizations and projects, and the availability of modern simulation tools and capable low-cost computers, will lead to an increase in the use of simulation in various decision-making activities that have been traditionally based on intuitive judgments.
5. Computer simulation is being incorporated in various curricula at colleges and universities. This increasing interest in simulation in educational institutions is likely to contribute in a significant way to the popularity and advancement of the field of simulation.

1.4 PLAN OF THE BOOK

This book is intended to cover the essential concepts related to the field of computer simulation with the major emphasis placed on discrete systems. An easy to use software tool that allows hands-on experimentation of the concepts covered in the book is provided. The general orientation of the book, however, is intended to be generic and independent of any simulation tool.

In Chapter 2 the definition of system and its related concepts with several easy to understand representative examples are provided. Systems classification with illustrative examples for various real-world systems under each class are presented. The chapter takes a generic approach to the presentation of the important system components and relations. The related issues are usually presented by other books in the context of a specific simulation language. These issues concern the meanings of entities, attributes, parameters, variables, system state, events, and so on. Models as representatives of systems for the purposes of communication and analysis are then described and classified. Analytical and experimental models are compared and contrasted using simple examples.

Chapter 3 concentrates on the experimental method of solving models of dynamic systems. A descriptive and a numerical example are given; they highlight the meaning and the process of simulation. Simplified manual methods for random variable generation that allow students to understand the details of examples are used. Generation of random numbers and random variables are then described briefly, and the method of constructing computer programs to handle the logic of dynamic discrete systems is demonstrated using a single-server queuing system example. A simplified description of the structure of simulation engines (event calendar, statistics collection, etc.) is provided. Gross flowcharts demonstrate the corresponding program structures. The related programs are listed in an appendix. A brief overview of the types of simulation software tools is then presented. The EZSIM simulation environment is introduced, and some details of the procedure to use the software are discussed using simple examples. The stages involved in a complete simulation study are then described. Certain sections of Chapter 3 may be skipped by those who are not interested in knowing the programming details related to simulation models.

Chapter 4 takes a generic approach to the description of common processes in discrete systems without specifically addressing any simulation tool. Knowledge of common processes in discrete systems greatly helps the simulation analyst in the translation of the system under study into the corresponding model by use of an available simulation tool. This knowledge also helps in comparing the capabilities of various simulation tools. The form of presentation in Chapter 4 is rather uncommon, since almost all simulation books that describe the common processes in discrete systems are based on a specific simulation language; they attempt to describe these processes only in connection with the capabilities of the language of their choice.

Chapter 5 provides comprehensive coverage of the capabilities of the EZSIM environment. The major functions of each modeling object provided by EZSIM are demonstrated. Several examples are used that help the user understand how modeling objects can be selected and set up to represent the common system processes.

Chapter 6 presents the issues related to the analysis of simulation input data. Methods of building histograms and fitting probability distributions to input data are demonstrated using numerical examples. The methods for generation of random numbers and random variates are then explained.

Chapter 7 elaborates on the topic of simulation output analysis. Establishing confidence intervals for the statistical estimators, setting the sample size, and determining the simulation run length are the major topics covered in this chapter. Terminating and nonterminating systems are defined, and several examples concerning output analysis for each type are provided. Other stages of the simulation process (including validation, experimentation, and documentation) are also discussed.

Chapter 8 presents several representative simulation application examples in the service and manufacturing industries. For each category, several examples that show the strength of simulation in the analysis and design of complex systems are provided. These examples also demonstrate various features of EZSIM.

Chapter 9 presents a broad overview of some representative simulation tools that are commercially available. A guideline for selection of suitable types of tools for specific situations is provided. The presentation includes a few representative modeling examples for some of the tools discussed.

Chapter 10 presents the new developments in simulation tools and discusses the future directions of simulation. Some concepts related to intelligent simulation environments, applications of machine learning in simulation, and object-oriented simulation are discussed.

1.5 CONCLUDING REMARKS

Despite its great strength, simulation is not without its pitfalls and drawbacks. Incorporation of inappropriate levels of complexity in the simulation model, lengthy development time, inherent inexactness of simulation results, misinterpretation of simulation output, and project budget overruns are likely to occur if careful measures are not taken into account when using simulation.

Besides the inherent appeal of simulation, which is largely due to its flexibility as compared with rigid mathematical models, the recent availability of numerous simulation tools adds to the appeal of simulation. Consequently, some users tend to resort to simulation for almost all problem scenarios even when other available techniques are more appropriate and perhaps easier to use. It must be emphasized that simulation modeling is to a large extent art rather than science. As is true in other forms of art, success is ultimately determined not by the sophistication of tools and techniques, but by how the tools and techniques are put to use.

CHAPTER 2

SYSTEMS AND MODELS

2.1 INTRODUCTION

In the introductory chapter we took the liberty of using some important terms such as *system* and *model* without elaborating on their definitions. This chapter presents the general meanings of these terms and many other important ones. It is important to note that the concepts presented in this chapter are rudimentary to understanding the systems approach to analysis and design in general, and the simulation approach in particular. Understanding these concepts should enhance the ability of the reader in gaining a systems orientation, which is helpful in all forms of systems analysis and design approaches. Realization of systems and models is essential to understanding simulation methodology.

2.2 SYSTEMS

There have been many technical definitions of the term *system*. For all practical purposes, however, a system may be viewed as a section of reality. Thus, everything may be thought of as a system: an atom is a system composed of protons, neutrons, and electrons; any object is a system of atoms; the world is a system of various objects in and on Earth; the Solar System is a system composed of the Sun, our Earth, and other planets; the Milky Way galaxy is a system composed of our Solar System and many other systems of stars and their orbiting planets; and finally, the universe is a system composed of the Milky Way and many other galaxies.

A common property of all of these physical systems is that they are composed of components that interact with one another. The nature of the interactions in these natural systems is determined by the physical laws that govern their behavior.

Like other physical systems, man-made systems are composed of components that interact with one another: a hammer is a system consisting of a metal end and a wooden handle that transfers human muscular force to the hammerhead; a car is a system consisting of chassis, engine, transmission, body, and interior elements; a computer is a system of processors, various kinds of memory modules, and input/output devices. All of the components in these systems interact with one another, either directly or indirectly. The nature of these interactions is determined by the physical laws of nature as well as man-made control schemes based on man-made rules.

Families, communities, countries, and their complex sociotechnical and socioeconomic organizations such as factories, banks, and governments are other forms of systems made up of many interacting components. These interactions are governed to a great extent by social behavior, which is based on human norms and values.

All of these systems may themselves be considered components of other, higher-level systems. In other words, every system is also a subsystem (with the possible exception of the universe itself). Because all of the components of a system interact with one another, this leads to the fact that every system is influenced by other systems. That is, no system is totally isolated from external influences.

A set of subsystems may be identified as a specific system and be hypothetically isolated from other systems only if it is the focus of a study. These isolating boundaries, however, may be drawn only with careful study of the external influences.

If external influences have such a great impact that they affect the behavior of the system in focus in major ways, then the boundary must be expanded to include those systems. For example, the design of a nuclear power plant must include the sociopolitical "subsystem" or the overall system will be seriously incomplete.

If the outside influences are still considerable, but their pattern and nature of behavior can be identified by use of a reliable observation method without the need to explore the internal structure of the outside systems creating the influences, then the influences may be treated as "inputs" to the system in focus. These inputs may be in the form of fixed values, tabulated information, or functional relationships.

Finally, if the external influences are minute and insignificant to the system in focus, then they may be ignored, at least initially. What influence is significant and what is not is a nontrivial question that can be answered only after the purpose of the study is established. If there is still a doubt about the relevance of certain inputs after establishing the purpose, subsequent testing and analysis may determine the sensitivity of system behavior to changes in those inputs. If these tests indicate that the system behavior is insensitive to some inputs, those inputs can be ignored.

Let us consider the problem of studying a traffic intersection for the purpose of finding the best traffic light timing for which the overall average waiting time of all cars that arrive at the intersection is minimized. Clearly, the behavior of this traffic intersection system (i.e., the focus of the study) is not independent of the traffic light timings of the adjacent intersections.

The analyst may choose to treat the car arrivals at the intersection as inputs to the system. He or she may then use some statistical observation methods to characterize these arrival patterns. This approach relieves the analyst from the need to study the internal behavior of the adjacent intersections.

If, because of the significant impact of the traffic light timing of an adjacent intersection, the effort to statistically characterize the car arrival patterns from that intersection fails, the analyst should expand the boundary of the system in focus to include that adjacent intersection. The advantage of this expansion of scope is the added realism about the behavior of the system. This expansion will result in more accurate analyses and may provide for studies that lead to better overall system performances (such as synchronization of several traffic lights). The disadvantage of expanding the system boundary is the added complexity that may sometimes make the system difficult, or even impossible, to study with the available analysis tools.

In many cases expansion of the system boundary is unnecessary, and some factors should simply be excluded. For example, in the case of the traffic intersection study let us assume that one of the streets leading to the intersection connects to a freeway at a distant location. Let us further assume that the freeway passes by an airport. Obviously, because of these interlinkages, the arrival pattern of the cars at the intersection is influenced to a certain extent by the flight schedules at the airport, but this influence is remote and small enough to justify its exclusion from the study.

It should be noted that what may be considered an insignificant influence to one observer may be quite important to another. Therefore, the two observers may choose different system boundaries, and even have different representations of what may seem to be the same system. For example, one may focus on a classroom with the purpose of studying the number and class (junior, sophomore, etc.) of students sitting in it and factors affecting these variables, but the other may focus on the same room to study the temperature changes in the room over a period of time. The system boundary, its components, and the relationships within it will be very different in the view of these analysts. Consequently, different observers of the same reality may see different things. A system is "in the eye of the beholder"; identification of its components, interactions, and boundaries depends on the beholder's purpose in looking at the system. Thus we get the following definition.

> A system is a section of reality that is the primary focus of a study and is composed of components that interact with one another according to certain rules within a boundary identified by the purpose of the study. A system can perform a function not performable by its individual components.

2.2.1 Classification of Systems

Parameters and variables are measures that characterize a system. Parameters are independent measures that configure the conditions of inputs and system structure. Parameters in man-made systems are directly controllable. Variables are measures that depend on parameters and other variables. For example, the weight and length of a pendulum are parameters; its speed within its oscillation range is a variable that depends on its weight, length, and other variables (such as its position). The number of machines in a shop is also a parameter (provided that it does not change during the course of study); the number of parts waiting to be processed and their waiting times are variables.

The set of values of some prescribed variables in a system at any point in time is called the *state* of the system at that point in time. Systems may be classified as static or dynamic. By definition, a static system is one whose state does not change over time. The state of a dynamic system does change over time. Note that, depending on the choice of the variables selected to represent the system state, a seemingly unique section of reality may be conceived of as a static system by one observer and as a dynamic system by another. A building structure, for example, may be viewed as a static system (if factors such as vibration, heat transfer with the environment, small expansions and contractions, etc. are excluded from the system boundary). An oscillating pendulum, a warm object cooling down, a traffic system with moving cars, a factory floor with moving parts, tools, and workers, an office environment with paper transactions, a communications network with message transactions, and a computer with data transactions, are all examples of dynamic systems.

Systems may be further classified by the way in which their variables (which are chosen to represent the system state) change over time.

If these system variables change continuously over time, the system is classified as a *continuous system*. If the system variables change in a discontinuous manner over time, the system is classified as a *discrete system*.

A warm object losing its heat to the environment may be viewed as a continuous system (if the amount of heat it contains is to represent its state) because the amount of heat in the object is a variable that continuously declines in value as time goes by. Likewise, a lake behind a dam may be viewed as a continuous system because its water level changes continuously over time. This level is a variable that fluctuates through the effects of the inflow rates of rivers feeding the lake and the outflow rates of the dam outlets, water evaporation, and penetration (see Fig. 2.1).

A bank that services arriving customers may be viewed as a discrete system if the number of customers in the bank represents the system state, and because this number changes only by discrete quantities (there is never a fraction of a customer in the bank), then the system state changes discretely over time. Here the system state is influenced by the arrival and departure rates of customers (see Fig. 2.2). Note that certain variables in this bank system (such as customer waiting times) are nondiscrete, but because they are not present in the set of variables representing the system state, the system is still viewed as discrete.

If some system variables change continuously and others change discretely, the system is classified as a *combined system*. In the example of the dam system, the total outflow of water normally changes in a continuous manner over time. The value of this outflow rate depends on the water pressure, which in turn depends on the water level of the lake behind the dam. If some of the dam outlets are instantaneously opened or closed, discrete changes will be imposed on the total outflow rate of water (see Fig. 2.3).

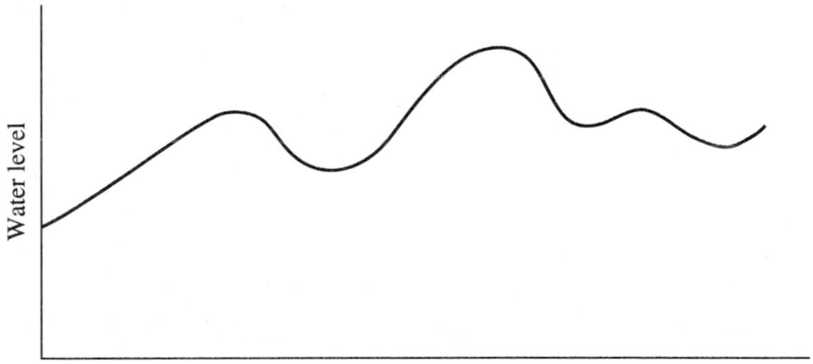

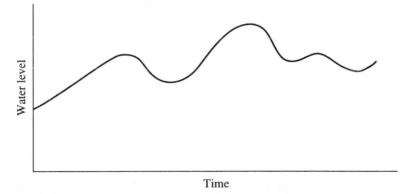

FIGURE 2.1
Continuous pattern of change in water level.

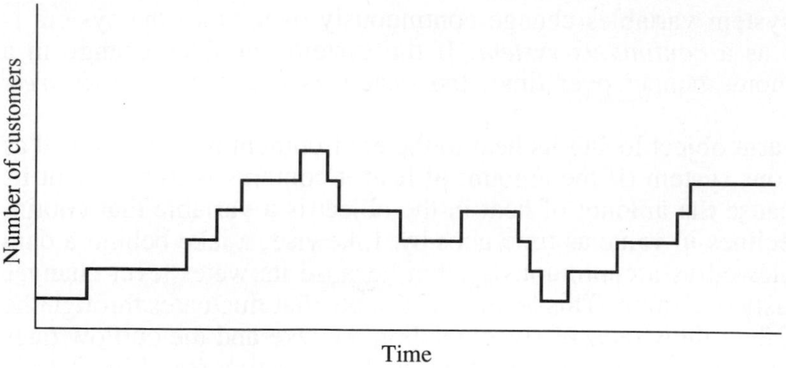

FIGURE 2.2
Discrete changes in the number of customers in a bank.

Whether systems are viewed as continuous or discrete, the time variable in their analysis is generally viewed as an independent variable that is usually continuous (exceptions are those rare situations in which discrete changes take place only at some specified points in time rather than at any point in time). All other variables in the system may depend on the time variable, other variables, or system parameters.

Figure 2.4 depicts the systems classification just explained. Let us reemphasize that the identification of a system and its boundaries, as well as the choice of viewing it as static, dynamic, continuous, discrete, or combined, all depend on the analyst's purpose in studying the system. Once the purpose of the study is established, the variable set representing the system state is carefully chosen. This then leads to the proper classification of the system in focus.

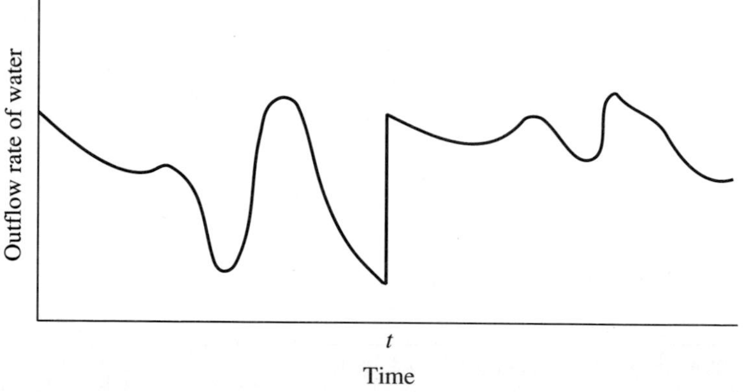

FIGURE 2.3
Opening of outlet at time t imposes discrete change on continuous flow.

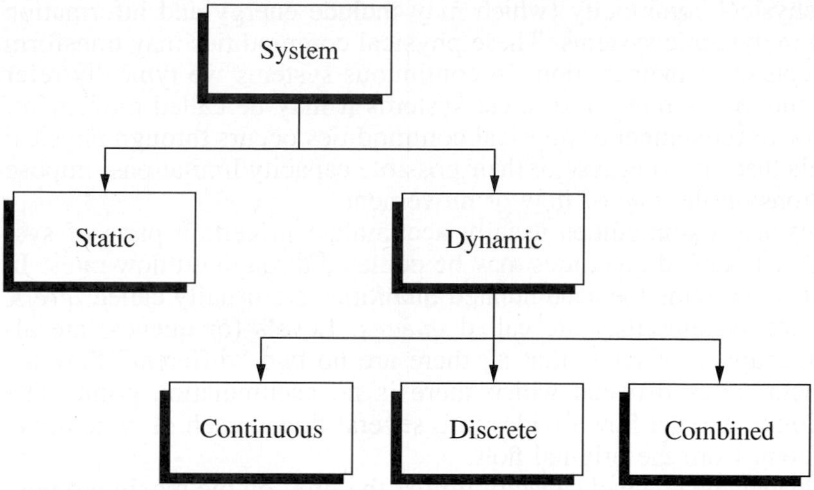

FIGURE 2.4
System classification chart.

Some preferred classifications for certain systems have been established. For example, most dynamic natural systems are best classified as continuous systems. Some man-made physical systems are best classified as combined systems (such as process industries, which include oil refineries, chemical plants, and the like). Finally, many man-made physical systems, such as discrete part manufacturing facilities, offices, hospitals, banks, traffic systems, and communications systems are best classified as discrete systems.

It should be noted at this point that, when viewed from a macro level, certain systems that otherwise may be classified as discrete are usually assumed (approximated) to be continuous. The motivation behind this assumption is the increased ease of representation and analysis (not that continuous systems are always easier to study). For example, in an urban planning and design project the analyst may view the flow of cars in freeways as a continuous commodity flow (like liquid in pipelines). Here, a great deal of modeling difficulty and excessive computational effort may be avoided by eliminating the need for tracing each individual car that moves through the freeway system. Some approximation errors should be expected, however, since the approach may well assume fractions of cars in addition to ignoring the details of behavior of individual cars that may affect the traffic flow.

2.2.2 Common Dynamic Systems Concepts

Whether they are viewed as continuous, discrete, or combined, all dynamic systems follow a common behavior. There is always motion of some physical commodity (which may include energy and information media) in dynamic systems. These physical commodities may transform in the course of their motion. In continuous systems we typically refer to this motion as flow; in discrete systems it may be called movement. The flow or movement of physical commodities occurs through physical channels that may (because of their possible capacity limitations) impose restrictions on the rate of flow or movement.

Physical commodities usually accumulate in certain parts of systems. Accumulated quantities may be depleted through outflow rates. In continuous systems the accumulated quantities are usually called *levels*; in discrete systems they are called *queues*. Levels (or queues) are always decouplers of rates; that is, there are no two "different" flow (or movement) rates between which there is no accumulation point. The exception is when a flow divides into several flows, each of which may be different from the original flow.

Besides the channel capacity limits, the limit on the maximum possible level of the accumulation point also places a natural limitation on the possible outflow rate of the commodity from the accumulation point. Capacities of channels and accumulation points are usually included in the set of system parameters (in certain situations they may be variables). Rate and level quantities may be considered as dependent variables of the system that may assume different values at different points in time.

To clarify these abstract concepts, let us use some examples. Water in an ecological system constantly flows through the channels of rivers and underground waterways and accumulates in lakes, seas, oceans, and underground reservoirs. Water may flow out of these reservoirs, still in the form of water, through other rivers and underground channels to flow into other accumulation points. Influenced by the heat of the sun and the internal heat of the earth, the water transforms into water vapor and flows out through the atmosphere to accumulate in pockets of clouds. The limits for the outflow rate and the accumulation level of water vapor are set by certain factors such as atmospheric humidity, temperature, and pressure (note that channel and accumulation point capacities may themselves be dependent variables rather than fixed parameters).

As another example, consider the movement of parts on a factory floor. The parts arrive at a certain rate at a machining station. They may then accumulate in a temporary storage area before being processed. The processing speed, the possibility of machine breakdown, and the possible unavailability of certain tools or operators at the station set a limitation on the departure rate of the parts. The departing parts are transformed to processed parts, and they may continue on to other stations for further processing and assembly that result in further transformations.

Figure 2.5 depicts a typical flow/accumulation structure. The rectangular elements in this figure represent accumulation points, and the

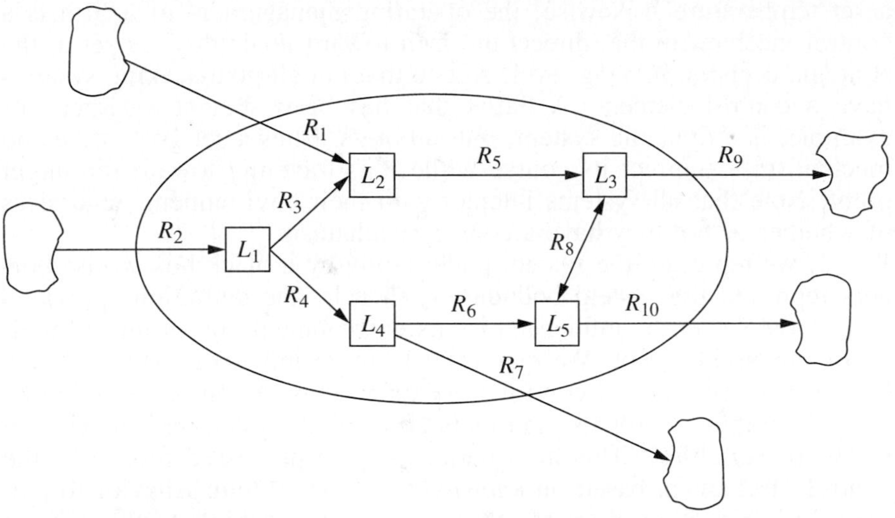

FIGURE 2.5
A typical flow/accumulation structure (dotted line shows system boundary). R—rate; L—level.

branches connecting the accumulation points represent channels. As shown in the figure, the commodity may take alternative channels to flow into and out of accumulation points.

We mentioned earlier that flow rate limitations are generally set by the channel capacity and the value of the level feeding the rate. In man-made systems, however, certain deliberate limitations may also be set on the rate of flow (or movement), the direction of flow, or choices of routing through the alternative channels of the system. This deliberately imposed limitation is the essence of control. Control is an important concept in systems analysis and design, and it deserves further elaboration.

2.2.2.1 CONTROL. The subject of control is important because it is related to many systems analysis and design studies. Traffic control, population control, production and inventory control, and pollution control are some of the common control problems. In many studies the objective is to design a control mechanism for the given systems that will attain the desired system performance. Simulation studies are no exception in this regard.

Control is a mechanism within all biological and some dynamic man-made systems that directs the system toward a set goal. An autopilot unit, for example, is a control mechanism that directs the aircraft system toward maintaining a set flight direction, altitude, and speed. An air conditioning system has a control mechanism that attempts to maintain a set temperature. Likewise, the operating management of a firm is a control mechanism that directs the firm toward goals that are set at the strategic decision-making level. Notice that not all goal-seeking systems have a control element. A bullet that has been shot at a target, for example, is a dynamic system, but although it has a set goal, it has no mechanism to correct its course while it is traveling toward the target point. Note that all systems interact with their environment, regardless of whether or not they have a control mechanism.

If we represent the system under study by a black box whose borders represent the system boundaries, then by the definitions provided earlier, the system is influenced by its environment and in return influences the environment. We may refer to these influences as inputs and outputs, respectively. A control mechanism may use information about the system inputs, outputs, and its internal state (system parameters and values of variables). This information may be processed further by the control mechanism based on *knowledge* of the system behavior to create new information (a set of other dependent variables) that is used to execute the control function.

Systems that encompass a control mechanism usually use either *feedback* or *feedforward* or both types of information. The information is received through the sensory component of the control mechanism. The feedback information relates to the current state of the system. The feedforward information relates to the current status of the environment within which the dynamic system operates.

Control mechanisms that are based only on feedback information compare the current state of the system with the goal state and take appropriate corrective actions. A disadvantage of this type of control is that compensating action is taken only after a disturbance has affected the system state. This means that an error must be present (i.e., the system state must deviate from the desired state) in order for the control action to be initiated. Although this may be acceptable in some situations, it may not be tolerated in some other scenarios.

In an air conditioning system, the control mechanism is based on feedback information only (i.e., the room temperature and not the outside temperature). The room temperature is sensed through the thermostat and is compared with the desirable goal temperature. The information regarding the error is then processed according to certain built-in knowledge about the specific air conditioning system (its BTU rating, air flow rate through fans, etc.) to determine the duration of time that the cooling (or heating) system is to remain on. In other words, the control mechanism determines the overall rate of heat exchange to maintain the target temperature for the room. The system actuators (furnace, compressor, fan, etc.) are then manipulated to execute the control commands. In the air conditioning system the knowledge about the system behavior may be hardwired in an electronic circuit board.

In feedforward control the environmental disturbances are measured before they have upset the system, and an anticipatory corrective action is taken. In the ideal case (not always possible), the corrective action compensates completely for the disturbance, thus preventing any deviation from the desired goal state of the system. However, the state of the system should be known in order to measure the extent of success of a feedforward control command. This means that for reliable and accurate control, feedback and feedforward control mechanisms should be implemented in a combined manner.

An aircraft autopilot system, for example, usually uses a combined control mechanism. In the control process the feedback information about the current state of the aircraft (its direction, altitude, speed, etc.) is taken by the control mechanism by means of sensors (altitude and velocity gauges, gyroscope, etc.) and is compared with the set goal values for these measures. The computed deviation from what is desired is used along with the feedforward information about the environment (direction and velocity of air, etc.) to actuate the control arms (rudder, aileron, throttle, etc.) in certain directions and by certain magnitudes to smoothly reach the goal state of the aircraft system. What helps the control mechanism in determining those certain directions and magnitudes is the knowledge about system behavior, which in this case may include facts and rules about fluid dynamics, thermodynamics, kinematics, and so on, represented in the form of mathematical equations in an on-board computer program. Evaluation of these equations for the sensed values representing the current system state and the values representing the goal state results in processed information that is useful in executing the control function. The resulting control commands of the autopilot unit influence factors such as rate of motion (speed), rate of ascension, rate of descension, and so on through manipulation of the system actuators.

In an organization, say a manufacturing firm, the set goal may be the achievement of a certain net revenue by a certain time. The control mechanism (management) senses the current net revenue through the accounting department and compares it with the set goal. Other related information such as current production rate, current accumulated inventory, current market demand rate, and time-phased availability of materials, people, and other resources may be sensed through other departments such as production, marketing, purchasing, and personnel. This information is then processed using knowledge about management techniques, which may be represented in the form of facts, rules, intuitive norms, management science mathematical relations, or simulation. The resulting management control commands (decisions) may then manipulate the production rate, the rate of hiring and layoff, the rate of advertising, and so on through various executives (actuators) to achieve the set system goal.

The above examples highlight the major elements of all control mechanisms: sensors, knowledge bases, information processors, and actuators. Figure 2.6 illustrates these elements in a combined feedforward/feedback control mechanism. The success of a control mechanism depends on how accurately the information is sensed, how intelligently the knowledge base is constructed, how quickly (timely) and accurately the information is processed, and how quickly and accurately the actuators execute the control commands.

It is now clear that control mechanisms operate on the basis of information. Knowledge is a logical integration of related information. Information is a useful set of some selected data. Data is representative of parameter and variable values. Not all data always constitutes useful information. To generate useful information, data may need to be filtered and structured to specific needs. For example, for a general store manager the details of what was sold at what hours of which days may be of much less use than a simple average daily revenue figure.

The term *optimal control* refers to the specific control condition for which the best system performance is attained. The measure of performance is generally expressed by a relationship known as *objective function*. The objective function represents the overall indicator of system performance that we desire to optimize. Among the performance objectives typically used in optimal control are cost minimization, profit

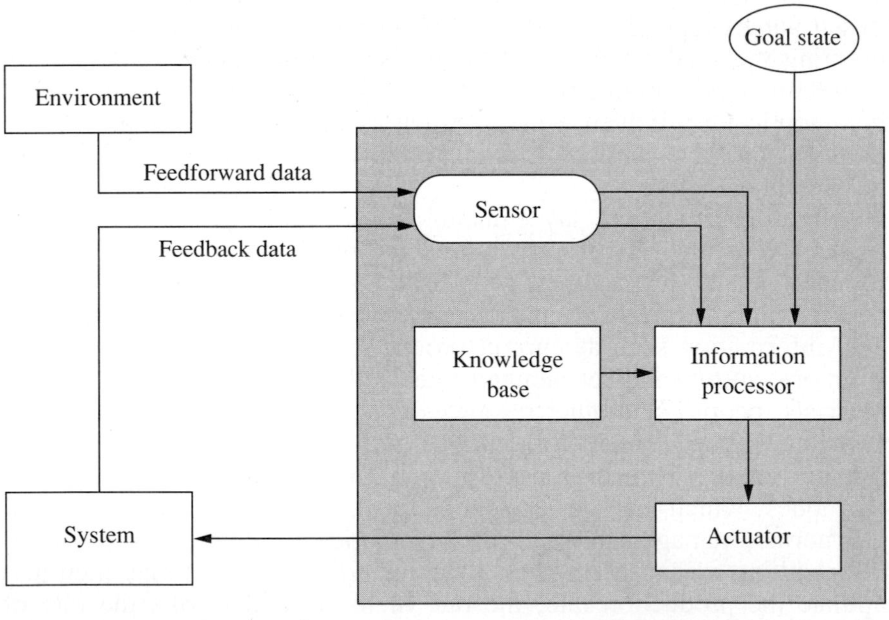

FIGURE 2.6
Structure of a combined feedforward/feedback control mechanism.

maximization, quality maximization, and error minimization. Generally there are constraints on the possible values of the variables that appear in the objective function. Identification of the values of these variables such that the objective function is optimized is the objective in designing an optimal control mechanism.

One noteworthy observation that could be made in the examples given in this section: the control mechanism always acts on the *rates* of commodity flow (or movement), on the basis of the information that may be gathered from rates, levels of accumulation points, parameter values, or any combination of these. Control never acts directly on the accumulation points because levels can be changed only through their inflow and outflow rates.

To clarify the relationship between the rate/level concept and the control mechanism, consider the simplified production control problem depicted in Fig. 2.7. The symbols on the channel lines represent the actuators for flow control (valves). Since there is a time lag for the production rate to keep up with the changing demand rate, and since, without inventory, high rates of demand would necessitate excessive production capacity (requiring a high number of machines, operators, etc.), the finished goods inventory is usually used to allow for relatively smooth changes in production rate while preventing out-of-stock situations. However, since inventories are costly to maintain, management usually has to set a desired inventory level considering the trade-off between the cost of carrying inventory and the out-of-stock costs represented in a logical relationship (objective function). A control mechanism is to be devised to maintain the goal of minimizing the overall operating cost. Assuming that it is not possible to control the demand rate (through means such as advertising), the only means of controlling the inventory level is through the production rate.

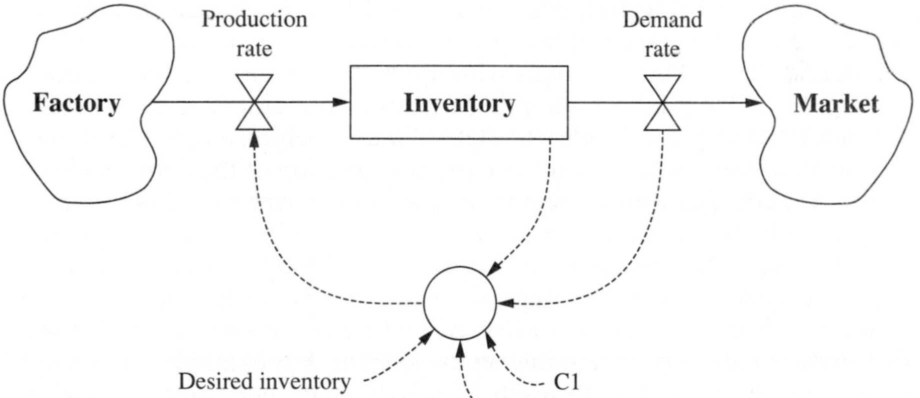

FIGURE 2.7
A level-rate representation of a production system.

Devising an effective control policy even for this simplified problem is not a trivial task, because it is difficult to predict the behavior of this dynamic system over time. Let us, therefore, arbitrarily assume that the management control policy is to set the daily production rate using the following simple relation, which is to be evaluated each day:

$$\text{Production rate for today} = (\text{C1}) \times (\text{Demand rate yesterday}) + (\text{C2}) \times (\text{Desired inventory} - \text{Inventory today})$$

C1, C2, and desired inventory in the above relationship are values that should be determined for designing and implementing an effective control policy. Note that the above control mechanism uses feedforward information originating from a rate (demand), feedback information originating from a level (inventory), and three parameters (C1, C2, and the desired inventory level). Processing of these information components results in new information that is used to manipulate the production rate. Figure 2.7 is a diagrammatic representation of this production control scenario. The material flow channels in this figure are shown as solid lines and the information channels are shown as dashed lines. In Chapter 8 an example of the application of simulation in production and inventory control is provided.

2.2.2.2 TRANSIENT AND STEADY STATE MODES. As mentioned, the state of a dynamic system changes over time. This change may be transient or steady. A dynamic system is said to be in the steady state mode when changes in the system state over time take place within a relatively fixed range (the range does not have to be small). This pattern of change is representative of a relative equilibrium that in many cases represents the system's behavior in the long run (typical day-to-day behavior). For example, an aircraft flying at its target altitude, target speed, and target direction may be said to be in the steady state mode. Notice here that the state of the aircraft system in this flight mode is not necessarily fixed, but it changes within a relatively narrow range around its targets, as the pilot (or autopilot) attempts to maintain the goal state. As another example, consider the state of a lake behind a dam. Assuming that several years have passed since the construction of the dam, the level of the lake changes within a relatively fixed range over various seasons of the year. The lake in this case may be said to be in its steady state mode.

The state of a dynamic system may go through transient changes when the system is in some atypical condition. Transient state modes are generally characteristics of starting or ending conditions, or conditions that impose radical perturbations on the system. For example, an aircraft taking off and ascending to reach its steady state mode may be said to be in a transient state. It also enters a transient state when it descends to land. Major altitude, direction, and speed changes imposed by the pilot or by other forces (storms, breakdowns, etc.) may also take the aircraft out of the steady state mode and into a transient mode. In the example of the lake, during the early years after the construction of the dam, the level of the lake goes through atypical changes (monotonically increasing) until it reaches the steady state, or equilibrium, mode at which typical ups and downs take place within a limited range. Unlike the changes in the transient state, which usually take a monotonically increasing or decreasing pattern over time, the system state changes during the steady state period are independent of time.

Studies are usually aimed at analyzing systems for their steady state behavior, provided that they have such a state. For example, when designing a dam, the lake level during the steady state mode is considered. Likewise, when deciding on the number of seats to be placed in the waiting area of a barber shop, the early moments after the shop opens (the empty state—atypical condition) are generally disregarded and the shop behavior after the initial cold start period has passed is studied. Some systems may not have a steady state; studying their transient state may be of interest.

2.2.2.3 FEATURES OF DISCRETE SYSTEMS. Since the emphasis of this book is on discrete system simulation, this section is written to highlight certain features of such systems. Recognition of these features will facilitate the understanding of simulation modeling concepts that will be presented in the following chapters.

The earlier discussions and examples regarding systems classification indicated that discrete systems differ from continuous systems in that their source of dynamism (the physical commodity) is in disjunctive form (unit parts, cars, people, etc.) rather than conjunctive form (liquid, gas, heat, etc.). We refer to each disjunctive unit of commodity as an *entity*. Therefore, entities are detached commodities whose motion in the system causes discrete changes in the state of the system. Entities do not have to be physical commodities. Messages and signals in a communication system, for example, may be considered as entities. It should be noted that discrete systems may have continuous variables. For example, the distance between car entities is a continuous variable. However, continuous variables do not appear in the set of variables that characterizes the state of a discrete system.

Entities may transform into other entities with varied characteristics (e.g., an unprocessed part becoming a processed one). They may also divide into a larger number of entities (e.g., the loaded-bus entity at the terminal station transforming into the empty-bus entity and several passenger entities). Finally, several entities may combine to form fewer entities (e.g., components assembled to form a product).

We mentioned that changes in the state of the system are caused by the motion of entities. This does not mean, however, that every motion of entities necessarily results in a system state change. To clarify this point, recall that what represents the state of a system is relative and depends on the analyst's viewpoint. For example, if the state of a room is to be represented only by the number of persons in the room, then it can only change (in a discrete manner) when someone either enters or leaves the room. The movement of people inside the room presumably will not affect the prescribed state of the system. The latter forms of entity motion are insignificant in the analyst's view and, although they may take place in the real world, the system state is assumed constant and unaffected by their occurence (recall that a system is a "section" of reality).

As another example, assume that an analyst is interested in studying both the number of customers inside a bank and the number that are in a line before the tellers' desk. The system state is represented by two different variables. When a customer enters the bank, the system state changes because the total number of customers in the bank changes. While the arriving customer is walking toward the waiting line, the system state is constant. Once the customer reaches the end of the line, the system state changes again since the number of customers in the line is another variable that represents the system state.

Notice that in the above example the system state changes at certain instances in time at which significant occurrences take place. Significant occurrences that result in state changes of discrete systems are called *events*. Furthermore, time-consuming actions not including waiting in queues (such as the customer walking toward the line or receiving service at the desk) in discrete systems are usually called *activities*. Therefore, in discrete systems the system state changes only at event times that take place at the beginning or end of activities.

In the course of their activities, entities may sometimes need to use and then release (or consume in whole or in part) certain *resources* that have limited availability. In these situations the entities may have to wait for their turn to use the resources. In the above example, the bank tellers are the resources that the entities need for processing. However, waiting for resources may not be the only delay type that an entity experiences. For instance, entities may have to pause and await permission to move on (like cars waiting for a stoplight to turn green).

Entities may take different routes through the various activities in which they may have to spend different times. The choice of routes that entities take, the time they spend in activities, the reason they wait, the resources they use, and the way they use them generally depend on certain conditions. For our bank example, the arriving entities may choose to join the shortest line (if there are two or more lines) and, depending on whether they need to make a deposit or withdrawal, their services may have different service times.

The recognition of entities and their routes, the nature of activities, the system resources, the various conditions governing the system behavior, the occurrences that lead to events, and the impact of the events on the system state are essential to the study of discrete systems.

2.2.2.4 MORE ABOUT ENTITIES. Discrete systems may have various types of entities. Furthermore, the same types of entities may have varying characteristics. For example, a public transportation system may include people as well as the buses that load, carry, and unload them. The people will have different characteristics such as their desired destination and the amount of money they can afford to spend. Likewise, buses may have different capacities, may be assigned to certain routes, and may be regular or express with different ticket prices. Entities may have various kinds of relations with one another. For example, the bus entity can carry several passenger entities, and a police entity can stop a bus entity.

Entities are characterized by their attributes. Each entity may have several attributes. An entity's collection of attributes is called its *attribute set*. Figure 2.8 shows the typical attribute sets for bank customer entities and for truck entities. Note that each attribute has an identifier (name, age, arrival time, etc.) and an instance (customer, 20, deposit, etc.).

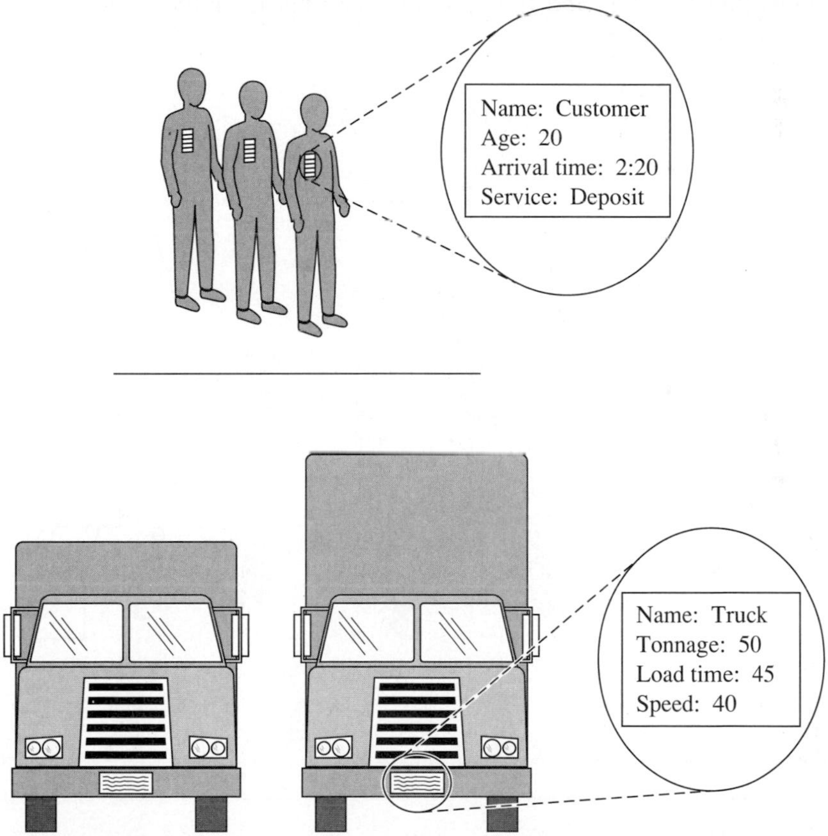

FIGURE 2.8
Sample attribute sets for bank customer entities and truck entities.

The attribute set of an entity is the information that is attached to the entity and found with the entity at its location. We can therefore consider attributes as local information. Entity attributes can be observed only where the entity is present. Certain other information, such as time, is global and can be observed from anywhere. For example, in a sports stadium the actual time of day and the score of each team may be displayed on a sign that everyone can see (global variables). The individual players (entities) carry attributes such as their number, the color of their team, the number of penalties counted against them, and so on, which have meaning only when expressed for a particular player. Thus, unlike the time variable, a single variable representing the number of penalty points cannot be shown on the stadium sign because there are as many such variables as there are players (see Fig. 2.9).

Typical global variables used in discrete system studies are time, entity counts (total number of entities of a certain type passed through a certain point in the system), the total number of entities in certain sections of the system (in each queue or in several queues, activities, and connecting routes), and the status of various resources (number in use, number available).

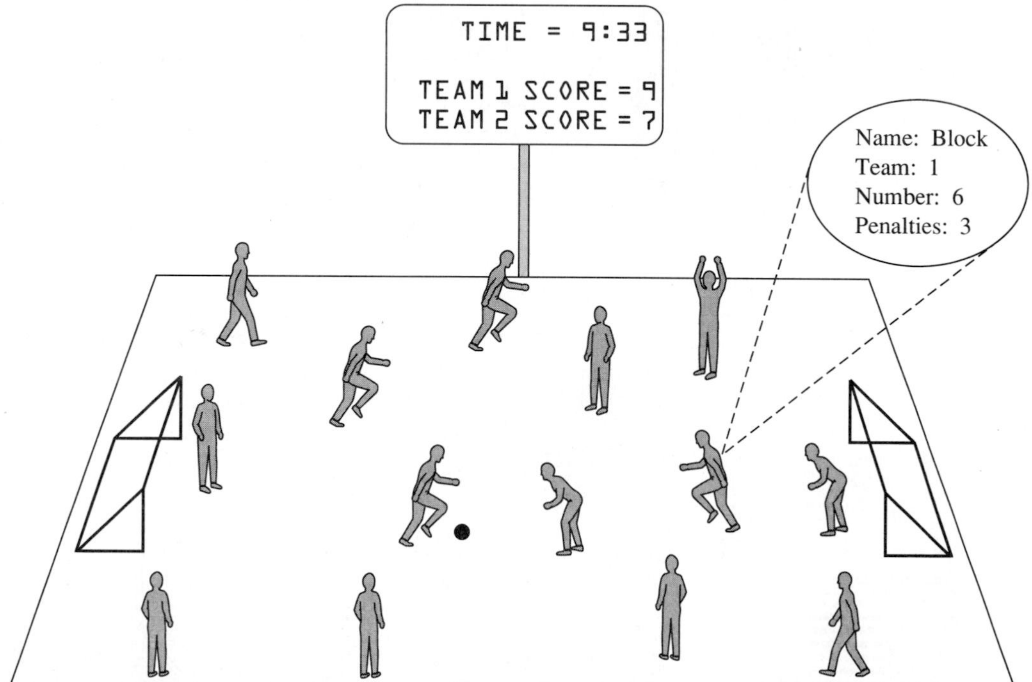

FIGURE 2.9
Local variables (attributes) are attached to the entities, but global variables are observable from any point in the system.

To summarize our discussion regarding the behavior of discrete dynamic systems, let us state that *entities* carrying *attributes* and taking on various *activities* and routes move through the system, use various *resources*, and create *events* that change the *state* of the system, while maintaining certain logical *relations*.

Before ending our discussion of common system concepts, let us list some brief definitions of the important terms related to these concepts:

System. A section of reality in the form of a set of components connected such that they can perform a function not performable by the individual components.

Parameters. Quantities in the system that do not change unless the analyst commands them to.

Variables. Quantities in the system that are determined by functional relationships and that change over time in dynamic systems.

System state. A snapshot of the system at any point in time characterized by the values of some selected variables in the system.

Events. Changes in the state of a discrete system.

Entities. Objects in the dynamic system whose motion within the system may result in the occurrence of events.

Attributes. Characteristics and properties that describe entities.

Relationships. Expressions of dependency between elements such as variables, parameters, and attributes of a system.

Activities. Time-consuming elements of a system whose starting and ending coincide with event occurrences.

Resources. Limited commodities that are used, consumed, or replenished by the entities.

Control. A mechanism that directs a dynamic system toward a set goal.

Transient state. Atypical conditions that impose time-dependent, radical perturbations on the system state.

Steady state. Typical conditions in which changes in the system state are within a fixed range and are independent of time.

2.3 MODELS

We mentioned earlier that a system is a section of reality that is the primary focus of a study. In order to communicate a system description or to analyze a system, one must first express the system in some form of representation. This representation is called a *model*. For example, a poet may see a tree (system) and represent it in words that describe the tree (verbal representation). A painter may look at the same tree but express it by a painting of the tree (pictorial representation). Although the painting and the poem are very different, they are models of the same system.

Models rarely convey all of the facts about the system they represent; therefore, they are in fact abstractions of systems. Certain representation forms, however, are more efficient at conveying certain facts. For example, the painting of the tree can better describe the colors of the scene, but the poem can better describe the motion of the leaves and the birds singing. Figure 2.10 demonstrates the relations of reality, system, and model. Notice that as indicated in the figure, what identifies a system is the purpose of the study, and once the system is identified, what can be used to identify a model among the candidate models is the strength of representation of various modeling alternatives.

Models are not only essential for communication; they are necessary for the analysis of systems. Our earlier examples showed that system studies may be aimed at understanding the behavior of existing systems or of hypothetical future systems. In general, studies require

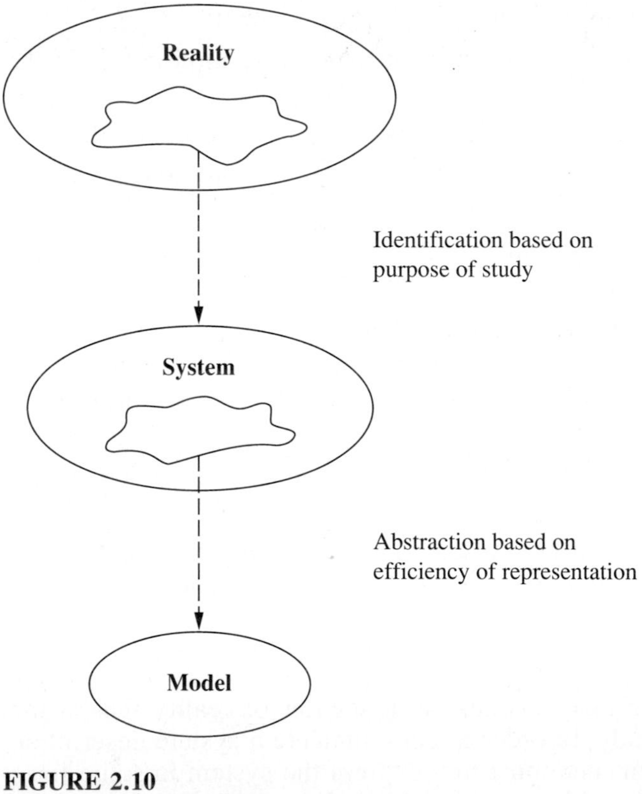

FIGURE 2.10
Relations of reality, system, and model.

experimentation and manipulations that are obviously impossible to implement on the nonexisting hypothetical systems. Also, in the case of the existing systems, this experimentation and manipulation may be very costly to implement on the system itself, since they may involve both time- and resource-consuming activities. They may also disturb the current system operation, and sometimes may result in the destruction of the system. Let us consider a few examples.

In the case of remodeling a living room, experimentation with various layouts when implemented on the system itself requires moving the objects in the room. Such experimentation, when applied to a factory floor, is even more dramatic since heavy machinery must be detached from its foundations and moved around in order to observe various alternative plant layouts.

When attempting to find the best timing of a traffic light for which the average waiting time of cars arriving at the intersection is minimized, one may actually attempt to manipulate the timing of the real traffic light for a certain length of time and observe the system behavior under various timing scenarios and hope to find the timing sequence that results in the best system performance. This approach demands the presence of the analyst (and/or operators and data gatherers) at the system site for a considerable length of time, and it may actually disturb the traffic flow at the intersection as well as several adjacent intersections.

When training and testing a technical flight crew to react properly to unexpected emergency situations (such as engine failure, extreme turbulence, loss of controls), use of the real aircraft system in a real flight may have grave consequences because as a result of a minor mistake, several lives and a multi-million-dollar aircraft may be destroyed in a training session.

A similar circumstance exists when designing a dam (a nonexisting system) for the purpose of improving irrigation and hydropower generation. If the actual construction takes place simultaneously with the design process, a multi-million-dollar project may be completed after several years only to find out that either there is an insufficient inflow of water to create any significant water level behind the dam to generate hydropower, or the accumulated water overflows the dam during certain seasons and floods the agricultural lands. To arrive at the best dam design using this approach, many actual dams would have to be constructed and observed in operation!

All of the above examples suggest that there should be a better way to perform systems analysis and design than experimenting with the system itself. To this end, models have been found to serve as more attractive means. A model is a representation of a system that can be experimented with and manipulated for the ultimate purpose of studying the system. In essence, models are used to do three things:

1. Study existing systems without disturbing their operation.
2. Study existing systems without destroying them.
3. Study nonexisting future systems.

Since various schemes may be used when representing a system, many kinds of models (which may not look alike) may represent the same system. Therefore, a valid model may or may not look like the system it represents, but it will contain all of the important facts about the system. For example, a circle may be defined as a system of points that have the same distance, or radius, from a point called the center. One representation of this static system is an equal-scale graphical model. Another representation could be a mathematical equation that places the center of the circle on the origin of an imaginary two-dimensional coordinate axis and specifies the circle as all points whose square of the *X* coordinate plus the square of the *Y* coordinate equals the square of the radius.

2.3.1 Classification of Models

Models can range from seemingly exact physical mock-ups of the system to abstract mathematical representations. Models of systems may be classified as being physical, graphical, or symbolic. All classes of systems (static, dynamic, continuous, discrete, and combined) may be represented by any of the model types.

Physical models, which are also called iconic models, may be to the same scale as the system itself. Examples of this sort of model are an aircraft cockpit model used for pilot training and an operational pilot factory built to study the manufacturing of a new product line prior to full-scale production. Physical models may also be of a smaller scale than the system that they represent. Examples are mock-ups of building structures used by architects and mock-ups of chemical plants. Some scaled-down models of three-dimensional systems may be two-dimensional (such as scaled templates used in plant layout designs). Scaled-up physical models of systems such as crystal and gene structures are also common.

Physical models do not always have to look like the system they represent. For example, since voltage changes across a capacitor coupled with an inductor follow the same pattern of potential energy exchange as an oscillating weight hung from a spring, the former electrical system is an analog of the latter mechanical system. Therefore, a capacitor-inductor coupling may be used to model a weight-spring coupling. This example serves to show that a common characteristic of all physical models is that they are systems themselves.

Graphical models may be two- or three-dimensional representations of systems. They may be static, such as drawings on paper, or dynamic, such as animated films and computer graphics. Graphic representations generally ease communications and enhance the understanding of abstract models (i.e., a picture is worth a thousand words). Therefore, many symbolic models are supported with graphics representations and interfaces. For example, EZSIM, the symbolic modeling tool with which the examples in the following chapters are implemented, takes advantage of the benefits of graphics.

Symbolic models are abstract representations of systems and as such they do not look like the systems that they represent. In many applications these models are a more effective means of system representation because of their ease of construction and manipulation. For example, a verbal description of a system is a symbolic model that could be expressed in any natural language (English, Chinese, etc.). A verbal model in certain applications is sufficient and much easier to build than a physical model of the system. Verbal models are common for certain applications but are cumbersome and inefficient to use when the system structure and the relationships within it become complex and overwhelming. Further abstractions may be required to better express the relationships in the system under study. Abstract and organized representation of the facts and rules within a system in either the form of structured seminatural language procedures, mathematical relationships, and/or computer programs are some commonly used forms of symbolic models. As the name implies, symbolic models use symbols to represent system components (parameters, variables, relationships, etc.).

2.3.2 Model Experimentation

Recall that the purpose of building a model is not merely to represent a system, but also to experiment with the model for various configurations and parameter values to ultimately draw some useful conclusions regarding the performance of the system under study. To these ends the model should be evaluated—or, as is commonly said, "solved"—to reveal the system behavior.

In the case of physical and graphical models, these solutions may be readily observed and measured. In the case of symbolic models, the solutions will be hidden in the abstract model structure. For example, in a graphical representation of the demand forecast for a company, the peak demand and its time of occurrence can be readily observed on the graph, but when the same relation of demand to time is represented mathematically (perhaps as a nonlinear function found through curve fitting to historical demand data), "looking" at the function representation does not reveal much information about the maximum point of the function. In the latter case the model has to be solved to obtain the desired information.

Solution methods for symbolic models are either analytical or experimental. The analytical method generally applies to mathematical models; the experimental method can apply to mathematical as well as some other forms of symbolic models. The analytical method requires the deductive reasoning of mathematical theories that apply to the problem at hand. This method usually results in precise, quickly obtained solutions. Analytical solutions are general, but an experimental method is specific and applies only to the given problem.

As an example of an analytical method application, consider the problem of finding the maximum point of the mathematical representation of the demand forecast mentioned above (see Fig. 2.11). To find the maximum of a function, we must refer to the calculus principles related to the concept of derivatives. According to these theories, the first derivative of a function of a variable (time, in our case) when evaluated for a certain point on the function provides the slope of the line that is tangent to the function at that point. Since at the maximum (or minimum) point the slope of this tangent line is zero (the line is horizontal), it follows that to find the maximum of the function the first derivative of the function should be set equal to zero, and the resulting equation should be solved to obtain the time value at the maximum (or minimum) point. Other mathematical theories are also available that can be used to solve these equations and to determine if the solution is a minimum, a maximum, or a deflection point.

As another example of an analytical method application, consider the problem of finding the area of a circle. Based on classical mathematical methods it was determined several centuries ago that the area of

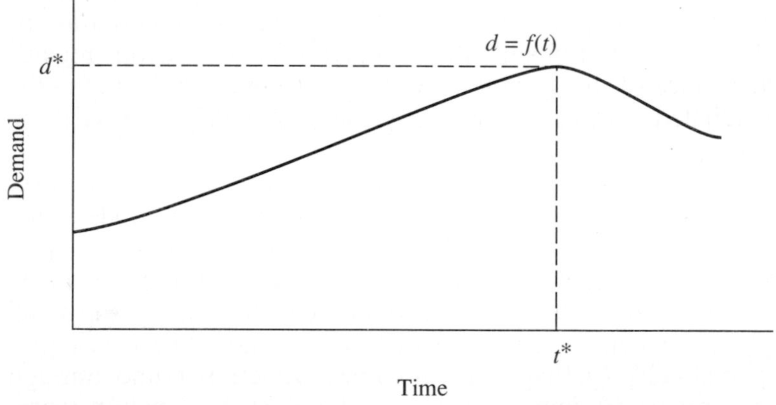

FIGURE 2.11
Functional relation of demand to time.

a circle is equal to a known constant (π) multiplied by the square of the radius. This formula provides fast and exact results, but most of its users take it for granted and are unaware of the logic behind its structure.

The major disadvantage of the analytical method to problem solving is that if the problem poses certain features not explained by the available mathematical theories (complex problem), the analyst has to develop and prove new theories. Since few analysts are mathematicians, they may be unaware of the available theories that apply to their problem, let alone be able to establish and prove new theories. Nevertheless, even expert mathematicians are unable to find attractive analytical solution methods for a wide range of large and/or complex problems.

The experimental method for problem solving generally uses simplified and at times commonsense approaches rather than sophisticated mathematical theories. The trade-off for this simplicity and ease of use is usually (not always) prolonged computation times and/or imprecise solutions. However, since most real-world problems are complex, experimental methods are often found to be the only feasible alternatives.

Instead of elaborating on the definition of experimental methods, let us demonstrate how they may be used on the two examples for which we discussed the application of the analytical method. Given the relation of demand as a function of time (an equation), to find the peak demand point without resorting to mathematical optimization principles, we can devise a simplistic approach based on numerical trial and error. We can start with an arbitrarily selected point in time, say t_1, and evaluate the demand function for this time value (see Fig. 2.12). We record this demand value, say d_1, and try another time value, say t_2. We then compare d_1 and d_2. If d_2 is larger than d_1, we may conclude that the peak demand point is closer to d_2 than it is to d_1. We subsequently

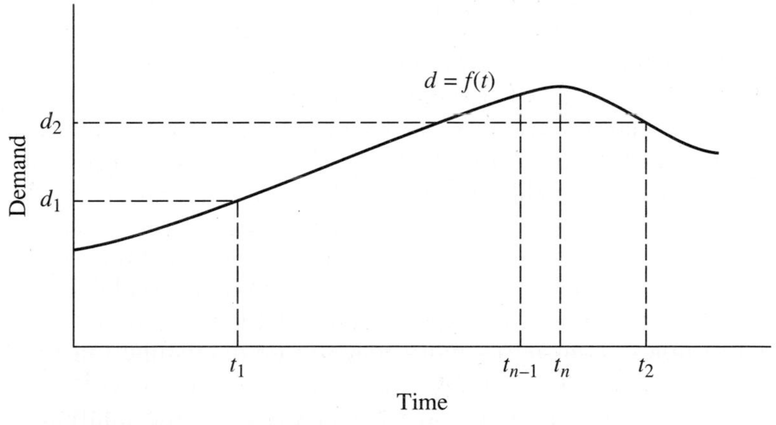

FIGURE 2.12
An experimental search method to find the peak demand value.

narrow our search neighborhood around t_2 and continue with our trial process until two subsequent trial points result in two demand values that are approximately (within our acceptable error range) equal to one another and accept the larger value as the maximum demand figure. In order to reduce the error in our solution, we would have to increase the number of trials (computations).

When finding the area of a circle, one experimental method is to fit several small and adjacent squares of equal size inside the circle. By counting the number of squares and multiplying this number by the area of one of the squares, we find an approximation to the area of the circle. We can reduce the error in this approximation by choosing smaller squares, but our computation effort will increase accordingly.

As another experimental approach to finding the area of a circle, consider drawing the circle on a square piece of cardboard with a known size, placing the cardboard on a flat surface, and dropping small beads from above in such a manner that the beads have a uniform chance of falling anywhere on the surface of the cardboard. We may then take the ratio of the number of beads that fall inside the circle to the total number of beads on the cardboard. This ratio would serve as a fair estimate of the ratio of the area of the circle to the area of the square-shaped cardboard. Knowing the ratio and the area of the square, we can then find the estimated area of the circle. Clearly, the goodness of this estimate depends on the number of beads that we drop on the cardboard. For example, if we drop only one bead and it falls outside of the circle, using this approach we would conclude that the area of the circle is zero! We can improve our estimate by increasing the size of our statistical sample (the number of beads that we drop on the cardboard).

The latter approach (which concerns a static problem) is based on the generation of random samples and is called the Monte Carlo technique. As will be seen in the next chapter, simulation uses the Monte Carlo technique for handling dynamic systems.

In the above examples we demonstrated the simplicity of the experimental method and its drawbacks, which are due to its inherent imprecision and computational involvement. When dealing with well-rounded problems such as finding the maximum of a unimodal function or finding the area of a known geometrical figure, experimental methods seem ungainly and cumbersome. What we did not demonstrate is the true potential of experimental methods in solving complex problems.

To highlight the potential of experimental methods in the context of the same examples, consider a more realistic demand function (see Fig. 2.13*a*) which is likely to be multimodal and a typical nongeometric contour (such as an aerial photograph of a lake). Here the analytical methods fail to find the global maximum (there are many local maxima) of the function because solving the nonlinear first derivative function

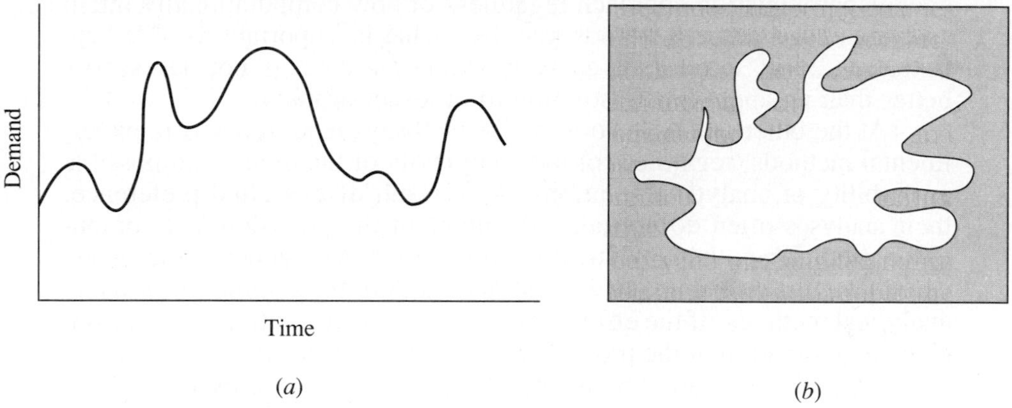

FIGURE 2.13
Complex but realistic scenarios. (*a*) Demand versus time. (*b*) Photograph of a lake.

becomes a complex task, and certain functions (such as functions with discontinuities) do not have defined derivatives. Likewise, since there is no geometrical formula for finding the area of a nongeometric contour, the analytical method does not apply to the shape shown in Fig. 2.13*b*. Nevertheless, the experimental methods explained above work just as well on these complex problems without the need for major modifications. (In the case of the demand function, the same algorithm, when tried with various starting time points, finds the global maximum.)

2.3.3 Some Remarks for Selection of a Modeling Method

In most cases the analyst's purpose is to find ways to improve the performance of an existing system or to design a new system. In these cases the model should undergo several parametric and/or structural configuration changes and be solved for each alternative scenario. As mentioned, this process is called model experimentation. If the model is too complex, the analyst will use experimental methods to solve for each model configuration. At times, however, experimentation may become too computationally intensive to use experimental methods. An approach generally taken in these situations is to approximate, if possible, the complex model with a simpler model that has an analytical solution. For example, the highly nonlinear demand function may be approximated with a quadratic curve (using a regression method), or the irregular figure contour may be approximated with a few arc pieces with known geometrical characteristics (such as portions of circles with various sizes).

Using the rationale that analytical methods are more precise than experimental methods, some analysts have a tendency to misuse the above approximation approach regardless of how computationally intensive the experimental methods may be. What is important to note here is that an exact solution to an approximate problem is not necessarily better than an approximate solution to an exact problem.

At the other extreme, some analysts always tend to resort to experimental methods, regardless of the complexity of the model and possible availability of analytical solutions. As a result of this blind preference, their analyses often do not take advantage of the precision and computational efficiency that analytical methods offer. As a general rule, when dealing with a problem, one should first explore the possibility of using analytical methods. If the effort to find an analytical solution (or a reasonable approximation of the problem by one that lends itself to an analytical solution) fails, then an experimental method should be considered.

2.3.4 Simulation—An Experimental Approach

The examples that we used to demonstrate the analytical and experimental methods concerned static systems; that is, no motion of a commodity was present. When dealing with dynamic systems, the structure of the models representing the systems generally becomes more complex. In the case of continuous systems, the analytical methods used for model experimentation employ advanced mathematical techniques such as differential equations of various orders that express continuous state changes over time.

In the case of discrete systems, especially when some probabilistic (stochastic) phenomena are present, analytical methods become very cumbersome to use. In such cases transition equations are used to express the nature of the transition of the system from one discrete state to another. Markovian chains and queuing theory (queues are accumulation points of discrete commodities, i.e., entities), developed in the field of operations research, are typical analytical methods used for constructing and experimenting with the mathematical models that represent discrete dynamic systems. These models, however, are restricted to some rather simple systems and are often limited to studying systems in their steady state modes. Several standard models (closed-form equations) have been developed that represent various, but a limited number of, queuing scenarios under certain stochastic effects. For example, models are available that determine the expected (average) length of queues, average server utilization (percent of the time the server is busy), average entity waiting time, and so on for a limited number of configurations of servers (parallel, serial, etc.) and queues (single queues or multiple queues before servers) and for limited patterns of entity arrival and service duration.

As shown in Fig. 2.14, simulation is an experimental modeling approach that applies to dynamic systems. Because of the limitations of

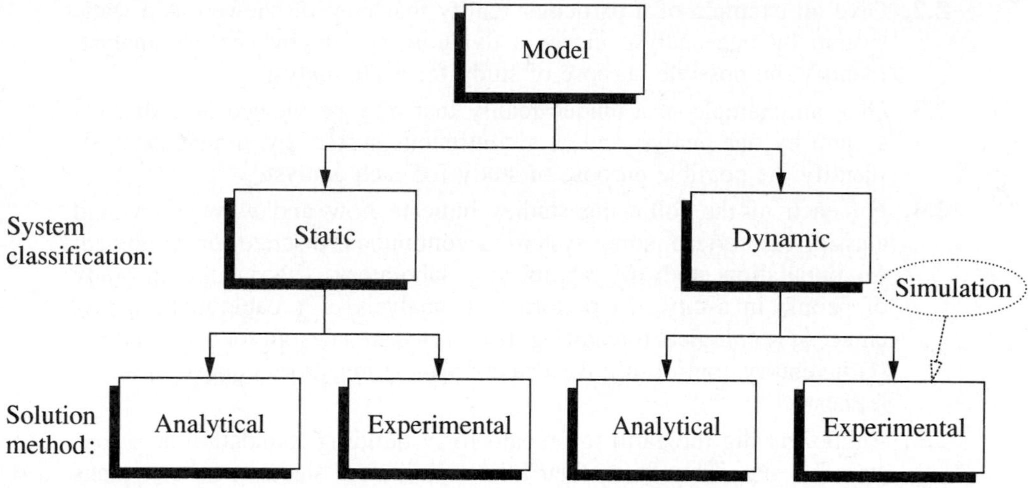

FIGURE 2.14
Illustration of the technical definition of simulation.

the analytical methods in handling complex dynamic systems, simulation has proven to be an excellent alternative in a variety of problem domains. The following chapters of the book concentrate on various issues related to the subject of simulation.

2.4 SUMMARY

In this chapter some of the important basic concepts related to systems have been presented. The materials that will be presented in Chapter 4 (under "Common Processes in Discrete Systems") are complementary to the contents of this chapter; however, those materials are not immediately presented here since the interim chapter facilitates their understanding. The description of models and their purpose, classifications, and types have also been presented in this chapter. Since simulation is an experimental approach, it is important to understand the capabilities and limitations of the experimental techniques as compared with analytical methods.

2.5 EXERCISES

2.1. For each of the following systems identify two possible purposes of study, and for each purpose identify the major outside influences and the major components that you would include within the system boundary: (*a*) a restaurant; (*b*) a traffic intersection; (*c*) a bank; (*d*) a small machine shop; (*e*) an airline company.

2.2. Give an example of a particular reality that may be viewed as a static system by one analyst and as a dynamic system by another analyst. Identify the possible purpose of study for each analyst.

2.3. Give an example of a unique reality that may be viewed as a discrete system by one analyst and as a continuous system by another analyst. Identify the possible purpose of study for each analyst.

2.4. For each of the following studies indicate how and why you would classify the corresponding system as continuous, discrete, or combined: (*a*) population study of rabbits in a laboratory; (*b*) population study of people in a city; (*c*) performance analysis of a cable-making machine; (*d*) ecological forecasting; (*e*) service area design for a gas station; (*f*) inventory control of a car dealership; (*g*) inventory control of an oil refinery.

2.5. Bulldozers dig the earth to prepare for a building foundation at a construction site. They dump their load on a limited size area where trucks of various capacities are loaded by several loaders. Load sizes of bulldozers, loaders, and trucks are different. Trucks travel back and forth between the construction site and a dump site. Trucks carrying foundation materials arrive twice a day, when the loader operators are called to operate the cranes that unload the materials. After unloading, the operators resume their loading operation. All equipment is leased at a certain hourly cost.

(*a*) Assuming that you are the construction manager, what would be your potential purposes of studying the above operation?

(*b*) Identify what you consider as entities, attributes, activities, resources, events, and possible accumulation points (queues) in the above system.

2.6. Give an example of a static system that can be approached by both analytical and experimental solution methods.

2.7. Modify your example in the above exercise such that analytical methods known to you cannot be used to solve the problem posed. What simplifying assumptions do you suggest that make an analytical method suitable for the modified system?

2.8. Repeat exercise 2.6 for a dynamic system (either discrete or continuous) of your choice.

2.9. Give a different realistic example for each of the following scenarios:

(*a*) An exact model of a system with an exact solution method

(*b*) An exact model of a system with an inexact solution method

(*c*) An inexact model of a system with an exact solution method

(*d*) An inexact model of a system with an inexact solution method

2.10. For each of the examples in exercise 2.9, alter the situation slightly such that one of the other three scenarios becomes more appropriate.

CHAPTER

3

SIMULATION

3.1 INTRODUCTION

Given the fact that most real-world dynamic systems are of a complex structure (e.g., a network of several queues), that the commodity movement patterns in them are based on some complicated conditions, and that their transient modes are sometimes of primary interest, analytical methods have proven to be of limited use in the majority of studies of dynamic systems. For example, a single traffic intersection with left turn signals and with typical, realistic arrival patterns of cars is impossible to study using queuing theory, unless simplifying assumptions (which are often unrealistic) are made. Under these conditions the experimental method is the only viable approach. As pointed out in Chapter 2, we refer to those experimental methods that are used in modeling and in model experimentation for dynamic systems as *simulation*.[1]

[1]It should be noted here that the word *simulation* is often used liberally by many to refer to a class of less-sophisticated analyses which involve simple experimentations (such as changing some entries in a spreadsheet to see the impact on the rest of the figures in the sheet). These "what-if" experimentations, which are typically applied to static problems, are not regarded in the technical communities as simulation.

To highlight the nature of the simulation approach to systems analysis and design, a nonnumerical example and a manual simulation of a numerical example that demonstrates the basic simulation procedure are presented in this chapter. Concepts related to computer implementation of simulation models are then presented. Readers not familiar with computer programming will still understand the related discussions. A general purpose simulation tool that does not require prior knowledge of programming is presented, and the general simulation process is explained.

3.2 A DEMONSTRATIVE EXAMPLE OF SIMULATION

Recall the case of studying a traffic intersection that was described earlier. The purpose of the study in this case is to find the traffic light timing for an existing intersection which minimizes the average waiting time of all cars that arrive at the intersection during a certain period (say, rush hour). In order to avoid disturbing the operation of the real intersection system through our analysis, we need to build a model of the system. To understand the basic components, the nature of the relations between the components, and the nature of the inputs to the system, we must first observe the real system. Let us assume that through observation of the intersection we find out that the streets leading to the intersection are both two-way streets. Let us further assume that the traffic light has left turn signals and that right turns are allowed only on a green signal. This completes our understanding of the system structure within the scope of our purpose.

To be able to re-create what takes place in reality, we need to record the arrival pattern of cars and their behavior once they arrive at the intersection (going straight, left, or right). To do this we may need to have several (say, four) assistants, each watching one of the four segments of the streets leading to the intersection. Using stop watches, these assistants should record the arrival time of each car (entity) and the direction that it goes upon leaving the intersection. Since we will be attempting to find a better traffic light timing than the existing one, there would be no need to record the actual timing of the traffic light (unless we want to compare the performance of the new traffic light timing with the performance of the current setting). This phase of the study is called data acquisition.

The next phase of the study is model building. To facilitate the understanding of the simulation concept, let us use a physical model of the intersection system. To build a physical model we may take a flat board, draw a two-dimensional picture of the intersection on it, and place a miniature traffic light in its center. Let us assume that our miniature traffic light is actually functional and is controlled by an adjustable timer. Next we need to bring several small model cars to the scene.

We should call upon our four assistants again and provide each of them with a number of model cars. Placing a clock (master simulation clock) where all the assistants can observe it, we then ask them each to look at their recorded data and move their model cars to the intersection as the time seen on the clock matches their recorded actual car arrival times. We also ask the assistants to observe the signals on the traffic light so that their cars may pass through the intersection only if the corresponding light is green. Otherwise they have to wait behind the stop lines (see Fig. 3.1).

As time goes by we may keep a record of the waiting times of the model cars. We continue this process until our assistants run out of data (the end of the simulation period). We then modify the traffic light timing, reset the clock to the beginning of the time of the actual data collection, and repeat the entire process. If we and our assistants are patient enough, we may repeat the above process several times, each time trying a different setting for the traffic light, and for each setting recording the average waiting time of the small cars (model experimentation). Note that in one of the experiments we can set the traffic light timing in our model to the current timing of the traffic light in the real system. The statistics collected from the model can then be compared

FIGURE 3.1
Physical modeling scenario of the traffic intersection simulation.

with the performance statistics of the current system. If the two statistics have reasonably close values, we then have a good reason to believe that our model is valid.

Using our recorded information, we can finally choose the timing that corresponds to the smallest average waiting time observed in our simulation (analysis of the output). We may then go to the actual intersection site and set the real traffic light timing to the best setting we found in our simulation (implementation). In this way we have managed to study the system and find its best performance condition without disturbing its operation during the course of our study. Notice that the simulation results in this example apply only to the period during which data acquisition was performed.

One important point to observe here is that the master simulation clock used in the above model need not be run at the same speed as the real time clock. For example, we may have our clock run ten times faster than the real time clock without affecting the validity of our study. This is like watching the system in fast motion (if our assistants can handle it!). Of course, the waiting times that we observe under this situation will be one-tenth of the actual waiting times of the cars, but since our study is relative, this has no impact on the results.

The above example, although rather unrealistic to implement, shows a number of important concepts about simulation. First, it illustrates the simulation process, which requires the definition of a purpose, a system study, data acquisition, model construction, model validation, model experimentation, and results implementation. Secondly, it serves to show the concept of simulation time and how it relates to the actual time; that is, the example shows that several hours of the actual system operation can be simulated in a few minutes. If models are represented in symbolic form and are manipulated by computers, then the simulation time can be several thousand times faster than the actual system operation time! Simulation may take longer than the actual time when simulating extremely fast systems (e.g., motion of electrons in a molecular structure). Finally, the above example demonstrates the difficulty involved in using nonsymbolic approaches to studying some dynamic systems.

3.3 A NUMERICAL EXAMPLE

As another example that highlights further concepts in simulation, consider the service terminal of a large air freight company that owns many cargo planes. The cargo planes are scheduled to arrive at the terminal, one at the beginning of each day, for possible maintenance operation. Each plane is inspected upon arrival. Assume that the inspection duration is negligible. Once inspected, the probability of finding an airplane in need of maintenance service is 0.5; that is, an average of 50 percent of the planes are found to need maintenance service. If a plane needs service, its maintenance operation may take either 0.5, 1, 1.5, 2, 2.5, or 3 days. The chance of the plane requiring any of these service times is 1/6.

The freight company currently utilizes one maintenance facility at the terminal. Each grounded airplane costs the company $5000 per day. The company management is interested in investigating the economic attractiveness of utilizing an additional maintenance facility at the terminal. Each facility costs the company $2500 per day to lease and operate. The purpose of the study, therefore, is to compare the operating costs of the system under one- and two-facility scenarios.

Note in this example that the details of the recorded data regarding planes that have needed maintenance service and their service durations are not used. Instead the data is presented in a summary form. That is, the data gathered about plane arrivals and service times has been statistically analyzed and summarized as probability distributions. There are certain advantages to this form of data representation. First, given this summary information, we are not restricted to a small number of observations (as was the case in the previous example). This allows us to extend the length of the simulation and observe the model behavior in much longer time spans. Second, we can experiment with various input parameters (various percentages of planes that need service and various patterns of service time) and check the model performance under each condition. But this form of representation requires a mechanism to generate random variates that represent the desired probabilistic behaviors. The details of the methods for summarizing input data as probability distributions and for the generation of desired random variations that represent those data will be given in a following chapter. For now we need to use a simple method to determine which arriving planes need service and what the required service time is if a plane is found to need service. The data in this example has been deliberately contrived so that the random behavior of the system may be re-created using a coin and a die. These simple physical random variate generators (sometimes called analog generators) emulate occurrences which have, respectively, 1/2 and 1/6 probabilities of taking place.

We can simulate the operation of the service terminal by tossing a coin for each plane arrival (taking place at the beginning of each day) and if a need for maintenance service is indicated, we can then toss a die to determine how long the service operation will take. Let us establish that if the coin toss indicates a head, we assume that the arriving plane needs service, and if the coin toss indicates a tail, the arriving plane does not need service. Moreover, let us establish that if the die toss indicates 1, 2, 3, 4, 5, or 6, the service duration will be either 0.5, 1, 1.5, 2, 2.5, or 3 days, respectively. Thus there are two probabilistic processes to be modeled: the probability that maintenance is needed and the duration of the maintenance operation.

The result of a ten-day simulation using the above procedure is shown in Table 3.1. Here, for each simulated day of operation a coin is tossed. If the result of the coin toss is a head, then a die is tossed to find the required service time. To understand the entries in the table it is best to first concentrate on the columns related to the one-facility scenario and then examine the two-facility scenario. Note that in the one-facility scenario some planes that need service have to wait until the facility becomes available. The second scenario, which uses two facilities, is capable of immediately accommodating all planes that need service. Figure 3.2 shows the same results in graphic form.

TABLE 3.1
Simulation of one- and two-facility scenarios

				One facility			Two facilities			
DP	CT	DT	NDS	SSD	ESD	NDD	SSD	ESD	NDD	FN
0	H	5	2.5	0	2.5	2.5	0	2.5	2.5	1
1	T									
2	H	4	2	2.5	4.5	2.5	2	4	2	2
3	T									
4	H	6	3	4.5	7.5	3.5	4	7	3	1
5	H	3	1.5	7.5	9	4	5	6.5	1.5	2
6	T									
7	H	4	2	9	11	4	7	9	2	1
8	H	2	1	11	12	4	8	9	1	2
9	T									
Total no. of days down						20.5			12.0	

Legend: DP = days past; CT = coin toss; DT = die toss; NDS = no. of days of service; SSD = start service day; ESD = end service day; NDD = no. of days down; FN = facility no.

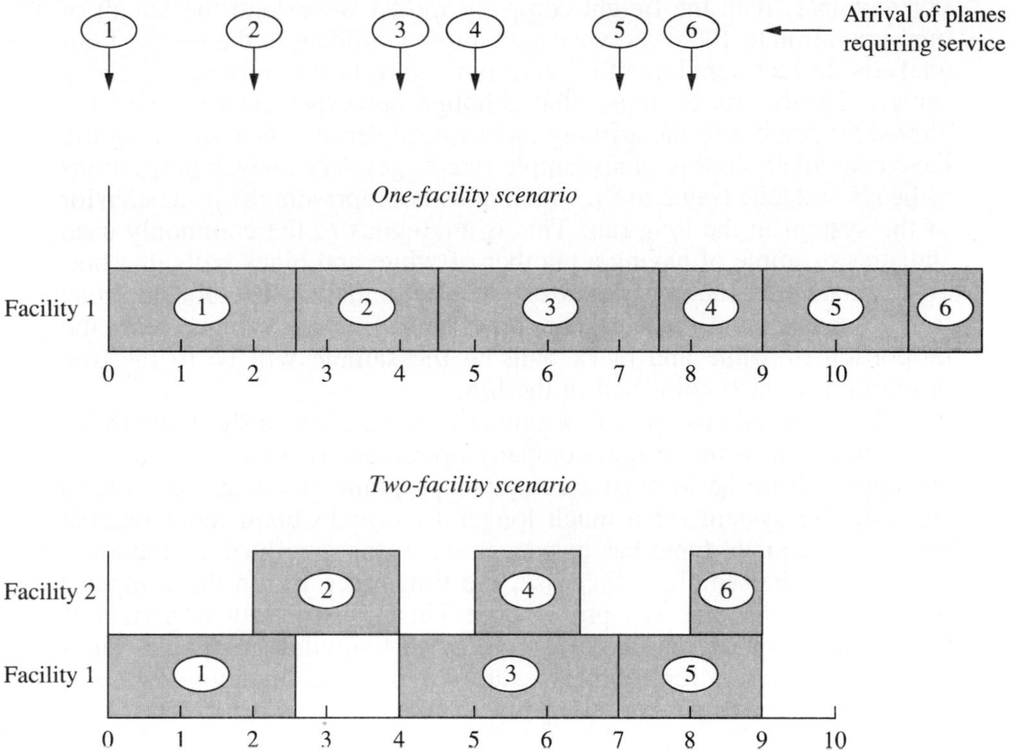

FIGURE 3.2
Graphical representation of the freight company simulation.

Based on our simulation analysis the total cost of operation of the service terminal may be found as follows:

$$\text{Total cost: One facility} = (10)(2{,}500) + (20.5)(5{,}000) = \$127{,}500$$

$$\text{Total cost: Two facilities} = (10)(5{,}000) + (12.0)(5{,}000) = \$110{,}000$$

This simulation analysis indicates that utilizing two facilities will save the company $17,500 for each ten days of operation ($52,500 per month). Notice that these results are only estimates found using a sampling procedure. The degree of confidence in their validity depends on the soundness of this statistical experiment performed using simulation. We will elaborate on the related issues in Chapter 7.

This example shows that simulation of stochastic systems is in fact a sampling process. Each replication indicates a sample of the system operation. As is the case for all other sampling methods, the larger the number of samples taken, the better the estimate of the system behavior.

For example, if in the freight company model we extend the length of our simulation to 100 days, we can be more confident in the result of our analysis. In fact, ten days of observation is too short to provide a reliable sample. Notice, for example, that although our experiment showed that almost 50 percent of the arriving planes need service (6 heads) it would have been likely in this small sample size to get very uneven proportions of heads and tails (say 2 to 8), which does not represent the true behavior of the system in the long run. This is analogous to the commonly used statistics example of having a number of white and black balls in a bin. To estimate the true proportion of each color ball in the bin one may take a sample from the bin. The larger the sample size, the closer the proportion of white and black balls in the sample will be to the true proportion of each color ball in the bin.

To avoid tedious manual computations, we chose a short length for our simulation of the freight company operation. If we could represent our approach in the form of a computer program, we would be able to simulate the system for a much longer time and obtain more reliable results. Of course, there has to be a limit on the length of a simulation run (size of the sample); otherwise, the time taken to run the computer program may become too prohibitive. This is especially important if the analysis requires day-to-day application of simulation (such as daily schedule generation for a factory), rather than a one-time application as was the case for the freight company.

3.4 COMPUTER REPRESENTATION OF SIMULATION MODELS

The manual simulation example presented in the last section serves to show the basic procedures required to perform a simulation study. As that example demonstrates, the simulation process may involve many computations. Accordingly, the cumbersome manual method that we used in that example is of little use when dealing with realistic problems that are often complex. Due to the magnitude of the computations involved in actual simulation problems, simulation became widespread only after computers became popular in scientific and business environments.

This section will briefly introduce the use of computers in the generation of random effects. In addition, the issues of coding the simulation model into a computer program that represents the logic of dynamic systems behavior and collecting and reporting the desired statistics on system performance are briefly addressed.

BASIC, a popular and easy to understand programming language, is used for the representative program discussed in this section. Another version of the same program is written in the C language. These programs are placed in Appendix A. Readers who are not familiar with computer programming should be able to understand the major structure and functions of the various modules contained in these thoroughly documented programs; therefore, they are encouraged to review the flowcharts and the programs. However, understanding these programs is not a requirement for understanding the rest of the book, since the book uses the EZSIM software, which does not need programming.

3.4.1 Creation of Random Effects by Computer

In the numerical example of the freight company, simple physical devices including a coin and a die, which have discrete outcomes (head or tail; 1,2,...,6) of equal probabilities of occurrence were used to re-create some random effects in the corresponding computer model. There are other physical devices that may be used to create certain discrete or continuous random outcomes. For example, a very balanced disk with its edge numbered within any range and with a low-friction shaft in its center (like a roulette wheel) may be devised. Each spinning of the disk will result in a reading by a fixed indicator arrow of a uniformly distributed random outcome. To make the disk useful for the creation of a uniformly distributed outcome within any given range, the disk edge may be numbered within the range of numbers 0 and 1. In this way, a simple numerical transformation of the indicator reading can provide the desired uniformly distributed random outcome for any parameter values (range) of the distribution. For example, if numbers uniformly distributed between the range of 60 and 100 are desired, the indicator reading may be scaled up by being multiplied by 40, the range, and shifted by 60, the lower bound of the range, to fall within the range. For example, if the aircraft maintenance time in the freight company example is distributed uniformly within the continuous range of 0.5 and 3 days, the following relationship may be used to create the corresponding durations:

$$X = 0.5 + r \cdot (3 - 0.5)$$

where r is the random outcome given by the disk indicator reading and X is the variable representing the random maintenance duration, which is distributed between 0.5 and 3 hours.

Note that using the balanced wheel, some nonuniform discrete probability distributions may be generated as well. For instance, in the freight terminal example, if the chance of an aircraft needing service upon arrival is 40 percent, all disk indicator readings below 0.4 could indicate a service requirement.

Also note that without the use of the 0–1 range numbering of the balanced wheel and the above transformation method, a different numbering of the wheel for each desired range of uniformly distributed random outcomes would be required. Random outcomes that are uniformly and independently distributed within the 0–1 range are called *random numbers*. As demonstrated above for the cases of uniformly distributed random outcomes with various ranges, the significance of random numbers is their utility in generating random outcomes with any probability distribution.

Because of the utility of random numbers, programming environments (e.g., BASIC, C, FORTRAN) offer a random number generator function. Each time they are called, these functions return a different random number. The following simple program written in BASIC to generate and display 100 random numbers may be used to observe the performance of a typical computer random number generator:

```
10 FOR I=100
20 PRINT RND(1)
30 NEXT I
```

The function RND(.) used in the above program is BASIC's random number generator; its function argument is the seed for the stream of random numbers generated by the function. Changing the seed results in a different sequence of random numbers in a new run (seed values other than 1 may need to be negative integers in some BASIC compilers).

The details of the mechanism for generation of random numbers by computers and the methods of using them for generation of random variates with various forms of distribution are presented in Chapter 6. The internal random number generators of BASIC and C are used in the programs listed in Appendix A, without elaborating on how the numbers are generated by the BASIC and C compilers.

3.4.2 Representation of Model Logics by Computer Programs

To represent the behavior of a dynamic system in a computer program, the relations within and between the system components must be known. This knowledge determines the way in which system variables should change over time. Given this knowledge, the simulation time may be advanced (according to a certain criterion) and for each updated value of time the updated values of system variables can be computed. In this way the changes in the state of the system over time may be traced and the necessary statistics gathered. Since simulation time plays a significant role in determining the flow of control in a simulation program, the indicator for it is usually called the master simulation clock (or simulation clock).

In a simulation of continuous systems where variables may change within any time interval, the simulation time is usually advanced in small increments, and at the end of each increment the values of time-dependent variables are recomputed. The treatment is different for discrete systems, since in these systems the variable values do not change between two consecutive events. Therefore, the simulation time in discrete systems is advanced from one event to the next, and at each event time the values of variables affected by the occurrence of the event are updated. The key to creating computer programs as simulation models for discrete systems, therefore, is in the recognition of all events that may take place and in the impact each event has on the system variables.

Generally, the occurrence times of all events taking place within a simulation time frame may not be known in advance. This is true since events are usually created by preceding events and according to some possibly complicated logic. For example, in a queuing system the event indicating departure from a service station depends on the arrival event times, the service time, and the status of queues. Hence, even though it may be possible to identify in advance all arrival events from start to end of the simulation (given the information about the interarrival times), it may not be as simple to predict all departure events in advance.

It should be pointed out that determining all event times (such as all arrival times) in advance and then going back in time to find the effects of these events on system variables is not a wise approach, since storing the information about events that are not to occur in the immediate future wastes computer memory. Efficient handling of events requires a limited time window that shows the most immediate upcoming events. The analogy of a car moving at night on a highway with its lights illuminating only a certain distance in front, to show the traffic signals and the upcoming road conditions only as far as necessary for safe driving, is an appropriate one.

In tracing the chain of events in a simulation program, the approach should identify all immediate future events caused by the event taking place at current simulation time, and it should place these events in a list of future events. The simulation time should then be advanced to the nearest event time in the latter list, and the changes affected by the corresponding event should be reflected on the system variables. After processing each event the information about it may be erased from the memory.

To highlight the major issues in simulation program development, let us consider a simple system: a single-server queuing scenario with uniformly distributed interarrival and service times. The associated computer program consists of the initialization routine, the input routine, the event timing routine, the arrival event routine, the departure event routine, the statistics routine, and the output routine.

As mentioned earlier, the realization of events and their effect on system variables constitutes an important step in devising a simulation program for discrete systems. In the example of the queuing system entity arrival, start of service and entity departure may be considered as events in the system. A program module may then be written for each event type to make the proper changes on the system variables as an event takes place. Recognition of interdependent events that take place at the same time could reduce the number of these event-related program modules. The start of the service event in this case is an example of a dependent event because it takes place either at the time of arrival of an entity to an empty system or at the end of service (departure) time of an entity when the queue is nonempty. Therefore, the start of service event logic may be implemented in the arrival and departure routines of the simulation program. The flowcharts in Figs. 3.3, 3.4, and 3.5 show the simple logic for the main routine and each event type, respectively.

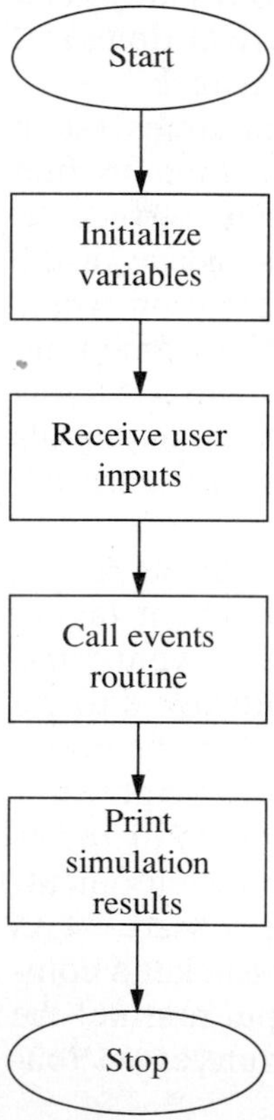

FIGURE 3.3
Flowchart of the main module in the queuing simulation program.

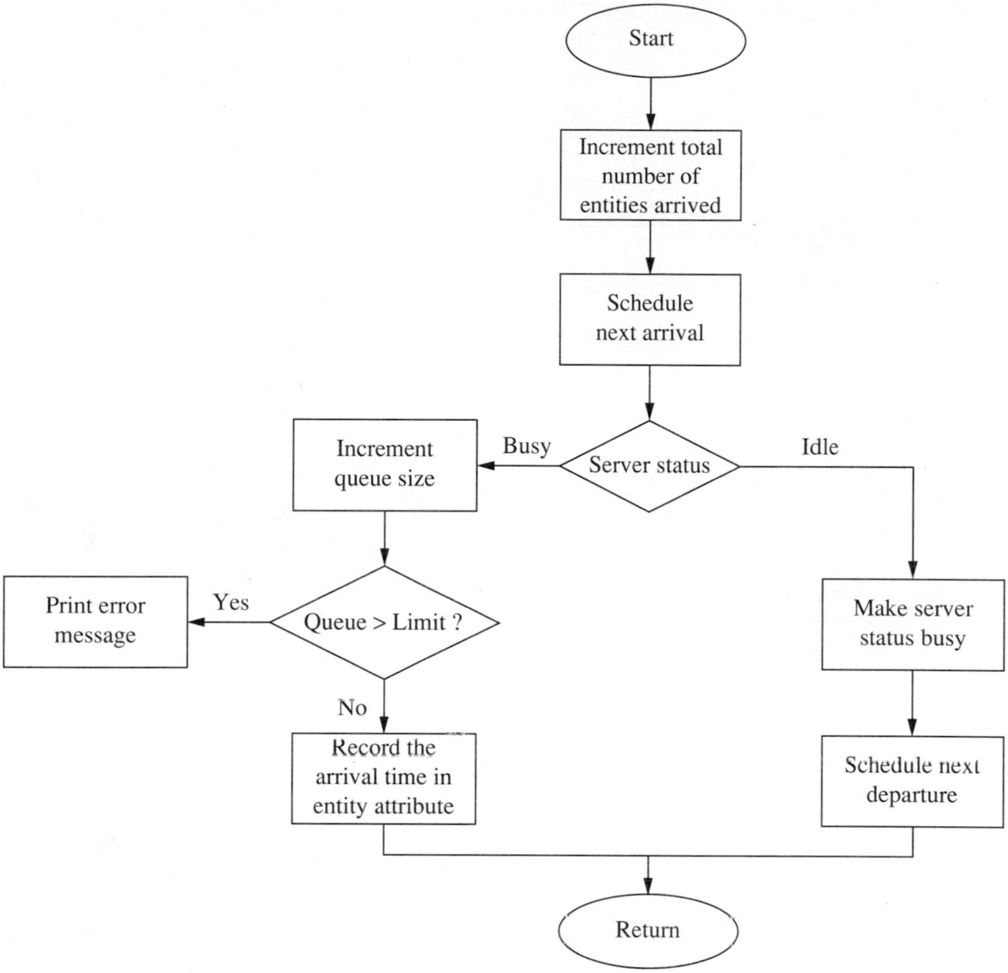

FIGURE 3.4
Flowchart of the arrival event module.

The program listings presented in Appendix A do not use specialized functions and are simple and generic; therefore, the programs may be easily translated to other languages such as FORTRAN and Pascal. A typical input parameter set and the related output are also shown in Fig. 3.6. As noted in the program, an arrival event always results in scheduling one future arrival event to take place at TNA (time of next arrival). This is known, since we are given the information about interarrival times. In other words, if we know the time of the current arrival we can determine the time of the next arrival by pulling a value from the corresponding

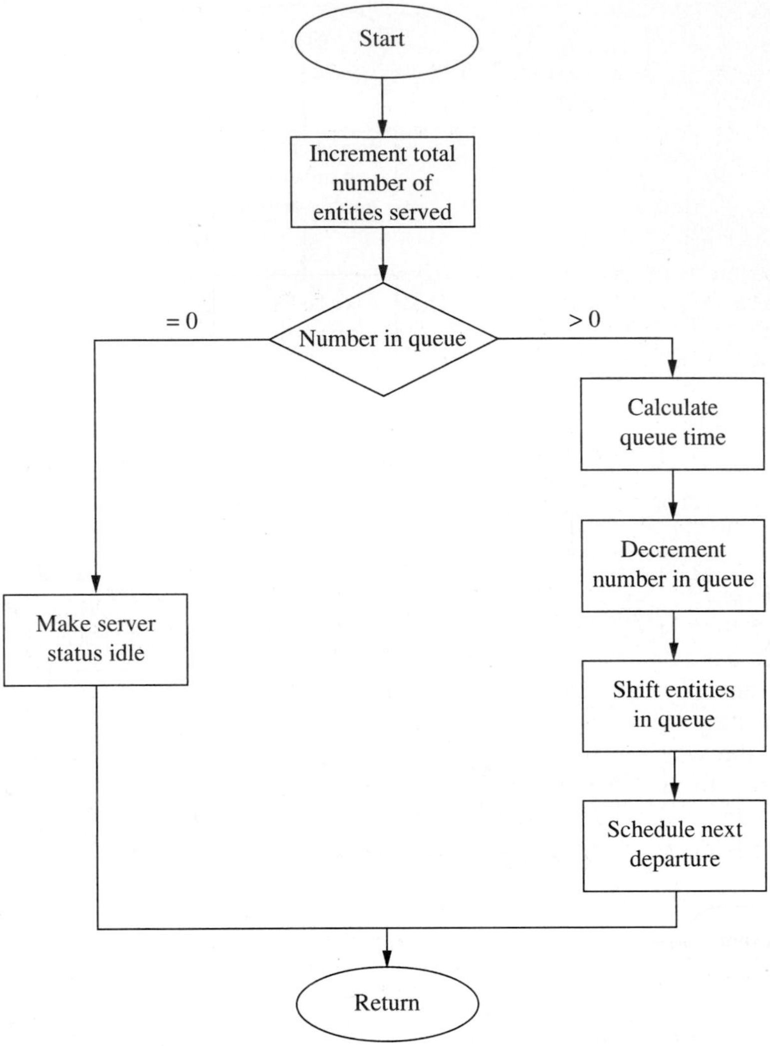

FIGURE 3.5
Flowchart of the departure event module.

probability distribution. An arrival event may result in scheduling a departure event at TND (time of next departure). This occurs only when an arriving entity finds the server idle. A departure event may also schedule another departure event. This happens when a departure event takes place from a system with a nonempty queue. From this discussion it follows that in our single server queuing system there may be a maximum of two imminent events in the list of future events, since there is always one arrival event and there is sometimes a departure event (see Fig. 3.7).

```
Input parameters:

    Length of simulation in time units? 1000
    Minimum interarrival time? 10
    Maximum interarrival time? 15
    Minimum service time? 9
    Maximum service time? 14

Simulation output:

    Number of entities entered=81
    Number of entities served=80
    Total time spent in queue=93.68
    Average (expected) waiting time in queue=1.17
    Average (expected) length of queue=0.41
    Average server utilization=57.91%
```

FIGURE 3.6
A typical input parameter set and outputs of the queuing simulation program.

In the example program a goal of simulation is to find the statistics on the time each entity spends in the queue and in the system. This is not possible unless each entity arrival time is recorded. This recorded time should be carried with the entity to the point of departure from the queue at which it may be compared with the current simulation time, TNOW, to provide the waiting time in the queue. Adding the entity service time to the queue time provides the entity system time. The procedure for recording the arrival times of the entities (an entity attribute) in the program uses an array called ATRIB. The entries in this array represent

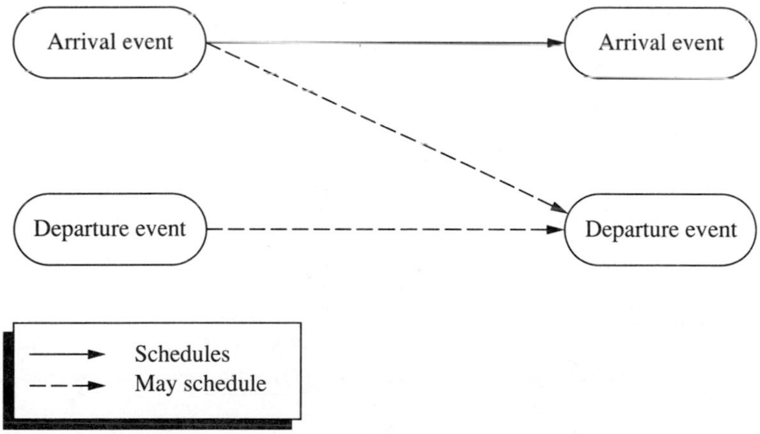

FIGURE 3.7
Events schedule other events to occur in future.

the entities in the queue. Upon arrival, the entity's arrival time is placed in the end of the array (see the Arrival routine). When an entity leaves the queue to enter the service station all entries in the array are moved one position forward (see the Departure routine).

The diagrams in Fig. 3.8 clarify the performance of the computer programs. They reflect the changes that take place in the queuing system at each event time. In these diagrams circles represent entities. The number inside the circle is the arrival time of the entity (entity's attribute). CLOCK is the simulation time (TNOW in the programs). "A" and "D" in the Event box represent whether the event is an arrival or a departure, respectively. TNA and TND are based on the interarrival and service times, which are created randomly. It is assumed that the first arrival takes place at time zero. The random interarrival times used in the diagrams are

3.9 2.5 3.6 2.5 1.9 3.3 3.7 3.4

The random service times are

1.8 1.5 1.1 4.8 3.8 2.2 3.5 3.7

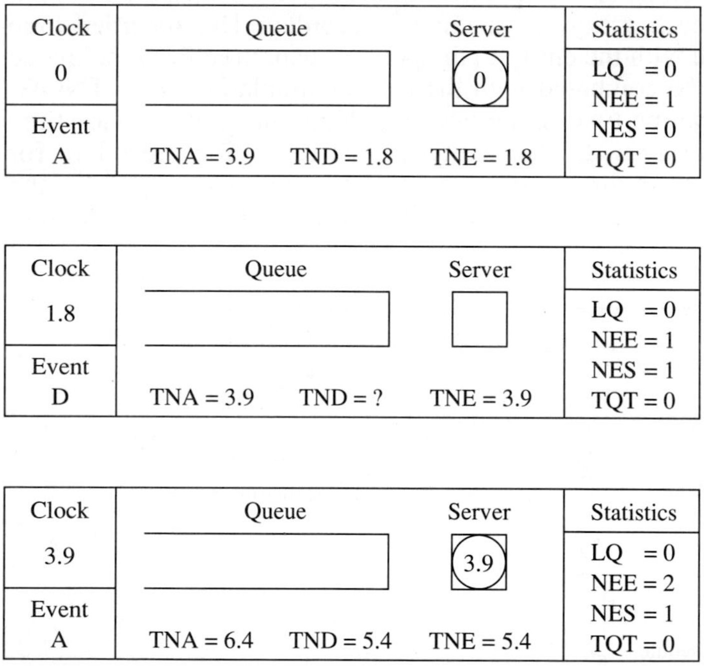

FIGURE 3.8
Diagrams showing changes in the queuing system at the event time.

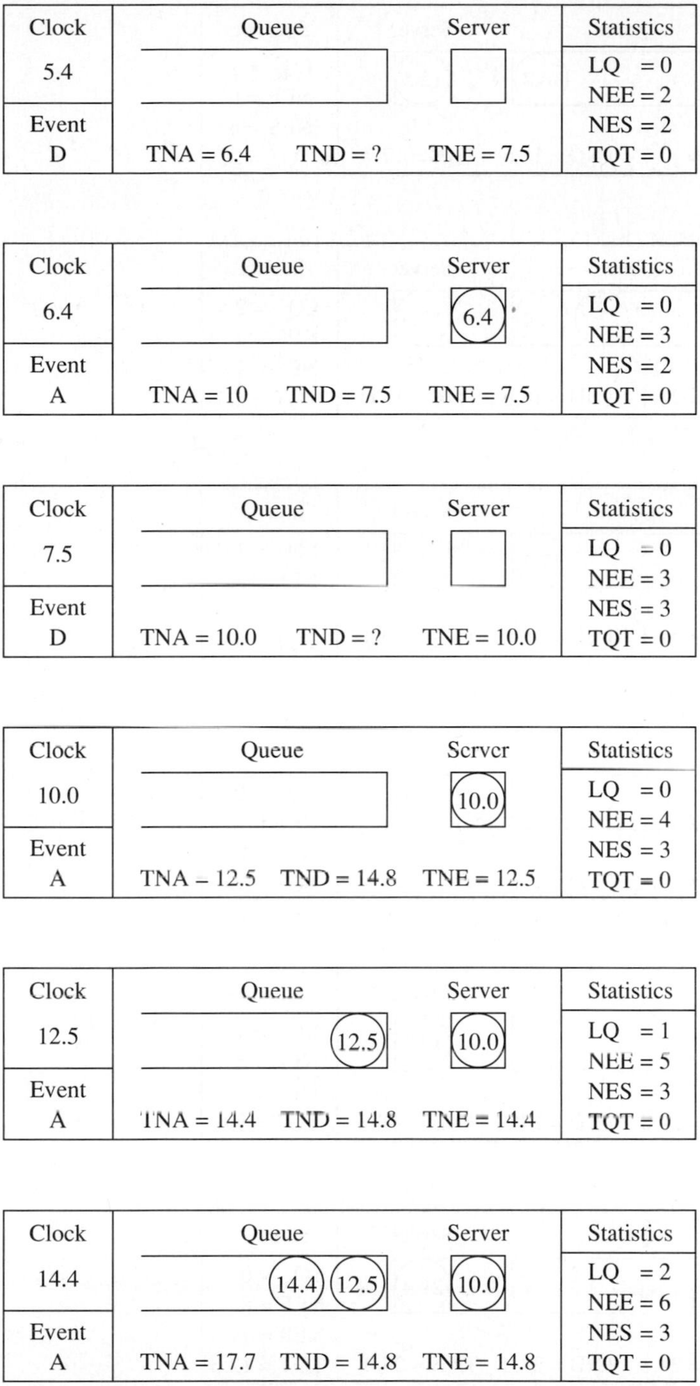

FIGURE 3.8 ***(continued)***

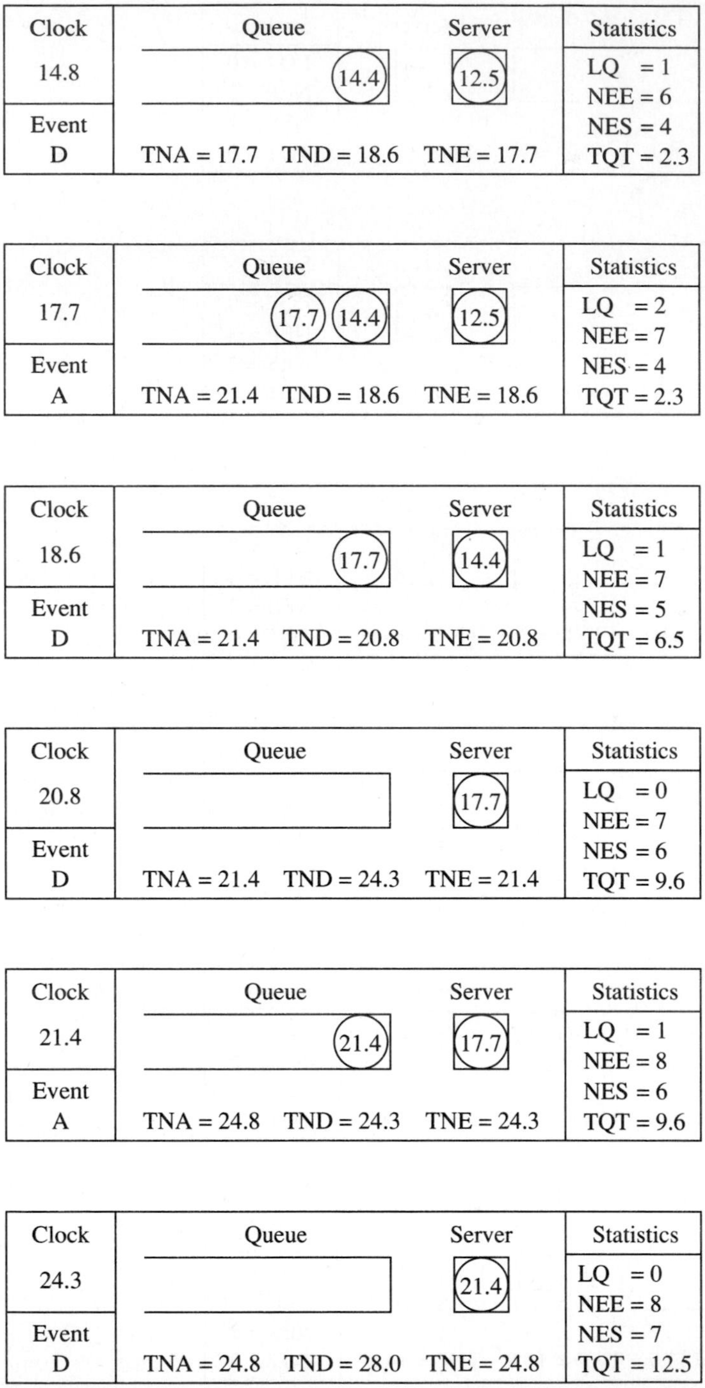

FIGURE 3.8 (*continued*)

TNE is the time of the next event, LQ is the length of the queue, NEE is the number of entities that have entered the system, NES is the number of entities served, and TQT is the total queue time (total time spent in the queue by all entities). Notice that TQT is updated only when an entity leaves the queue to start service.

3.4.3 General Structure of Simulation Programs

The structure of the simple program provided in this chapter represents the general structure of all event-based simulation programs. The following discussions regarding the general structure of simulation programs may also clarify certain points about the specific example program discussed in the preceding section. Generally, the following modules constitute a discrete event simulation program written in a procedure-oriented language (general programming language):

Main routine transfers control between the major modules of the program.

Initialization routine initializes all variables and clears the statistical data that may have been gathered in a previous run.

Events timing routine locates the most imminent future event, advances the simulation clock to the time of the event, and calls the corresponding event-processing routine.

Future events list (calendar) contains the list of the unprocessed future events. Other information, such as the attributes of entities causing the event, may also be stored in this structure. Note that in the example program there are a maximum of two future events possible at any point in time. Consequently, there is no need for making an elaborate future events list for this case. If, for example, there were many parallel servers in our queuing system, there would have been a possibility of several upcoming departure events which would have needed to be scheduled (at the start of service times) in the future events list. Since events and their related information constantly appear and disappear in the future events list, a dynamic structure for this list that efficiently utilizes the computer memory is essential for realistic applications. The combination of the event timing routine and the event calendar are often called the *simulation engine*.

Event processing routines are individual modules each representing an event in the system. Each module, when called by the events timing routine, makes the proper changes (caused by the event) to the system variables and possibly schedules an event of the same or different type in the future events list.

Library routines include a module for pseudorandom number generation and several modules for random variates with various distribution types.

Statistics routine collects and processes certain statistics that are specified by the user. These statistics may deal with information concerning the length of queues, the entity waiting time in queues, utilization of facilities, and utilization of resources. The general quantities desired in statistics reports are the mean, standard deviation, minimum, maximum, and last value observed at the end of a simulation. Graphs of frequency histograms and/or plots of variable values over a specified time period may be constructed in the statistics routine.

Output routine gathers the values collected by the statistics routine and may perform some operations on these values to create measures such as overall averages. It prints the results and graphs in readable report formats.

3.5 STATISTICS IN SIMULATION

Statistics in dynamic systems are generally of two types: observation-based statistics and time-based statistics. Observation-based statistics, as the name implies, are collected at certain observation points. For example, the time spent in the system is observed at the entity departure point and not at any other time. Generally, all statistics that are of "time" dimension (waiting time in queue, entity traversal time between two points in the system, interdeparture time, etc.) are based on observation, and all other types of statistics are time-based.

The basic parameters, mean and variance, for an observation-based statistic (e.g., waiting time) for n observed values are as follows:

$$\overline{W} = \frac{1}{n}\sum_{i=1}^{n} W_i$$

$$\text{Var}_W = \frac{1}{n-1}\sum_{i=1}^{n}(W_i - \overline{W})^2$$

As an example of observation-based statistics consider the example shown in Fig. 3.8. The average waiting time in queue is $12.5/7 = 1.79$. This statistic is based on the observed waiting times of 7 entities that completed service.

Time-based (or time-averaged) statistics usually apply to quantities that do not represent "time." For example, the value of a variable representing the number of entities in a certain section of the system and the number of entities in a queue are represented as time-based statistics. A reliable value of these statistics may not be obtained if their levels are observed only at some epochs of time. The correct approach weighs each value taken by these variables against the time that each value persists. For example, looking at a queue at two points in time and observing

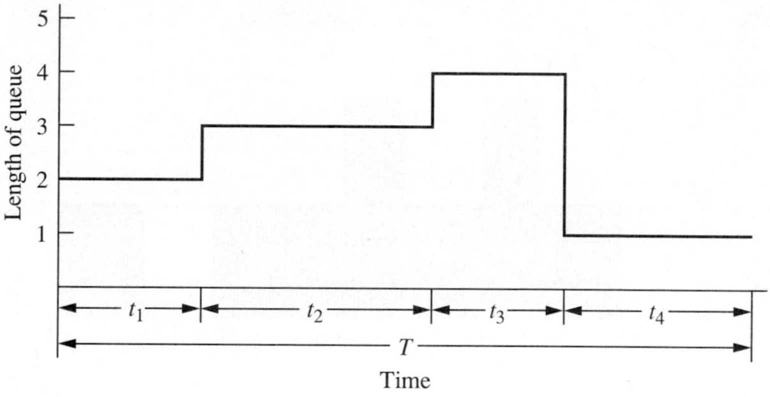

FIGURE 3.9
Changes in the length of a queue over T time units.

queue lengths of zero and ten, and subsequently concluding that the average queue length is five, would be an incorrect assessment of the true statistic; the zero length may have persisted for only one minute, but the length of ten may have persisted for many hours. The correct way to collect statistics is to monitor the queue length for each value that it takes and compute the time-weighted average of these values. Figure 3.9 illustrates a typical change in the number of entities in a queue over T time units. In this figure t_i is the length of time between the $(i-1)$th and the ith event, and L_i is the queue length at the ith event time. As this figure shows, each queue length L_i is persistent for t_i time units. The mean and variance of the queue length are found using the following equations:

$$\overline{L} = \frac{1}{T}\sum_{i=1}^{n} L_i t_i$$

$$\text{Var}_L = \frac{1}{T}\sum_{i=1}^{n} (L_i - \overline{L})^2 t_i$$

Figure 3.10 shows the pattern of changes in the size of the queue in the example shown in Fig. 3.8. For this example the average queue length as a time-based statistic may be calculated in the following manner:

$$\begin{aligned}\text{Average queue length} &= [(12.5-0)(0) + (14.4-12.5)(1)\\ &\quad + (14.8-14.4)(2) + (17.7-14.8)(1)\\ &\quad + (18.6-17.7)(2) + (20.8-18.6)(1)\\ &\quad + (21.4-20.8)(0) + (24.3-21.4)(1)]/24.3\\ &= 0.51 \text{ entities}\end{aligned}$$

In a simulation program the sum of the statistics and the sum of their square values may be updated each time the variable changes to compute

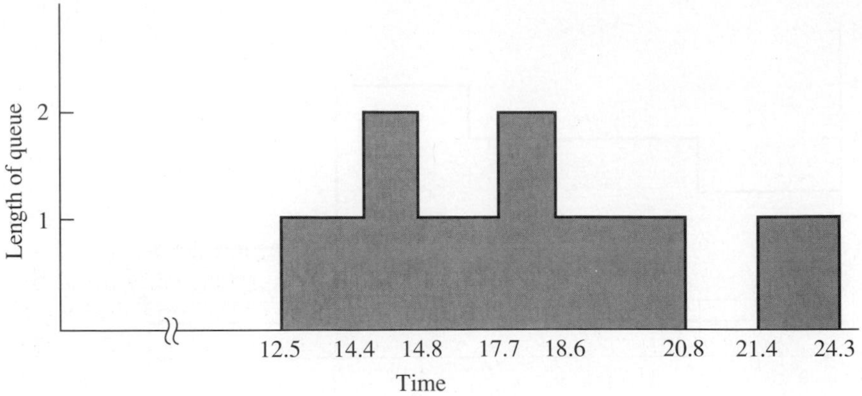

FIGURE 3.10
Pattern of changes in the queue size of the example in Fig. 3.8.

the mean and variance of each statistic at the end of the simulation. For a variable, say X, the following relationships are used to compute observation-based statistics:

$$SUMX_i = SUMX_{i-1} + X_i, \qquad SUMX_0 = 0$$
$$SSQX_i = SSQX_{i-1} + X_i^2, \qquad SSQX_0 = 0$$

The equivalent equations for time-based statistics are

$$SUMX_i = SUMX_{i-1} + X_i t_i, \qquad SUMX_0 = 0$$
$$SSQX_i = SSQX_{i-1} + X_i^2 t_i, \qquad SSQX_0 = 0$$

An efficient way to collect time-based statistics on a variable is to update the statistics each time the variable changes. A simpler but less efficient approach is to update the statistics at each event time regardless of the change in the variable. The latter approach is used in the example program, where the value of the statistics is multiplied by the quantity (TNOW−TLE) which is the equivalent of t_i in the above formulas (TLE is the time of the last event). In either case the statistics are weighted by the time segment between the current and the last observations. The cumulative total of all weighted values is then divided by the total length of time over which the statistics are collected.

Facility utilization statistics are best computed using a binary variable that takes on values of 0 or 1 for the idle and busy server conditions, respectively. Collecting time-based statistics on this variable would then provide the proportion of time that the facility is busy. Average utilization is at most equal to 1 for a facility with a single server. For a facility with multiple parallel servers the average utilization may be as high as the number of servers at the facility. For the freight company example the average utilization for a 10-day simulation under the first scenario is 1 (see Fig. 3.2). For the second scenario of this example we have

$$\text{Average utilization for the first server (facility)} = (2.5 + 3 + 2)/10 = 0.75$$

$$\text{Average utilization for the second server} = (2 + 1.5 + 1)/10 = 0.45$$

$$\text{Overall facility utilization (expected number of servers in use)} = 0.75 + 0.45 = 1.2 \text{ servers}$$

Facility "busy time" statistics are observation-based statistics on the duration of uninterrupted busy status. Each observed statistic is the time segment starting from each idle-to-busy transition and ending at the next busy-to-idle transition (at which the data point is observed). Facility "idle time" statistics are collected in the same manner. Consider for instance the second scenario of the example depicted by Fig. 3.2. Average busy time for each facility (server) is calculated as follows:

$$\text{Average busy time of server 1} = (2.5 + 5)/2 = 3.25 \text{ days}$$

$$\text{Average busy time of server 2} = (2 + 1.5 + 1)/3 = 1.5 \text{ days}$$

$$\text{Average busy time of overall facility} = (3.25 + 1.5)/2 = 2.375 \text{ days}$$

Average idle times can be found in the same manner:

$$\text{Average idle time of server 1} = (1.5 + 1)/2 = 1.25 \text{ days}$$

$$\text{Average idle time of server 2} = (2 + 1 + 1.5 + 1)/4 = 1.35 \text{ days}$$

$$\text{Average idle time of overall facility} = (1.25 + 1.35)/2 = 1.3 \text{ days}$$

Statistics on other types of resources may be collected in much the same way as the facilities statistics. A resource unit is considered busy if it is in use and idle if it is in its base stock or in transit to the point of use.

3.6 SIMULATION SOFTWARE TOOLS

The practice of computer simulation started with programming in generalized computer languages, mainly in procedure-oriented languages such as FORTRAN and PL1. Commonly used data structures, such as event calendars and queues, and commonly experienced processes, such as usage of facilities and material accumulation and flows, resulted in the creation of program modules that were used frequently in various applications. The need for the integration of these modules into an overall software system was soon realized, resulting in the emergence of simulation languages. Available simulation tools for discrete systems are now classified into three categories: (1) general purpose event-oriented languages, (2) general purpose process-oriented languages, and (3) special purpose simulation environments.

Event-oriented languages are based on the discrete event programming concepts presented earlier in this chapter. These languages provide the simulation engine and other standard modules. The user has to provide the individual event-processing routines, so he or she must have some programming skill. Because of the user's control of the contents of individual event routines, these languages are generally more flexible than process-oriented languages in representing complex and specialized processes.

Process-oriented languages are generally easier to use. These languages attempt to provide a set of modules, each representing a commonly occurring process (usually in queuing systems). For example, one module may create entities with the desired timing, another module may represent a queue to which entities join, and facilities may be represented by yet another module. Model building is then reduced to properly sequencing the calls to these modules. A flowchart, block diagram, or network representation of the model prior to writing the program statements is usually helpful. Since not all possible processes may be predicted and incorporated in a simulation language, some process-oriented languages have built-in capabilities that allow sufficient modeling flexibility, and others allow for linkages with user-written procedure-oriented modules that handle specialized processes. Therefore, a process-oriented language may also provide for event-oriented modeling. The advantage of event orientation is the flexibility that it provides. This flexibility may be needed when modeling uncommon processes. The combined event/process-oriented language allows the user to benefit from the advantage of each type of orientation. When modeling a system using the combined language, the process orientation of the language is used to the maximum extent possible (for simplicity and speed of modeling). If there are certain sections of the system that may not be represented by the available modules, specialized routines may be written and linked to the model using the event-orientation capability of the language.

Special purpose simulation environments are a special class of process-oriented software that are dedicated to certain application domains, such as manufacturing, communication networks, traffic flow, and scheduling. Since they generally do not require programming, they are called "environments" rather than languages.

Chapter 9 briefly presents some representative simulation software systems for each of the above categories. The necessary considerations for selection of a simulation software tool are also presented in that chapter. The next section introduces EZSIM, which is a new general purpose simulation tool. This tool is used in the simulation examples given throughout the rest of the book.

3.7 EZSIM—A GENERAL PURPOSE DISCRETE SYSTEM SIMULATION TOOL

The emergence of new software technologies through progress in computer science provides for creation of software systems with easier man/machine interfaces. When utilized in simulation, these technologies will help bring simulation directly to the disposal of end users with various technical backgrounds and computer programming skills. In the future, this factor will have a profound impact on the practice of simulation. The availability of modern simulation tools is especially important to work environments with a limited number of technical professionals, as these tools demand far less knowledge and technical sophistication of their users. Expert simulation analysts can also benefit from the increased model building and verification speed that the new environments offer.

The design and development of EZSIM has been an attempt to address the above issues. The following sections present the EZSIM environment and some simple examples that are comparable with the queuing models discussed earlier in this chapter. Only a small subset of the EZSIM capabilities are used in these examples. Following the Chapter 4 presentation of generic concepts regarding the common processes in discrete systems, the remaining features and capabilities of the software will be presented in Chapter 5 with numerous modeling examples. EZSIM has been designed by the author over the last several years. The effectiveness of EZSIM has been tested in various educational courses and realistic industrial projects.

EZSIM is a general purpose process-oriented simulation modeling tool for discrete systems involving entity flow. The software design allows for maximum modeling flexibility within the process-orientation frame. Models in EZSIM are represented in network form. Each node in the model network represents a process, and branches show the entity path from one node to another. The choice of processes considered in EZSIM is consistent with the general concepts presented in Chapter 4. The system was designed with the following objectives in mind:

1. To bring the power of simulation analysis to users with no detailed knowledge of simulation modeling and/or computer programming.

2. To detect and prevent logical modeling errors, to the greatest extent possible, at the early stages of the model building process.

3. To ease the process of learning and teaching the simulation modeling concepts through a pleasant and user-friendly environment, that does not demand the user to know obscure syntactic formats.

4. To help expert simulation model builders and analysts by saving model development time and minimizing logical and syntactic errors, which are frequently made in the conventional simulation model building and computer programming practices.

EZSIM is available in different operational modes. The stand-alone mode provides an easy to use modeling tool, a discrete event simulation engine, an output module, and a real-time graphic animation capability. These all exist in a single integrated environment. In program generator modes, EZSIM performs as a generic front-end environment capable of generating high-level codes for several popular simulation languages. This modeling environment is similar to that of the stand-alone mode.

EZSIM allows its users to quickly and easily build a model of the system under study, run the model in either the batch mode or animation mode, verify the model, and observe the desired statistics. The users can quickly change model parameters or configuration and run the model several times in a single session.

Numerous windows, menus, and context-sensitive help prompts are available throughout the above stages. EZSIM uses a combination of user interfaces that are graphics-based and menu-driven. The modeling approach in EZSIM enables the user to concentrate on the system structure and high-level dependencies while the system checks, to the best extent possible, the integrity of the model structure as it is being constructed by the user. Detailed information such as nodal parameters and branching conditions may be specified afterwards.

3.7.1 Procedure for Using EZSIM

This section is a brief presentation of EZSIM capabilities. Many detailed capabilities not covered here can be easily understood through interaction with the numerous context-sensitive help windows and menus of EZSIM. Figure 3.11 shows the flowchart of the procedure for using the stand-alone mode of EZSIM, and the following sections address the major components of this structure. The presentation will be based on the features of the EZSIM stand-alone mode. Most of the modeling features for the program generator modes of EZSIM are similar to those of the stand-alone mode. In a later section the differences between the two modes will be highlighted. The following stages are involved in a complete stand-alone EZSIM session:

- Model network construction
- Nodal parameter specification
- Model initialization
- Desired statistics specification
- Execution in batch or in real-time with animation

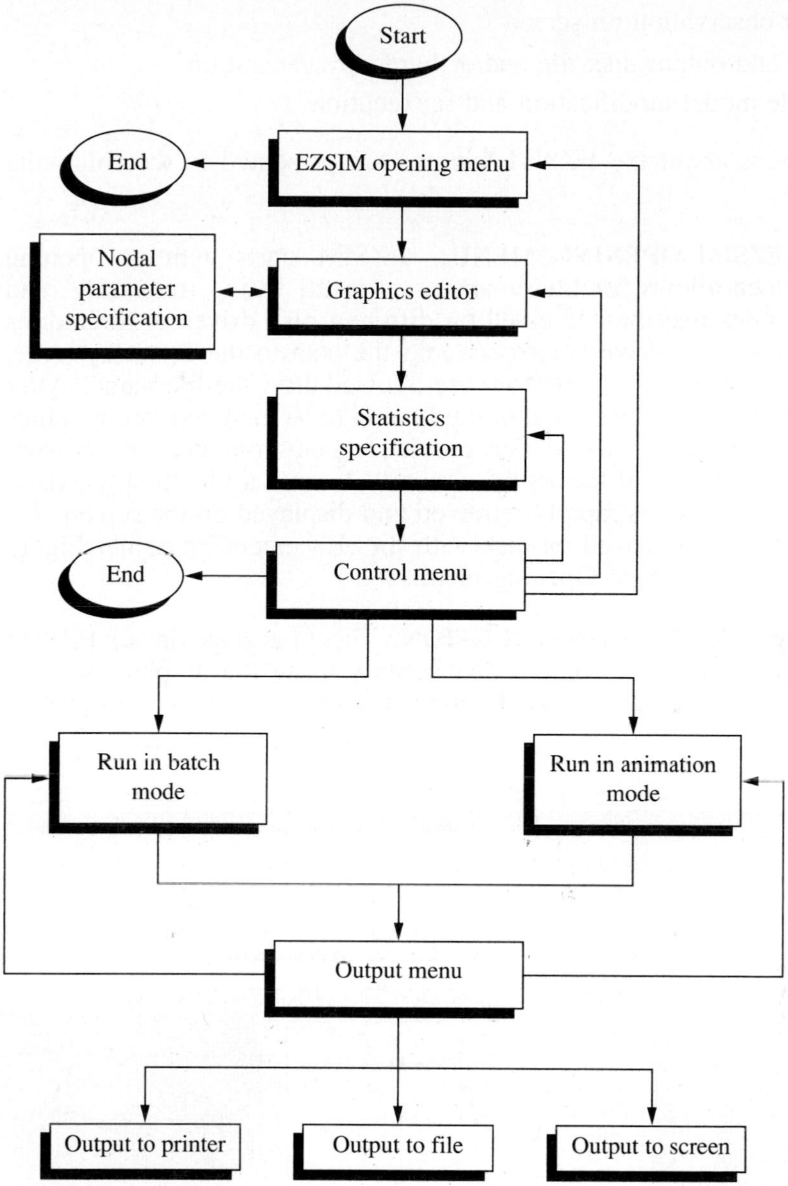

FIGURE 3.11
Flowchart of the procedure for using the stand-alone EZSIM environment.

- Output observation on screen
- Model and output disk file and/or hardcopy generation
- Possible model modification and reexecution

More details about the EZSIM features are presented in the following sections.

3.7.1.1 EZSIM OPENING MENU. EZSIM starts with its opening menu, which allows for file creation, retrieval, listing, renaming, and deletion. Files may be addressed on different disk drives or directories by identifying the drive letter code and the path to the directory (e.g., A:\EZ\MYFILE). No extensions are allowed after the file name. After the user specifies the file name of the system to be modeled, the graphics screen of EZSIM is invoked. This provides an environment for construction or modification of the model network. In case a file already exists, the current network is rapidly retrieved and displayed on the screen. All EZSIM models are saved on disk with the .EZ extension name. Figure 3.12 shows the EZSIM Opening menu.

3.7.1.2 NETWORK CONSTRUCTION. The first stage in an EZSIM session is the network construction activity using the graphics screen. At the beginning of this stage the user is asked to identify the optional

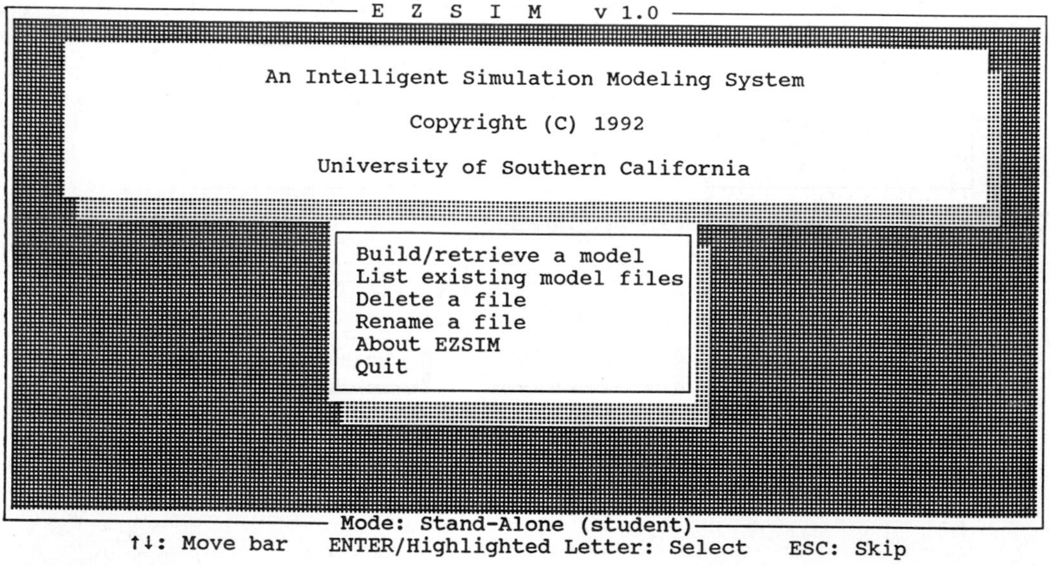

FIGURE 3.12
EZSIM Opening menu.

project header information. This information includes the project name, the date, and the analyst's name. Hitting the H key while in the graphics mode shows the help screen. All graphics control parameters and node definitions can be reviewed on this screen. To view the definition of each of the nodes, the corresponding key indicated by the highlighted uppercase letters in the node name should be pressed. For example, pressing the S key will provide the definition of the SOURCE node. A node definition window can be seen alternatively by moving the scroll bar on the node using the arrow keys or a mouse and pressing the return key. Hitting the Esc key while in the help mode returns the control to the graphics screen. While in the graphics screen, pressing the node code key places the corresponding node at the cursor position. The nodal abbreviations can be seen in a column on the right side of the graphics screen by pressing the F1 key.

After placing each node in the network, the assigned name for the node is asked for. Nodes can be connected by placing the cursor (either by arrow keys or by a mouse) on a node and pressing the - key. This node is then connected automatically to the next node on which the - key is pressed. An on-line check is made to make sure that connections are allowable. EZSIM analyzes each node and its possible relations to other nodes in the network. It then determines what components are logically needed to be added to the model. It also detects logical errors at the early stages of model construction. For example, connecting a QUEUE node to a TERMINATE node results in an EZSIM prompt indicating that such a connection is meaningless and is not allowed (reasons for queue formation are indicated in Chapter 4).

Note that when the user specifies the source and destination nodes for connection, EZSIM automatically determines the routing for the connection path and performs the connection. This automatic procedure speeds the network fabrication process. In complex models, an area of the network may become congested with several nodes and connection lines, where the inputs and outputs to and from nodes may not be clearly identifiable. In this case, one may specifically choose to see the inputs and/or outputs of a selected node. The desired branches can be seen by placing the cursor on the node and hitting Alt I, which flashes all the input lines to the node in a different color. Hitting Alt O flashes all the output lines. Pressing the space bar key stops the flashing action.

The EZSIM network may be as large as the host computer memory allows. If the network is too large to fit in the screen it may be scrolled up, down, left, or right. The network navigation gauge on the upper left corner of the screen indicates the portion of the network that is not shown on the screen. The flashing arrows on the upper left corner of the screen indicate the position of those network portions with respect to the visible screen area. This capability is especially useful when making disjoint networks, as one may otherwise lose some of the disjoint network positions when reviewing or constructing a large model.

The following is a summary of features related to network construction in EZSIM:

- To see the node descriptions and brief graphics commands, hit H while in network construction mode.
- To see the brief node codes on the graphics screen, hit the F1 key.
- To place a node on the screen, place the cursor on the desired location and hit the letter code for the node. Locations too close to a current node or branch are not accepted. Use the arrow keys or a mouse for cursor movement.
- To connect two nodes, place the cursor on the origin node and hit the - key. Then move the cursor to the destination node and hit the - key again.
- To specify the parameters for a node, place the cursor on the node and hit the Enter key. To change the values entered, use the backspace or delete keys.
- To flash the input or output branches, place the cursor on the desired node and hit the Alt I or Alt O keys. Hit the space bar to unflash the branches.
- To delete a node, place the cursor on it and hit the F2 key.
- To delete a branch, place the cursor on the origin node and hit F3. Then move the cursor to the destination node and hit F3 again.
- To move a node with its connecting branches, place the cursor on it and hit the F4 key. Then move the cursor to the desired new location and hit F4 again. This places the node at the new location and reroutes the connected branches accordingly.

The network construction component of EZSIM submits a global representation of the system structure to a rule-based procedure that interactively asks for the complementary information needed to progressively construct a complete system representation and form an internally complete model.

3.7.1.3 NODAL PARAMETER SPECIFICATION AND DECLARATION. Generally, there are some parameters or other declarations that need to be specified for each node of the network. The user may choose any node in the model network by placing the cursor on the node and hitting the return key. This invokes a dedicated window for node parametric and declarative information identification. A series of relevant questions (accompanied with lists of potentially valid answers and context-sensitive help prompts) and menus of choices are generated at each level of inquiry related to the parameter identification of each selected node. For example, several standard probability distribution functions are provided in a menu that appears automatically whenever the need arises. Nodes may be selected in any desired sequence. Input information for a node may affect the conditions of other related nodes. In such a case EZSIM brings the relation to the user's attention and asks for clarification.

Each node that is placed on the screen is identified by its type code and its user-given name surrounded by a single-line rectangle. As one completely identifies this information for a given node, its surrounding rectangle becomes a double line. A question mark on the lower right corner of the screen appears as soon as the first node is placed on the screen. This is to indicate that there are some nodes for which the parametric information is not yet specified. This is useful when the network is larger than the screen, because it signals the existence of single-line rectangles in invisible parts of the network.

Identification of some of the parametric and declarative (variables and attributes) information in some nodes may not be possible, however, if the identification process for another related node is incomplete. This has been a deliberate design consideration in EZSIM to help the user avoid overlooking some important declarations needed to complete the model. EZSIM signals such situations and helps the user in creating a complete and well-rounded model. Any nodal identification process may be temporarily deferred to a later time if the Esc key is pressed during the process. This scheme allows the user to first concentrate on the global architecture of the model and then elaborate on details.

After a node selection, a dedicated window that guides the user in identifying each of the parameters relevant to the node in question is shown. For each item in the list, there is usually a help window, which is shown in the bottom box of the screen. Hitting the return key after entering each piece of information takes the cursor to the next question. In certain cases, the nature of the questions asked in each succeeding stage depends on the choice of response to the current question. In most instances the possible choices of answer are given in selection lists in which the desired answer can be selected using a scroll bar and hitting the return key; in many instances a default choice for the answer is given within <> characters. In these cases hitting the return key without entering anything results in the selection of the default value. Note that the backspace or delete keys may be used to change a previously given answer to the node parameter identification question.

After completing the answers to all of the questions regarding a node, control returns to the graphics screen for selection of the next node or further network construction and editing. If an attempt is made to exit the network mode (by pressing the Esc key) while the parameter identification or network construction processes are incomplete, a pop-up window on the bottom right corner of the screen warns the user about the situation. This window gives the choices of either returning to the network, saving the incomplete model for completion at some future time, or abandoning the model without saving it.

Hitting the Esc key after the completion of the nodal parameter identification stage starts the system initialization stage. This initializing stage occurs only if there are quantities (such as user variables and resources) that need to be specified. It should be pointed out that the model construction process can be aborted at any time by pressing Ctrl C, which causes a return to the Opening menu.

3.7.1.4 STATISTICS SPECIFICATION. After completing the nodal parameter specification stage, the statistics screen is displayed. Here the user may specify the desired statistics by any identifier name that is to appear in the simulation output. A selection menu allows the user to choose the desired statistics from the following list of options:

- Arrival time of first entity at a node
- All times of entity arrivals at a node
- Traversal time between two nodes
- Time between arrivals
- Entity count at a node
- Any user-defined variable
- An expression

Once a selection is made, the user is asked to identify where in the system the selection applies. For example, if the time to traverse is the option selected, the user is asked to identify the two points in the system between which the traverse time statistics is desired. In this case a listing of all nodes appears in a window from which the user may select the two nodes.

For some statistics, the identification of whether the statistics are to be considered observation-based or time-based is required. Several pages of statistics menus may be used for a model. Hitting the Esc key while in the statistics screen will bring up the Control menu.

It should be mentioned that EZSIM automatically collects and reports some frequently needed statistics. These include the statistics on all types of queues, facilities, and resources specified in the model.

The statistics in the output are identified under the associated node or resource names given at the modeling stage.

3.7.1.5 EZSIM CONTROL MENU/MODEL EXECUTION. After the above stages are complete, the EZSIM Control menu is displayed. The Control menu is the gateway to any of the preceding stages of model building and to the execution and output generation stages. Return to the model network or to the statistics menu for possible model modifications can be achieved by a keystroke. As stated before, the model may be executed in either batch mode or animation mode. If the batch mode is selected from the Control menu, the desired length of simulation in time units must be identified. The length of the transient (cold start) period after which the desired statistics are to be collected should also be given to EZSIM. The simulation process then starts and the current simulation time, as it progresses, is displayed. The user may momentarily stop the execution and observe the outputs at any time during the execution process. The execution may be resumed after the output is observed. Figure 3.13 shows the EZSIM Control menu.

If the selected execution mode is animation, the user is asked to identify the animated entities by his or her choice of symbols and their associated colors. The animation may be observed in either the continuous mode or the step mode. The control then returns to the model network and the motion of entities through the branches and nodes of the network model is displayed. Networks that are larger than the screen

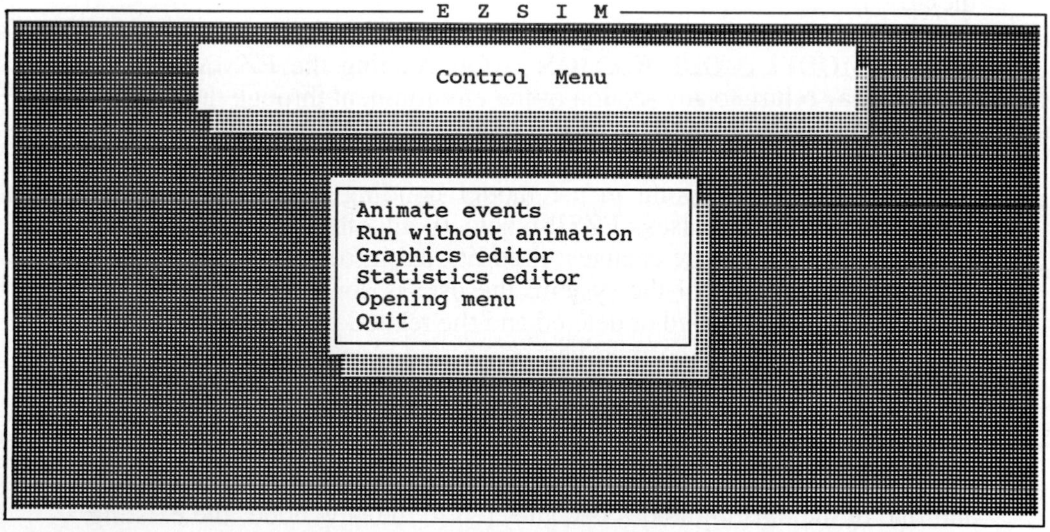

FIGURE 3.13
EZSIM Control menu.

may be scrolled in any desired direction. While the animation progresses, the current simulation time is shown on the upper right corner of the screen. The movement speed of entity symbols in the animated model can be changed by a keystroke.

The network animation capability of EZSIM provides a simple means for model verification. An animation can be set up and performed in a matter of seconds. The motion of entities as they go through various branches and nodes according to the modeling logic, the status of the nodes, and the changing values of resource levels and user variables can be observed during the course of animation. The animation process may be interrupted at any point to review the simulation output. The user may resume the animation process after the desired output is observed.

3.7.1.6 EZSIM OUTPUT MENU. The Output menu allows for the observation of the selected statistics and graphs on the screen, during or at the end of either the batch or the animation execution mode. The entire simulation output and the network model with its related information may be saved on the disk or sent to the printer by means of the related selections in the Output menu. Resumption of either batch execution (if interrupted before the end of the specified simulation time) or animation execution may be made through the output menu. Also, the batch execution time can be extended by the desired time length by selection of the associated option from the Output menu. (Note that extension of the simulation time applies only to the batch mode execution.) The Output menu also has a gateway to the Control menu, from which other parts of the system may be accessed. Figure 3.14 shows the EZSIM Output menu.

3.7.1.7 MODEL MODIFICATION. After running the EZSIM model, the user may return to any section of the environment through the Control menu. One of the major capabilities of EZSIM is its ease of modifying the model network and the related modeling information. Modifications may be made at any point in the model building process or after its completion. In such cases, EZSIM detects all other components that may be affected by these changes; if additional information is needed to maintain the integrity of the system, the user is consulted.

Nodes may be added or deleted and the related parametric or declarative information may be modified as desired. Every time a change is made to the structure of the network, the affected nodes will switch to a single-line rectangle representation, meaning that the parametric and/or declarative information regarding the nodes affected by the changes needs to be modified. For example, if a node is added to the network and one of the existing nodes is connected to the new node, the existing node may change to a single-line rectangle representation and a question mark will appear on the lower right of the screen. The new branching

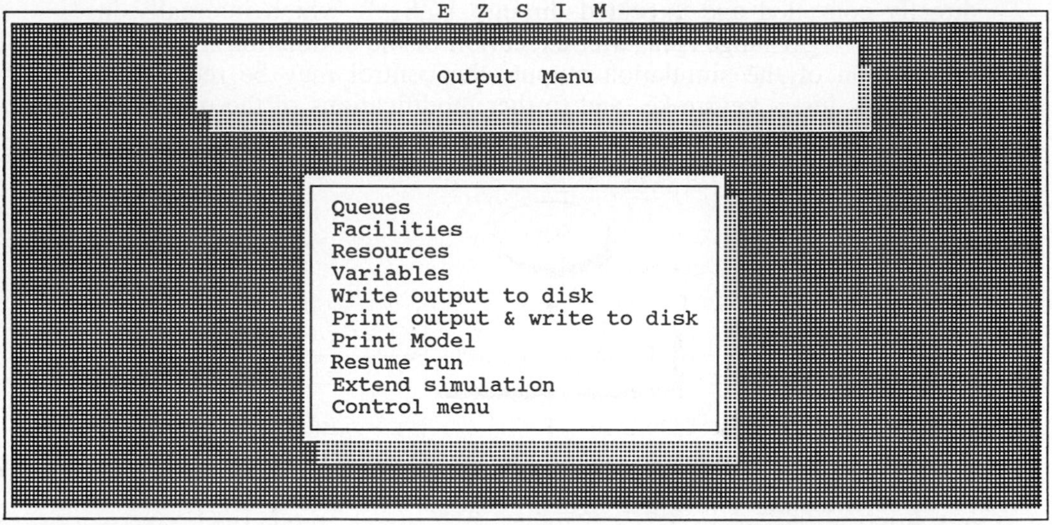

FIGURE 3.14
EZSIM Output menu.

information for the node affected by the change should be subsequently identified.

A realistic simulation analysis always requires several modifications to the original model in order to test various system configurations and the associated performances. This process may be very cumbersome using the conventional approach to simulation modeling. EZSIM makes this task as easy as possible and provides measures for automatic monitoring of the system structure. EZSIM attempts to prevent the user from making logical and structural mistakes as well as syntactic errors throughout the model construction and modification processes.

3.7.2 EZSIM Program Generator Mode

As mentioned before, most of the modeling features in the program generator mode of EZSIM are the same as the ones in the stand-alone mode. Since EZSIM, as a target language code generator, is restricted to the capabilities of its target languages, certain necessary conformations with the structure of these languages impose some requirements that make the system different than its stand-alone mode. Currently SLAM, SIMAN, and SIMNET simulation languages are supported. The source codes generated by EZSIM for the target languages are in ASCII format. Therefore, they can be modified if needed using standard editors or word processors. The source codes generated by EZSIM may be directly compiled and executed through EZSIM, which internally invokes the target language. After execution of the simulation model and observation of the simulation output, the control may be returned to EZSIM by just a keystroke, and further modifications of the model can be performed. Figure 3.15 shows the procedure flowchart of EZSIM as a

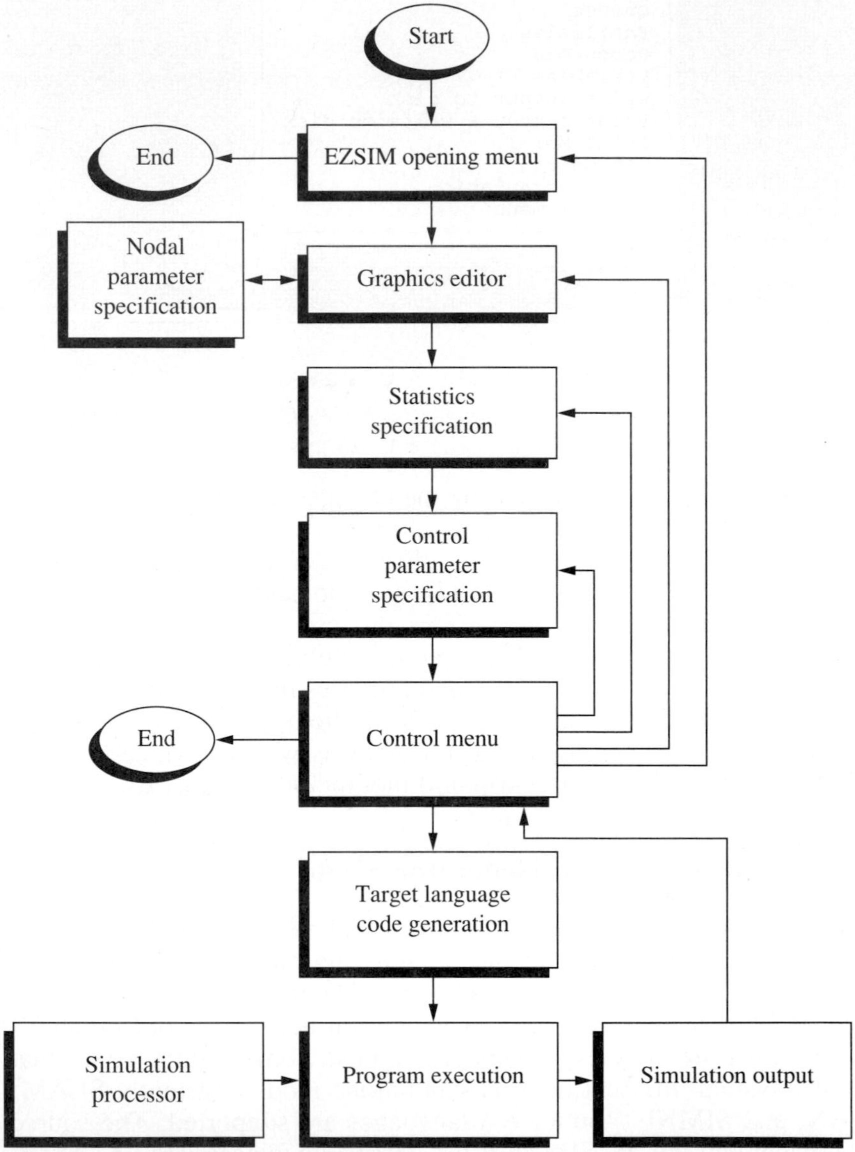

FIGURE 3.15
Flowchart of the procedure for using the program generator mode of the EZSIM environment.

code generator, and the following sections present the related components.

3.7.2.1 CONTROL PARAMETERS SPECIFICATION. Some global simulation information required by the target language is identified using the Control Parameters Specification menu. Some default values for many of the parameters in this menu are provided. After completion of this stage, the EZSIM Control menu will appear. This menu allows the user to return to either of the above stages or go on to view and/or run the automatically generated program of the model.

At the final stage, an interactive session prompts the user for general control parameter specifications (run length, project name, etc.). After this session, EZSIM creates the equivalent target language program and writes it on the disk in a format conforming to that required by the target language.

3.7.2.2 EXECUTION OF THE EZSIM GENERATED CODE. One of the options on the EZSIM Control menu is the listing of the target language source code for the model constructed using the EZSIM environment. The target language source code is readily executable if its processor is available under the active directory. To execute the code, the Run option on the EZSIM Control menu may be chosen. From this point on, the target language processor takes over and runs the simulation program. The user-defined names for nodes and requested statistics will appear in the target language output, which may be saved on the disk by the language processor. To review the output, the related choice from the Control menu may be selected.

Note that EZSIM resides in memory while the target language processor is loaded and executed. Any other memory-resident program which is active concurrently may disrupt the operation of EZSIM or the target language processor. Users must make sure that these applications are disabled while running EZSIM. For large models it is recommended to quit the EZSIM environment to free the memory that it occupies and run the generated target language source code in the same way that it is run without EZSIM. This will require a few simple commands to be entered at the DOS prompts. The target language operations manual should be consulted for this purpose.

The target language code may be modified outside of the EZSIM environment, if needed, using any editor that handles ASCII files. The EDLIN program in DOS and several word processors and programming language editors allow for the creation and editing of ASCII files. This form of editing may be needed to use certain capabilities of the target language that may not be incorporated in EZSIM. Attempt has been made in the design of EZSIM to minimize this requirement.

Note that EZSIM saves its version of model representation under the assigned file name with the extension name .EZ. The target language related files are given extension names that are in conformance with the language specifications. For example, the SLAM source code is given the extension name .DAT and the SLAM output is assumed to have the extension name .OUT.

3.7.3 Some Design Considerations

EZSIM is designed to run on IBM PC XT, AT, PS/2, and compatible systems. To run even on the lowest end hardware platforms, and to avoid the incompatibility problems of the numerous graphics adaptor cards and screens utilized under DOS, the current implementation of EZSIM employs only the text screen and IBM's text mode graphics primitives. This makes EZSIM portable and its network graphics very fast. Future versions of EZSIM will also be available under Microsoft Windows and Unix operating systems. Since most of the internal operations of EZSIM are based on symbolic manipulations, EZSIM was first prototyped and tested in the LISP environment and then was translated to C for increased speed and enhanced portability.

3.7.4 Demonstration of the EZSIM Application Procedure

In this section two simple examples are given to familiarize the reader with the EZSIM environment in the stand-alone mode. In the first example only three nodes are used (SOURCE, DELAY, and TERMINATE). The second example presents a queuing system simulation model and introduces two additional nodes (FACILITY and QUEUE). Following is a brief description (complete descriptions of node features will be provided in Chapter 5) of each of the nodes used in the examples:

The SOURCE node creates entities. Every entity that the SOURCE node creates may have a name, which becomes an attribute for the entity (called NAME). If no name is specified for the entity it will be shown as a ? in animation. It is possible to specify the total number of entities to be created by a SOURCE node.

The DELAY node is used for creating a delay that corresponds to the traversal time of the entity from one node to another. The traversal time is selected from the standard time window (described below) and may be represented by various statistical distributions, an attribute value, a user variable value, or an expression.

The TERMINATE node ends the path of entities. Each TERMINATE node can end the entire simulation if the entity count specified in it is reached.

The FACILITY node acts as a server. Entities remain in the node for the duration of their service. When the node is occupied, the arriving entities have to wait until the node is free. FACILITY node must be preceded by a QUEUE node. The entity leaving the facility node may be given a new name. This will allow for showing the unprocessed and processed parts, for example, with different symbols in the animation mode. Multiple parallel servers may be specified for a given facility.

The QUEUE node represents buffers before FACILITY nodes and is always succeeded by a FACILITY node. A QUEUE may have capacity limitation, various priority disciplines such as first-in-first-out (FIFO), last-in-first-out (LIFO), and so on.

Throughout EZSIM, various time elements such as time of first entity creation, time between creations, and delay times in DELAY and FACILITY nodes, have a standard window that allows for specification of time. This window offers several options. When the option of statistical distribution is specified, a window listing several distribution names allows for the selection of the desired distribution. After the distribution is selected, the user is prompted with messages asking the user to identify the values of each of the parameters of the distribution and the choice of the random number seed (selection of the default seed is recommended until the related concepts given in Chapter 6 are studied).

Example 3.1. Suppose that cars arrive at the entrance of a freeway segment with an exponentially distributed interarrival time with a mean of 3 minutes. The travel time through the freeway segment is distributed normally with a mean of 15 minutes and a standard deviation of 2 minutes. We are interested in observing, using animation, the changes in the total number of cars on the freeway during a selected time period.

The network model for this example is shown in Fig. 3.16. To create this network start EZSIM, select the Build/Retrieve option of the Opening menu and give a name, say EX1, to your model. In the graphics screen hit S to place a SOURCE node at the cursor position. You will be asked to provide a name for the node. The maximum number of characters

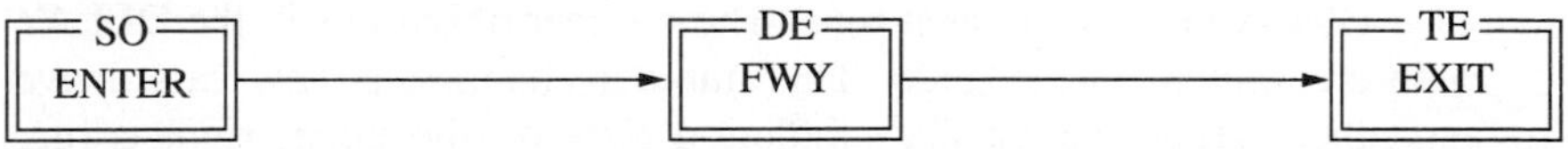

FIGURE 3.16
EZSIM network model of the freeway example.

for the node name is 6. Your given name, say ENTER, will appear inside the node. Move the cursor to the right, allowing sufficient distance from the source node, and hit D to place a DELAY node at the cursor position. In the same manner place a TERMINATE node in the proper location. Now place the cursor in the SOURCE node and hit -. Place the cursor in the DELAY node and hit - again. The two nodes will be connected by a line. In the same manner connect the DELAY node to the TERMINATE node. Hit H for help at any stage of network construction to see the operation codes that you may need (undoing the last action, moving a node, etc.).

To complete your simulation model you should identify the parametric information for each node. You may select the nodes at any sequence by placing the cursor on them and hitting the return key to observe the associated parameters window. You may start with the SOURCE node in this example. When you select this node a window appears. The first question asked in this window is the entity name. You may give the name CAR to the entities generated at the node. Set the first creation time at zero by hitting the zero and then the return key at the time window prompt. To identify the intercreation time select the Statistical Distribution option of the time window and select the exponential distribution. Enter 3 for the mean and select the default random number seed. Hit the return key at all other question prompts to select the default values. In the same manner, select the normal distribution for time at the DELAY node, and enter the parameters of 15 and 2. Choose the default option for termination count at the TERMINATE node. At this point the question mark at the lower right corner of the screen disappears, meaning that the model is complete. Hitting the Esc key takes you to the Statistics menu. To see the function of this menu, choose a statistics name, say TRIP. This statistic is to denote the car trip time from the entrance to the exit points of the freeway. The statistics choice on the menu that pops up is, therefore, "Traversal time between two nodes." EZSIM automatically assigns the observation-based type statistics to your choice, and it then asks you to point at the two node names between which you want the traversal time statistics. After selecting these two nodes select the No default for the question regarding graph, and hit the Esc key to go to the Control menu. At the Control menu choose the Animate option, which prompts you to select a symbol for the entity called CAR. You may choose any character by hitting the related key. You are then given the choices of foreground and background colors. Simply move the arrow to the desired choice and hit the return key. You may now run your model in continuous animation mode by hitting the C key.

Observe the model as it runs. The number of entities in the DELAY node are shown on the node. This number changes as new cars arrive and others leave. Watch the simulation time on the upper right corner of the screen as it advances from one event time to the next. You may speed the animation up or down using the PgUp and PgDn keys. Hitting the arrow keys moves the entire network and the animated entities. If you hit the Esc key during animation you will see the Output menu. Select the Variables option on this menu to see the statistics called TRIP, which you specified earlier. To go back to animation select the Resume option of the Output menu.

Select the Control menu option of the Output menu, and go from there to graphics to modify your model. Select the SOURCE node and modify the number of entity creations from infinity to a limited number, say twenty. Animate the new model and observe its operation. Experiment with giving a limited number for the termination count as well. After completing this example, you will be prepared to use EZSIM to experiment with larger models which may involve various types of processes.

Example 3.2. In this example a queuing system similar to the earlier examples given in this chapter is studied. Assume that customers arrive at a facility for service. The customer interarrival time is distributed uniformly between 10 and 20 minutes. The customer service time is distributed uniformly between 8 and 15 minutes. Assume that the buffer before the facility can accommodate a practically unlimited number of people. Let's assume that we are interested in simulating the system for 500 arriving entities or 10,000 minutes of operation, whichever happens first, to find the queue and facility statistics.

Figure 3.17 shows the network diagram of the system. Hitting Q and F keys places QUEUE and FACILITY nodes, respectively, on the graphics screen. The parameters for the SOURCE and TERMINATE nodes are selected in the same manner as in the previous example. At the SOURCE node the maximum number of creations is set at 500. The name CUST may be given to the entity at the SOURCE node. At the parameter specification stage for the QUEUE node, default values should be selected for all fields in the related window. For the FACILITY node, one server with the specified service duration should be identified. Default values for other options should be selected. A new name, say SCUST (to represent serviced customer), may be given to the entity leaving the FACILITY node. To do so, a Y response should be given to the question "Would you like to change the entity name?" in the FACILITY window. The name SCUST may then be entered.

After the network construction stage, the Statistics menu appears. Since in this example we are only interested in the statistics on the queue

FIGURE 3.17
Demonstration of queue and facility nodes.

and the facility, which are automatically collected and reported by EZSIM, the Statistics menu may be exited by hitting the Esc key. In the Output menu the model may be observed in the animation mode. Specify entity symbols C and S for unserved and served customers, respectively. Note that in the course of animation numbers in the lower side of the QUEUE and FACILITY nodes represent the current number of entities in the corresponding node. The model may be run in the batch mode as well. The simulation length of 10,000 minutes may be specified after selecting the "Run in batch mode" option of the Control menu. A transient period length may be specified after which EZSIM starts collecting the statistics. The default value of zero may be selected for the transient period at this stage.

Figure 3.18 shows the simulation output for this model. Notice that the simulation in this case has ended prior to the 10,000 minute time limit. This is because all 500 entities entered the system prior to this time. The output reflects various statistics such as minimum, maximum, and the last number of entities observed in the queue, mean and standard deviation (STD) values for queue length, waiting time in queue and server utilization, and mean uninterrupted busy and idle times for the facility. To create this output, the Print Output option is chosen in the Output menu after the execution is completed. Selecting the Print Model option in the Output menu results in printing the network and the list of information regarding each node in the network model.

```
                    *** E Z S I M STATISTICAL REPORT ***

Simulation Project: EXAMPLE 2
Analyst: BK
Date: 15/1/92
Disk file name: CH3EX2.OUT

Current Time: 7553.900  Transient Period: 0.000

Q U E U E S:
NAME       MIN/MAX/LAST             MEAN     STD      MEAN    STD
           LENGTH                   LENGTH   LENGTH   DELAY   DELAY
-----------------------------------------------------------------------
LINE        0/ 1/ 0                 0.03     0.17     0.47    1.12


F A C I L I T I E S:
NAME       NBR    MIN/MAX/LAST      MEAN     STD      MEAN    MEAN
           SRVRS  UTILIZATION       UTLZ     UTLZ     IDLE    BUSY
-----------------------------------------------------------------------
SERVER     1      0/ 1/ 0           0.77     0.42     3.47    11.64
```

FIGURE 3.18
Output of the queuing system simulation.

3.8 THE SIMULATION PROCESS

The stages involved in simulation studies have been partly shown in the illustrative examples given in this chapter. The flowchart in Figure 3.19 shows the tasks involved in a complete simulation process and the related sequences. Generally, the following stages may be identified as the main tasks of a simulation analyst to conduct a complete and successful simulation study:

System realization. Definition of the purpose of the study and subsequent abstraction of the reality leading to system identification (identification of the system boundary and components and the nature of the relationships among the components). This abstraction process is largely an art.

Data acquisition. Identification, collection, and proper representation of the data, describing outside influences and behavior of some internal processes.

Model construction. The abstraction of the system into a representative model that to the best extent possible and desirable mimics the system behavior. Modeling tools such as simulation programming environments may be used at this stage; therefore, the simulation software selection task may be included in the model construction phase.

Verification. The process of establishing that the computer implementation of the model is error-free.

Validation. The process of establishing that the model and the data correctly represent the important aspects of the system (an error-free computer program does not always represent a valid model).

Experimentation. The process of devising relevant and efficient experimental conditions under which the model behavior is examined. This stage may require parametric changes as well as structural changes to the model; that is, model reconstruction may be performed at this stage.

Analysis. The process of making sense out of the simulation output data for each of the experimental scenarios, and comparison of scenarios.

Documentation. The process of describing the problem and the methodology used to address it with minimal use of technical simulation jargon, and the translation and summary of the results of the simulation output analysis into a useful information format that relieves the end user of the task of digging the facts out of computer printouts. The documentation process may include recommendations for the implementation stage.

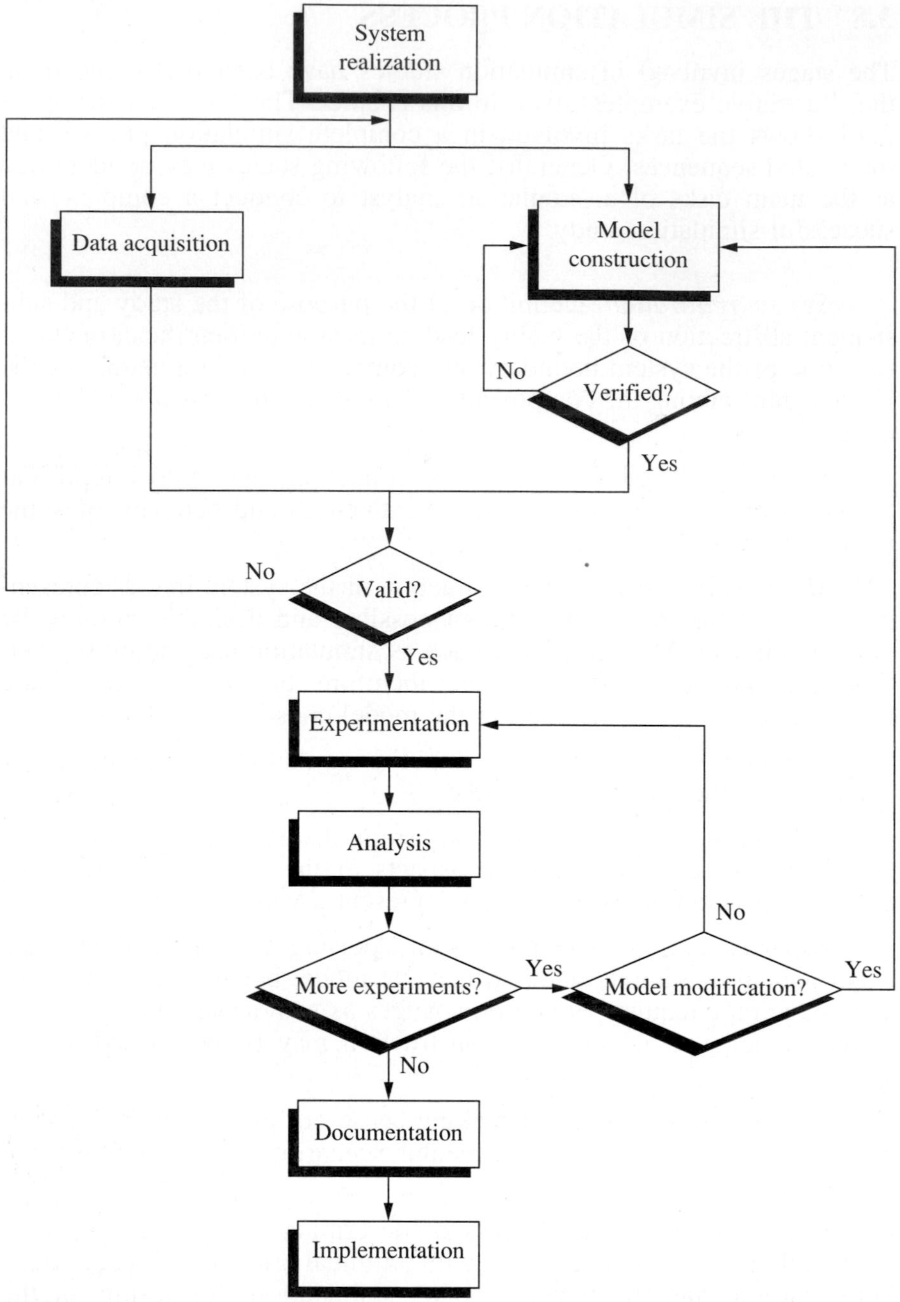

FIGURE 3.19
The simulation process.

Implementation. The process of making decisions that lead to changes in an existing system or to the construction of a new system on the basis of the simulation study. This stage may or may not involve the simulation analyst.

3.9 SUMMARY

A numerical example of simulation using the manual approach followed by a simplified queuing system and the corresponding computer simulation program were presented in this chapter. The major components of an event-oriented simulation program were described, and the available simulation software tools were categorized and an introduction was provided to EZSIM which demonstrates the simplicity of using a process-oriented simulation tool as opposed to writing event-oriented programs in procedure-oriented computer languages. Although the details of simulation modeling for various application examples have not yet been given, it is hoped that the materials presented thus far provide a basic and broad understanding of the activities and tools involved in the dynamic field of computer simulation.

The remaining chapters of the book will elaborate on several stages of the simulation process using the terms and concepts presented here. Other features of EZSIM will be introduced, and the software will be implemented as a teaching tool on a number of selected examples from various domains of application to prepare the reader for actual use of simulation in various realistic systems analysis and design projects.

3.10 EXERCISES

3.1. Extend the manual simulation of the freight company example to 20 days. Compare your results with the ones obtained in the 10-day simulation. You may use the given data for the first 10 days of operation.

3.2. In the freight company example assume that planes arrive every half day. Using the same probability for service requirement and for service duration, simulate the system for 10 days of operation and various number of facilities (1, 2, 3, 4, etc.) to find the lowest cost alternative.

3.3. Customers arrive at a bank with a time between arrivals that may be either 1, 2, 3, 4, 5, or 6 minutes at equal chances of occurrence. The service time by the bank teller may be either 2 or 2.5 minutes. Using a die and a coin, generate customer arrival and service times, and simulate the system for 20 minutes of its operation. Plot the number in the system for each minute of system operation, and find the average customer waiting time and the teller utilization (percentage of time the teller is busy).

3.4. An operator should complete four remaining jobs to end his work day. Each job takes either 5 or 7 minutes to complete with equal chances of occurrence. A completed job may be either accepted or rejected at equal chances. If a job is rejected it is sent to the operator for rework. Each rework takes 3 minutes, and the reworked job is always accepted. Simulate this system once, using a coin for generating the operation times, to find the remaining time of the operator at his work place. Repeat the simulation process four additional times. Using the results of your five simulations, find the expected (average) value for the operator's remaining time.

3.5. A device has two major components that operate in parallel. The device fails if both components fail. The probability of each component failing after each hour of operation is 0.5. This probability is independent of the total number of hours of component operation. Using a coin, simulate the system operation to find the time of device failure. Repeat the process 10 times to find the average operating time before device failure.

3.6. In a battle three tanks A, B, and C shoot at one another. After each shot it takes 7 seconds to reload the tank gun. In each round of shooting a tank chooses to aim at one of the other two tanks with an equal probability. The chance of a hit is 1/6. Assume each tank has an unlimited source of ammunition. Simulate the combat five times using a coin and a die to find the duration of the battle. Assume that during the first 7 seconds of the combat A shoots first, then B shoots (if not hit by A), and then C shoots (if not hit by A or B).

3.7. Modify the relations for generating interarrival and service times in the single-server queue simulation program to generate the random patterns specified in the freight company example. For various lengths of simulation, compare your program results with the manual simulation results.

Hint: To generate discrete random outcomes you may first generate a random number and, through a series of if-then statements, select an outcome based on the value of the random number (e.g., if $R < 0.5$, a repair is needed; if R< 1/6, the repair time is 0.5 days). Compare your computer simulation results with those of the manual method.

3.8. Modify the single-server queue simulation program to incorporate each of the following cases (apply each modification to the original program):

(*a*) The queue capacity is limited to 10. If an arriving entity finds the queue to be full to capacity it leaves the system (balks). You are to determine the total number of entities that balk, and the average time between balk events.

(*b*) There are two parallel servers instead of one. Entities use the first server if both servers are available. Determine the average server utilization.

(*c*) There are two queues before a single server. Arriving entities choose the shortest queue. The server selects entities from the longest queue to serve. When both queues have the same length the first queue is chosen by arriving entities and the server. The statistics of each queue is to be found.

(*d*) Entities waiting in the queue leave the system if their waiting period exceeds 3 time units.

3.9. Compare the results of the original simulation program (as given in the appendix) for the single server queuing system with the output of an equivalent EZSIM model. Try run lengths of 100, 1000, 5000, and 10,000 time units. Assume uniformly distributed interarrival times between 5 and 10 minutes and uniformly distributed service times between 3 and 9 minutes.

3.10. Compare the results of the program that you wrote for part (b) of Exercise 3.8 with the output results of an equivalent EZSIM model. Assume that interarrival times are uniformly distributed between 8 and 12 minutes and each server's service time is distributed uniformly between 10 and 14 minutes. Choose a simulation time of 500 time units.

3.11. Run the single-server queuing program with the original parameter values for run lengths of 20 to 100 with increments of 10; 100 to 1000 with increments of 100; and 1000 to 10,000 with increments of 1000. Plot the averages of waiting times, queue lengths, and server utilization values that you obtain from each run on three different graphs. Describe your assessment of the results.

3.12. Modify the single-server queuing program given in Appendix A such that the minimum, maximum, and standard deviation values for the three statistics (waiting time in queue, length of queue, and server utilization) are also calculated and printed by the program.

CHAPTER 4

COMMON PROCESSES IN DISCRETE SYSTEMS

4.1 INTRODUCTION

In Chapter 2 some of the common features of discrete systems were described. In this chapter we extend those discussions to provide a comprehensive view of the common processes and properties of general concern in the study of discrete systems. The discussions in this chapter will be general in nature. Therefore, no specific simulation language will be addressed in connection with the topics presented. The advantage of this approach is that it provides the reader with a generic knowledge of the nature of behavior of discrete systems and the desirable capabilities of a modeling tool that an analyst may use for describing such behaviors in a simulation model.

A general knowledge of the common processes in systems greatly enhances the analyst's ability in the tasks of system identification (identification of components, relationships, and boundary) and of system abstraction in the form of a simulation model. Moreover, possessing the generic knowledge presented in this chapter, one should be able to understand the major features of all of the available simulation languages with minimal effort. The presentation attempts to provide various aspects of each common process, which may or may not be incorporated into a selected simulation modeling tool. This knowledge should also serve as a criterion for comparison of the strengths and weaknesses of various simulation modeling tools for various application scenarios.

Before proceeding with the rest of this chapter, it is recommended that the reader review the section on the common features of discrete systems presented in Chapter 2.

4.2 A VIEW OF COMMON PROCESSES IN DISCRETE SYSTEMS

Process-oriented simulation modeling tools attempt to provide prewritten and precompiled program modules that represent commonly occurring phenomena (processes) that take place in various realistic scenarios, thereby relieving the model builders of the tedious task of frequently reconstructing these program segments for use in their various simulation projects. The task of simulation programming then reduces to appropriately selecting and logically linking these ready-made modules, and specifying the related parametric values.

What may be viewed as a common process is very subjective and specific to the application domain. Hence, not all general purpose process-oriented simulation languages present the same set of modules as a unique repertoire of common processes. Nevertheless, there is a relatively large overlap in the offerings of all of these languages that includes a number of nearly similar processes. The basis for this similarity is the movement/accumulation concept of the physical commodity, which is universal and applies to all application domains in which entity motions are present. Moreover, certain occurrences are found to take place frequently in the domains in which simulation analysis is often used (manufacturing, office operations, traffic flow, transportation, construction management, etc.). Therefore, various languages attempt to include some capabilities that apply to the specifics of these application domains. The process-oriented simulation languages may differ in their syntactic structure (such as names used to represent each common process, assumed parameters and their default values, and the format of their program statements), and in the capabilities of the modules representing the real-world processes considered in the language.

The organization of the following sections will be based on the entity (commodity) movement pattern through various processes and selection of alternative routes, and on the conditions for entity accumulation in queues. A commonly used form of representation of entity flow by process-oriented languages is the network representation. A network is a collection of nodes that are interconnected by branches. In simulation models, a node may represent a common process that some entities go through.

4.2.1 Entity Movement

As explained in Chapter 2, dynamism in the discrete systems is caused by the movement of entities which results in the occurrence of events that change the system state over time. In most systems the entities enter the system as inputs through the system boundary, move between the components within the system, and perhaps leave the system boundary in the form of output from the system. In a few other cases the system may initially contain some entities that move through the system, circulate between the components of the system, and perhaps leave (in whole or in part) the system boundary. Alternatively, some entities may never leave the system boundary (e.g., pallets in a warehouse, loaders and other moving construction equipment in a construction site).

4.2.1.1 ENTITY CREATION. To create entities as inputs to the system (like customers arriving at a bank, or cars arriving at an intersection) a common process acting as an entity source should be considered. The module representing this process is called source, create, generate, and so on by different simulation languages. The desirable properties of this module are as follows.

The module should allow for setting the *time of creation of the first entity.* This value may be a constant or a random variable.

The module should provide for the selection of the *time between entity arrivals,* which may be a constant or a random variable. The module should provide a random variate generator that provides samples from several theoretical probability distribution functions. The user should be able to easily select the desired choice by merely specifying the name of the distribution and the values of its parameters. Provisions should also be provided for customized probability distribution functions.

The module should allow for the *assignment of a name* to the entities created. This name will be assigned to the entity as an attribute. This information may be used in later stages in the determination of entity routing through various branches on the basis of the entity type and for ease of specialized treatment of different entities that enter the same segment of the system.

It would be desirable for the module to allow for the *batch creation* of entities, that is, the creation of more than one entity at the end of each intercreation time interval.

The module should provide for *stopping the creation of entities after a certain number* have been created. Among other advantages, this capability provides control of the simulation process length by means of the number of entities rather than by time.

The module should allow for stopping the creation of entities *after a certain amount of time* has elapsed.

As mentioned earlier, entities may be initially contained in a system rather than enter the system from the outside. Usually in such cases the entities are initially located at the accumulation points. Modules that correspond to the accumulation points (queues) should provide for initialization of the desired number of entities in each of these system components. They should also allow for setting of the attributes of such entities as the user initializes them in the system.

4.2.1.2 ENTITY TERMINATION. In most systems entities that enter the system also leave it. The module that represents this entity departure or disappearance is called terminate, sink, depart, out, and so on by various simulation modeling languages. Although the entities that leave the system are of no further concern to the analyst, the entity termination process is essential for clearing these entities and their attribute information from the computer memory. Otherwise a memory overflow may be experienced during the course of the simulation. A desirable property of the module representing the termination process is the ability to specify the number of entity terminations required to end the simulation process.

The choice of controlling the entity creations in the source module by number or by time, and the number of entity terminations needed at the termination module to end the simulation, provides the analyst with the option of ending the simulation when the system is in either the empty or nonempty state. For example, if a particular simulation model uses one source module and one termination module, then by specifying a certain maximum number of creations at the source, say N entities, and not specifying a termination count limit at the termination point, the source module stops creating entities when N entities have been created, but the simulation continues until all of the entities clear the system through the termination module. In this case the simulation ends when the system is in an empty state. Likewise, if the source module is to stop its entity creations at a certain time, and no number is specified in the termination process, then the simulation continues until all entities clear the system. This feature is helpful in simulating scenarios such as a bank that closes its doors to arriving customers at a certain time but continues to serve the customers that are already in the bank. The bank operation then ends when all customers in the bank have been served.

Alternatively, if the number of entities specified in the source module is greater than that specified in the termination module, then at the end of the simulation there will be some entities left in the system.

4.2.1.3 ENTITY TRAVERSAL. Entities traverse between system components through branches that connect system components. The entity motion through branches may take time, in which case the entity will be delayed in arriving at its destination point. Following is a list of desirable features for the module representing entity traversal:

A desirable property of the entity traversal module is the ability to specify the *delay time* as a constant, a random variable, a user variable, or an expression combining several of these elements. Also, since various entities may traverse through the same branch at different speeds (which may be represented by one of the entity attributes), the traversal module should allow for making the delay time a function of an entity attribute.

Another desirable capability of the module is the ability to *signal the end of the traversal time* with the occurrence of an event. This should allow for those types of entity traversals whose completion may not be possible to schedule in advance.

In many practical situations branches through which the entities traverse have "practically" unlimited *capacity,* that is, an unlimited number of entities can traverse through the same branch simultaneously. Examples of this are customers traversing between the bank entrance and the end of the waiting line and airplanes flying between two cities. In certain other situations the branch capacity may be limited. Examples of this are cars moving on a narrow street and material-handling equipment moving on a factory floor. A capability within the traversal module itself or another means of controlling the capacity of branches should allow for these provisions.

Since at times entities need *carriers* to transport them from their sources to their destinations (such as people getting into cars and buses, trucks carrying construction materials, and parts being moved by the material-handling equipment), it is desirable to provide the modeler with the ability to assign carriers to entities and their routes. More information about entity carriers will be given in succeeding sections.

It should be noted that the possible material-handling scenarios could be very complicated and dependent on technology. This is especially true in the manufacturing domain. In this domain conveyors may be used on which all materials move at the same synchronized speed. Overhead cranes may be used that have acceleration, deceleration, movement of the main crane, trolley motion, and the vertical raising and lowering of picking devices. Automatically guided vehicle (AGV) devices may be used, which may have allocated or random routes with routes crossing one another. In the cases of multiple cranes and AGVs, the capacity of the branch sets limits on the number and speed of the entities traversing through it. Also in these situations the speeds of the entities ahead limit the speeds of the entities behind. Because of these complications, specialized simulation tools that are dedicated to the manufacturing domain are generally easier to use than the general purpose simulation modeling tools, some of which may provide no provisions for modeling these complex material-handling systems.

4.2.1.4 ROUTE (BRANCH) SELECTION. In many practical situations entities have alternative routes that they may take. Examples are cars arriving at an intersection, arriving customers choosing different waiting lines to join, and the customer at the beginning of a waiting line choosing different servers if more than one server becomes available. The desirable properties of the module representing the selection process are as follows:

The module should allow for the *probabilistic selection* of branches. That is, it should be possible to assign each alternative branch a probability of selection by the arriving entity. An application example of this type of selection is the case of the parts leaving an inspection station at a factory. A certain percentage of the parts may proceed to the next operation, a certain percentage may be sent for rework, and the rest of the parts may be scrapped. The percentage values here indicate the probabilities associated with each alternative branch that an inspected part may take.

The module should allow for the *conditional selection* of alternative branches; that is, it should be possible to assign a condition (a rule) to each alternative branch. A branch is then taken by the arriving entity if the associated condition is met. The conditions may be based on the values of the global variables, the entity attribute values, or a complex expression. An example for the global variable case is the routing of arriving cars in one direction prior to a certain time of day and then changing directions after that hour. An example of the attribute value case is the routing of the customers to different servers depending on the type of service required (the service type required being represented by an entity attribute). An example of the expression case is the selection of an alternative branch if the length of every possible queue to join is larger than a certain maximum limit.

Since *selection from different queues* (waiting lines) is a commonly occurring process, it is desirable to have provisions for some specialized conditional branching on the basis of the status of the alternative queues that an arriving entity may join (shortest length, largest remaining capacity, cyclic or random selection, etc.). Another common process is the selection of which queue to serve if the server has the option of

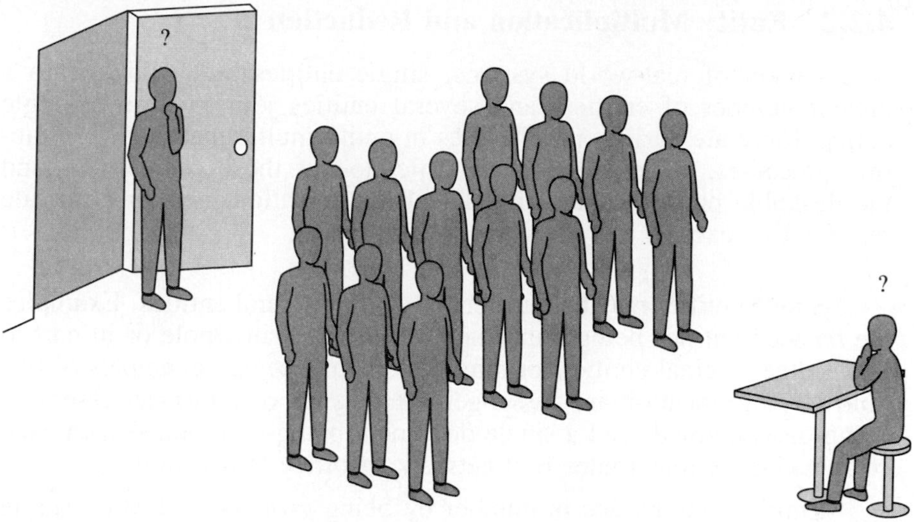

FIGURE 4.1
Selection of a queue to join and selection of a queue to serve.

serving more than one queue. Certain criteria such as selecting from the longest queue, selection of one entity from each queue (i.e., in assembly operations), cyclic selection, or random selection are typical. Figure 4.1 shows a scenario in which both of these queue selection types apply.

Another common process is the *selection of a server* from among multiple servers with varying characteristics (see Fig. 4.2). Possible criteria here may be the selection of the server with the longest idle time, cyclic selection, or random selection.

FIGURE 4.2
Selection from several servers.

4.2.2 Entity Multiplication and Reduction

In a variety of real-world systems, single entities break down into a greater number of entities, and several entities join to form a single entity. There are various possibilities in entity multiplication and reduction processes. Following is a classification of these possibilities and the desirable properties associated with a simulation modeling module representing each:

Single entities may be multiplied to form several entities. Examples are batched entities being unbatched (ungrouped) in whole or in part to release the original entities forming the batch. Specific examples of this kind of multiplication are passengers getting off of a bus, an assembly being disassembled, and a single document being reproduced with each copy taking various routes in a network of office transactions.

Entities may reduce in number by being *grouped* to form a single entity that may not be broken down at a later stage into the original entities forming the group. Here, the original entities lose their identities (attributes) and the grouped entity may take on the attributes of one of the original entities or a combination of the attributes of all of the original entities. For example, five bulldozer loads of dirt may make one truckload of dirt in which the individual bulldozer loads are not separable (and do not need to be).

Entities may be grouped to form a batch that will be *unbatched* at a later stage. The group will move as a single entity with assignable attributes. The original entities, however, preserve their identities and may be ungrouped later, each taking a different route in the system. Examples are passenger entities getting into a bus and parts being placed in a bin for transportation or storage. As the above examples demonstrate, carriers may sometimes be needed to move the grouped entities. There may be various loading conditions for carriers (they may be filled to capacity, take whatever is available, wait for a certain time to depart, etc.), carriers may have various departure priorities, and they may have different characteristics (capacities, routes, speeds, etc.).

Usually there are certain *grouping conditions* that apply to either of the above grouping scenarios. These conditions are generally based on either the number of original entities or some combination of values of a certain attribute of these entities. For example, a tour bus may leave whenever there are forty passengers available, or a truck may leave whenever the total weight of the available load reaches its remaining capacity. In the former case the total number of the original entities is the condition for grouping; in the latter case the total of individual weights, which may be stored in a certain attribute of the original entities, sets the grouping condition.

Finally, entities may be *assembled* to form a single entity. Unlike the above grouping cases, in which various entities coming from various routes may satisfy the grouping condition, specific entities each arriving from a certain route are usually needed to complete an assembly operation. Examples are components assembled to form a subassembly and specific documents needed to form a package to be mailed.

4.2.3 Entity Use of Resources

As explained in Chapter 2, resources are limited commodities that entities may use for certain durations of time. Generally, entities may physically move to certain fixed locations at which some resources reside in order to use the resources. They may also receive certain other kinds of resources that are made available to them wherever the entity happens to be. The latter types of resources may be carried by the entity, used, or consumed in whole or in part, in which case a portion of the resource may be released to the base stock of the resource at some later stage. Because of the differences between the two types of resources, many simulation languages prefer to handle them differently. We will refer to the first type of resource with the special name of *facility* and will call the second type *resource*. Following are the descriptions of common properties of each resource type:

Facilities usually represent servers, operators, workstations, or machines with which certain *service times* are associated. The duration of service could have properties similar to that of the entity traversal time described earlier (fixed, random variable, etc.). Facilities may have several homogeneous servers working in parallel. A common property of a facility is its possible breakdown or idle time allowance, which may be random or scheduled. Also as mentioned before, certain selection criteria may apply when selecting from among several nonhomogeneous facilities or when selecting an entity to serve from multiple parallel queues before a facility that just became idle. Finally, entities may concurrently need other types of resources while receiving service at the facility (such as tools or machinists for a machining operation).

Resources generally have a *base stock* that initially contains a known number of resource units. Entities can request a certain number of resource units. There may be a delay involved in receiving the resource. The resources seized by the entity may be kept while the entity goes through various processes (e.g., delay process). They may be returned to base stock after use, in which case there may be a return time involved. Entities may have different priorities in accessing the resources. At times a higher priority entity may preempt a lower priority entity that is in possession of the needed resource. The time during which an entity uses a resource generally need not be explicitly identified since the entities possessing the resources can go through several time-consuming activities of the system (such as being delayed in traversal activities and spending time in facilities).

Figure 4.3 demonstrates two scenarios for modeling a facility with two parallel servers. In the first modeling scenario a facility is used for which two parallel servers are defined. The arriving entities

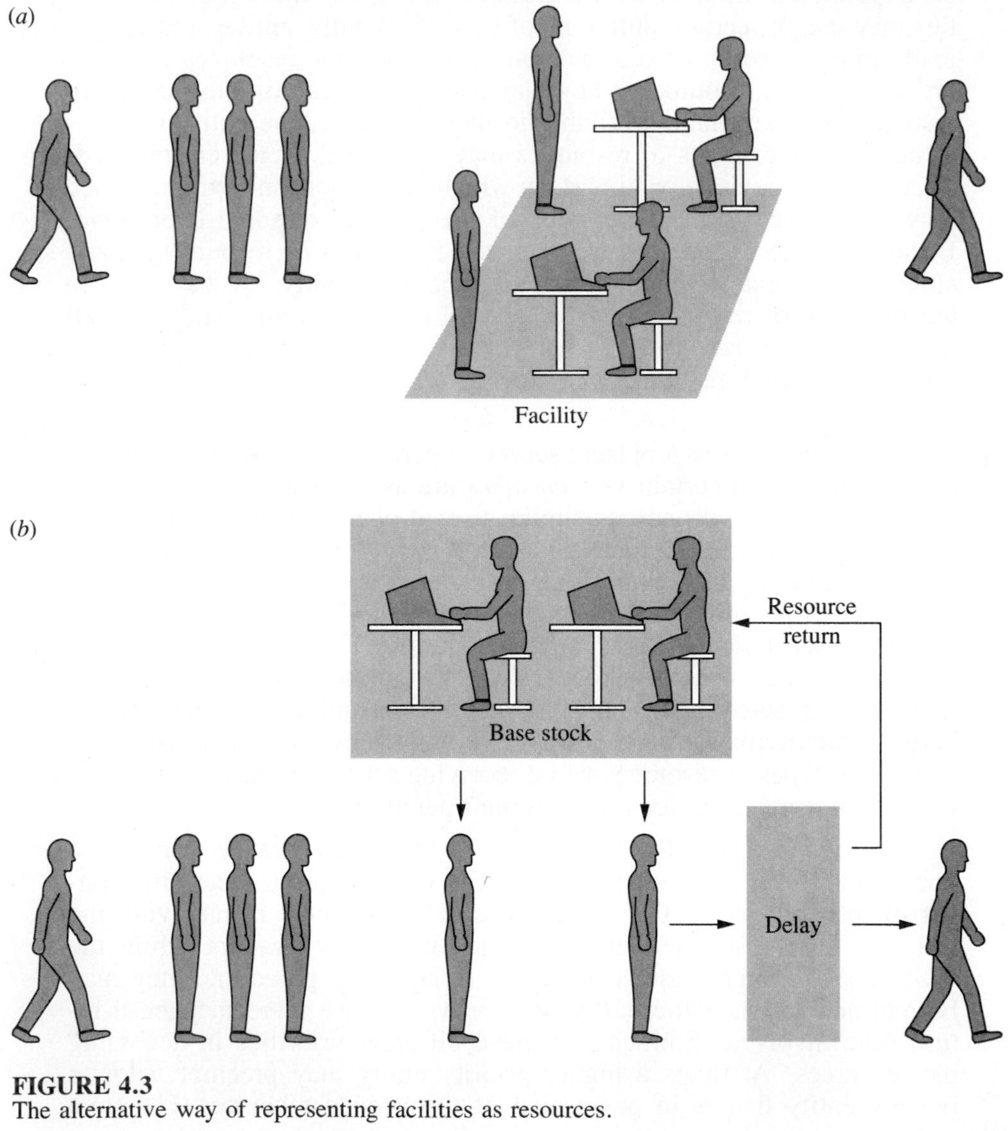

FIGURE 4.3
The alternative way of representing facilities as resources.

receive service at the facility by one of the servers. The entity is delayed at the facility for the duration of service. In the second modeling scenario a base stock of a resource is defined. The base stock is initialized to contain two units of the resource. Arriving entities await a resource unit once they enter the system. After seizing the resource, the entity and its seized resource are delayed for the duration of service. After this delay period the resource is released to its base stock.

An important remark should be made here regarding the modeler's choice of identifying a system component as either an entity or a resource. Generally speaking, when a system component is represented as an entity the modeler has the choice of identifying specific routes for entity movement. This possibility does not exist for resources that are retrieved from their base stock, used, and returned to the base stock. The routes for resource movement are unidentifiable unless an entity is specified to carry the resource through some specific routes. As an example, consider the possibility of representing a piece of material-handling equipment as a resource. In such a case the idle carrier may be assumed to be at some unidentified location (base stock) for which the travel time to the requesting entity's location may be specified with a given delay, but the exact route for the carrier traversal may not be identified. If, on the other hand, the carrier is defined as an entity, one may specify its exact routing to reach the entity requesting the carrier. Material-handling equipment in some simulation languages is treated as a resource. Depending on the specific scenarios and the information desired from the simulation study, this may be considered a limitation! Another distinction between entities and resources is that each individual entity may be traced through the system, but resources of the same type cannot be characterized individually (resource units may not assume attributes!). This leads to the fact that the individual resource units in the base stock are indistinguishable. That is, one may not know, for example, which specific resource unit arrived last to the base stock.

Unlike the resource base stocks that hold a single number representing the quantity of available resources, entity accumulation points (queues) are more complex in structure since they must keep track of the information about each individual entity (entity attributes and entity order in the queue) contained in them. An advantage of using resources instead of entities (when possible) is that resources demand less computer memory.

Let us consider an example that clarifies the above point. Suppose that in a manufacturing operation components A and B are assembled together with screws. One alternative is to represent components and screws as entities, as shown in Figure 4.4*a*. However, knowing that screws are shop-floor usage (SFU) materials and balks of them are usually made available at assembly stations, and that they are not sent to the station individually via a particular route, it is more efficient to define

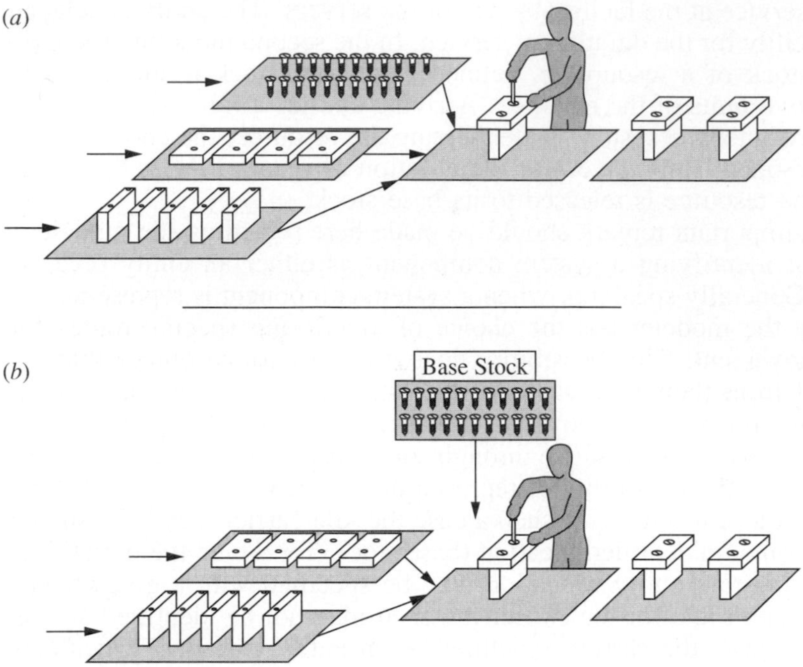

FIGURE 4.4
The alternative way in which some entities are represented as resources.

them as resources with a given base stock. These resource units may then be seized at the assembly station for completing the assembly. Figure 4.4*b* shows this alternative scenario. Notice that unlike the previous example, in this example resources are not returned to the base stock. Also notice that in this example several resource units (not necessarily one) may be seized for each operation.

4.2.4 Entity Accumulation

As mentioned in Chapter 2, throughout the course of their movements in the system, entities may wait at certain accumulation points for a variety of reasons. In discrete systems these accumulation points are generally called *queues*. We will first identify the circumstances that necessitate queue formations and then list the properties that potentially apply to all queues, regardless of the type of situation that necessitates their existence. Following are the processes in which queues are usually formed:

The most common situation for queue formation occurs *when entities require the use of facilities*. Facilities may be occupied or temporarily unavailable for reasons other than that they are busy processing entities (breakdown, scheduled maintenance, operator break, etc.). The arriving entities that find the needed facility unavailable or occupied by other entities have to wait in a queue. There may be multiple (parallel) queues before a facility. For example, in an assembly operation each of several components leaving their machining operations or the warehouse have to wait in their respective queues before the assembly station. The facility would then need to receive a component from each queue in order to perform the assembly. Note that the availability of the facility in this case is a necessary, but not a sufficient, condition for entity departure from the queue. The availability of other matching entities is the other condition. As explained, various other selection rules apply to facilities that have multiple queues.

Queue formation may take place when arriving entities *request a certain resource* (other than facility). If the resource base stock becomes empty, the entities have to queue up until the requested resource becomes available. Each entity may demand one or more resource units. Therefore, an entity that has only a portion of its needed resource available must wait in the queue until its desired number of resources is provided.

Entity grouping is another condition that necessitates queue formation. When a certain number of entities is needed to form a group (in either batched or combined forms) all entities that arrive earlier than the last one (which makes the required number) have to wait in a queue before departing in the form of a single, grouped entity. Also, when carriers are used to move a group of entities, the arriving entities need to queue up and await the carrier arrival.

In certain situations entities have to *await the arrival of other entities* with the same value of a specified attribute at some points in the system before they can proceed. Once the *matching* entities are all available at their assigned queues, all matching entities leave simultaneously, each taking its own route. Notice that the main difference between the assembly process and the match process is that in the assembly process the incoming entities combine to form one entity, but in the match process entities are only delayed in queues to be synchronized by their matching entities. In the matching instance each entity may take a separate route and move at a different speed. As an example of queue formation for entity matching, consider the case of unloading a truck delivering soft drink boxes to a grocery store. A number of employees (entities) take the boxes from the truck, carry them to the storage room entrance, and hand them to other employees who take the boxes and stack them up in the storage room. In this situation, if either of the employees delivering or receiving the boxes is not available, the other has to wait at the storage room entrance. Once the delivery is made employees take their own routes.

Finally, entities at certain points in the system may need to *await permission* to proceed. The permission may be issued by other entities at remote locations in the system. Examples are cars waiting at an intersection for a green light, customers waiting for the opening of the entrance to an exhibit area, and documents awaiting an approval signature before proceeding and going through other office transactions. The departure point for queues formed for the above types of reasons is usually referred to as a *gate* by several simulation languages. Associated with gates are switches that can open and close the gates. Switches may be activated either by the entities going through the gate or by other entities traversing in locations distant from the gate. Although it is possible to use switches as timers to open and close a gate with a desired frequency, a desirable feature of the module representing the gate process is a self-timer for the gate that provides for its automatic opening and closing with certain durations. This feature eases the modeling task as it eliminates a need for the creation of imaginary entities to activate the gate switches. Application examples for this feature are the timing of traffic lights and the opening and closing of the entrance to a movie theater at the beginning of each show.

The following general properties may be associated with all queues:

Queues may have an unlimited or limited *waiting area capacity* for accommodating the incoming entities. Examples of limited areas are the number of seats available in a barbershop waiting area, the space available for cars before a drive-in bank teller, the buffer between two consecutive machining operations, the space in a warehouse, the computer memory buffer for incoming programs (jobs), and so on. When an arriving entity faces a full queue, it may balk and return to the queue after a certain time, or it may go to another part of the system. Alternatively, the arriving entity may block the movement of the entities behind it, but this can only happen when preceding entities can be accommodated by a system component. A common blocking process is in serially positioned machining facilities with limited buffers in between. When a buffer becomes full the entity in the preceding facility may be forced to remain in the facility until space becomes available, even if its service at the facility is completed. This situation results in facility blockage.

Queues may have various *priority disciplines* for ordering entities. First-in-first-out, or FIFO, which applies to situations such as customers in a bank line, and last-in-first-out, or LIFO, which applies to cases such as parts stacked up before a machining operation, are two common priority disciplines in queues. A variety of other disciplines that may be based on the value of an entity attribute are also common. Examples of these disciplines are situations in which older customers or jobs with shorter processing times are to be served first. The customer age or job process time may be represented by an attribute. The priority discipline in these examples will be on the basis of the highest or the lowest value of the given attribute, respectively.

Entities joining a queue may have a *limited acceptable waiting time* after which they may choose to abandon the queue and go to another point in the system. Examples are callers to a utility company's telephone service system who may hang up after being on hold for a certain time, customers waiting in a line who decide to leave because of the slow movement of the line, and a communications switching system selecting a new channel after a certain delay in order to route a transaction that is awaiting a free channel.

Queues may have specific *initial conditions*. They may be initially empty or nonempty. This is a common phenomenon in cyclic systems that contain a limited number of entities that leave the queue, go to other parts of the system, and return to the queue to start the entire process anew. An example of this process is the case of trucks that wait to be loaded at a mining area, travel to the dumping point, and return to the original queue for reloading. Note that queues that are initially nonempty can act as entity sources. It is desirable to have the option of defining and setting the attribute values for the entities initially available at a queue.

4.2.5 Auxiliary Operations

Thus far, our discussion of common processes in discrete systems has considered those occurrences that correspond to physical phenomena readily observable in real-world scenarios. For each of the processes described, one may construct or use a modeling module (usually called an object, a primitive, or a construct) that represents the process. When one constructs a model of a system there are, however, certain auxiliary operations that may be needed, which may or may not correspond to any physical phenomenon. Following is a description of some of the most common auxiliary operations.

4.2.5.1 ASSIGNMENT OPERATIONS. In many situations a modeler may need to define some global variables or entity attributes and assign, or occasionally change, the values of these variables and attributes. For example, to trace the total number of entities in a section of the system, one may define a global variable and increment it every time an entity enters the section and decrement it whenever an entity leaves the section. Likewise, an entity attribute representing a characteristic, say weight, may change at some point in the system (such as the weight of a truck that has dumped its load). A change may also be made to the current level of a resource, or a new resource level may be set at a certain time. The values that are affected may be changed by constant or random amounts or be derived from complex expressions. It is often desirable to use if-then-else conditions at a single assignment module to avoid excessive conditional branches. This capability would be useful in situations when an entity must be directed to one of several alternative assignment points on the basis of a given set of conditions. Generally, the more capabilities built into the assignment module of a process-oriented simulation tool, the more flexible the tool becomes for modeling a variety of situations.

4.2.5.2 FILE OPERATIONS. In certain cases needs arise for transferring, deleting, or copying entities already in queues. We refer to these transactions as *file operations*. The choice of the term *file* is due to the fact that queues are complex data structures in which entities with various attributes are filed in a certain order for future processing. The indicator for selection of entities in files is usually the entity rank in the file (first, last, or *n*th). A specified entity attribute having a specific value is another common indicator for selection of entities from files. Customers leaving a longer queue to join a shorter one and a postal worker withdrawing third-class mail in the queue of incoming mail to be processed are examples of file operations.

4.2.5.3 HANDLING AND UTILIZATION OF VARIABLES. As mentioned in Chapter 2, there are typically two kinds of global variables that are frequently used in systems simulation studies. These are *user variables* and *system variables*. User variables, as the name implies, are explicitly defined by the user for specific purposes. For example, as shown in Fig. 4.5, a user variable may be defined to keep track of the total number of entities in a certain section of a system. In this case, each time an entity enters the section it increments the variable, and each time an entity leaves the system it decrements the variable (assuming that no multiplication or reduction of entities takes place within the considered boundary). Collection of time-based statistics on the variable can provide the desired information.

System variables are those global variables that are automatically created by the simulation tool, and their number and type depend on the specific model used. The most common system variables are the current simulation time, the current number of entities in various queue files, the current number of entities in facilities, the current number of resources in use, and the like. System variables are frequently used when specifying conditions, durations, and various forms of expressions, and at various assignment stages for setting user variables and entity

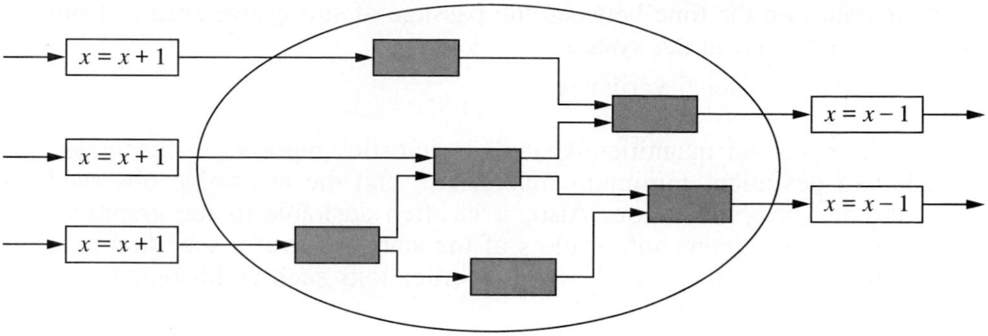

FIGURE 4.5
User variable x is used to keep track of the total number of entities in a certain section of a system.

attributes. A simulation tool should provide access to these variables when requested by the user.

4.2.5.4 SYSTEM INITIALIZATION. User variables, resource levels, status of gates, and the possible number of existing entities in queues and their corresponding attribute values should be identified at the beginning of a simulation. A simulation tool should allow for initialization of these quantities prior to the actual model execution.

4.2.5.5 STATISTICS SPECIFICATION. Another auxiliary operation in a simulation model is the specification of the desired statistics. A simulation model, no matter how sophisticated and precise it is, may be of no use if it does not provide the desired information about various aspects of system performance. Certain statistics that are commonly needed by users deal with information concerning the length of queues, the entity waiting time in queues, utilization of facilities, and utilization of resources. Because of the high frequency of user needs for these statistics, it is desirable for a simulation tool to automatically generate them. There are, however, other kinds of case-specific statistics that should be left for the user to specify. These are generally one of the following types:

- Statistics on the arrival times of entities to a point in the system
- Statistics on the number of entities passing through a certain point in the system (entity count)
- Statistics on traversal time of entities between two points in the system, the most common type being the system time, or the total time spent in the system (from creation to termination)

- Statistics on the time between the passage of successive entities from a certain point in the system
- Statistics on global variables

The general quantities desired in statistics reports are the mean, standard deviation, minimum, maximum, and the last value observed at the end of a simulation. Also, it is often desirable to see graphs of frequency histograms and/or plots of the statistics over a specified time period and with the desired graph specifications such as histogram cell sizes or plotting time intervals.

As explained in Chapter 3, statistics in dynamic systems are generally of two types: observation-based and time-based. A desirable feature of a simulation tool is the provision for specifying various statistics and their types. The simulation tool should properly track, compute, and output the specified statistics in readable formats.

Since in many instances analysts need to study their system performance under steady state conditions, it is desirable to disregard the statistics during the transient state (cold start period) of systems under simulation study. The simulation tool should allow the user to define the estimated length of the transient period after which the desired statistics are to be collected. The capability to automatically perform multiple runs with each run starting with a possibly different set of initial conditions (different values for parameters, initial levels of resources, initial values of user variables, etc.) is also a desirable feature in a simulation tool since it can save some model experimentation time. This feature may not be crucial if the tool provides facilities for rapid model modification and execution (i.e., if operations such as source code modification, link, and compilation are not required for each change in the initial conditions).

4.2.5.6 TRACING ENTITIES IN THE MODEL. Simulation models are usually hard to debug for error correction, model verification, and some basic aspects of model validation. Consequently, to help facilitate the debugging process a tracing capability that shows the position of some specified entities at various stages of model execution is usually provided by simulation languages. A model that runs without error is not necessarily a correct model, since some misrepresentation of reality in the model may exist. A trace capability that prints the name of the position of the specified entities at various event times, or preferably an animation capability that shows the motion of entities in the model network or flow diagram itself, is an essential feature of a good simulation tool.

4.2.5.7 SCENE ANIMATION. Scene animation is a capability that can enhance the presentation of a simulation model. A scene animation generally shows a background with some fixed components (e.g., machines in a factory simulation, or sky, hills, ponds, and runways in a flight simulation), and some moving icons representing entities. Certain graphics primitives such as bar and pie charts, level meters, and dials may also be used in animation to show some statistics of interest. Some simulation tools offer concurrent animation; others offer playback animation. A concurrent type animation can run simultaneously with the simulation model, but the playback type can run only after the simulation model has been run and its trace results of certain events within a desired time segment are stored on a disk. The animation module in this case reads the data from the disk and changes one animation frame to the other in accordance with the changes in the data.

Animation usually shows a short period of system operation, so it is not a proper means for performance analysis of the simulated system. Many events usually take place in a simulation run of a typical length that may not be entirely captured in a few animation frames. In addition, creation of scene animation usually takes a considerable effort, so scene animation is rarely used for debugging and trace purposes. Because of the difficulty involved in building scene animation and the limitation of its applications, most simulation studies do not use scene animation.

An important application of the concurrent type scene animation is in operator training (e.g., pilot training using flight simulation). Another important application of this type of animation is in computer games, which are gaining in popularity.

The desirable properties of a scene animation package include ease of use (preferably without programming), high-resolution graphics support, a library of animation icons, high-speed operation, smooth motion of moving images, and true three-dimensional support. The latter property is rarely incorporated in commercial simulation packages. Most animation software systems are two-dimensional, even though some of their application instances may appear to be three-dimensional. True three-dimensional systems are mostly based on wire-frame representations of solid geometry (a rather crude representation). Some advanced animation systems that run on high-end hardware platforms use more sophisticated solid representations (e.g., constructive solid geometry or boundary representation used in some supercomputer-based three-dimensional animated movies and virtual reality applications). Scene animation is used primarily for presentation purposes, but once developed, the animation work can be used for model verification and validation purposes. However, using this type of animation only for the latter purposes is not advised, since it is usually very time-consuming to develop scene animation.

4.3 SUMMARY

Knowledge of common processes in discrete systems greatly helps the simulation analyst in the translation of the system under study into the corresponding model by use of an available simulation tool. This knowledge also helps in comparing the capabilities of various simulation tools. In this chapter we have taken a generic approach to the description of these common processes without specifically addressing any simulation tool. In the next chapter we will introduce the EZSIM simulation environment and the details of the modeling capabilities that it provides with reference to each of the common processes described in this chapter.

4.4 EXERCISES

4.1. Identify the common processes involved in the scenario given in Exercise 2.5.

4.2. Unit loads arrive at an air terminal to be transported by cargo planes of various capacities, which arrive at the terminal with certain interarrival times. Loaders that occasionally break down, operated by operators who take scheduled breaks, place the unit loads in the planes once the planes arrive. The system components in this example (loads, planes, loaders, and operators) may be represented alternatively by either entities, resources, or facilities. How would you represent these components in your simulation model?

4.3. Describe the common processes involved in modeling each of the following systems:
(*a*) A traffic intersection with left turn signals
(*b*) A factory floor with parts arriving in kits and being routed to various assembly stations to form various products
(*c*) A computer system serving several remote users who log in at random times
(*d*) The rides and the shows in an amusement park
(*e*) An office system of your choice with its typical transactions

4.4. What specific statistics do you think are desirable in the simulation of each of the scenarios listed in Exercise 4.3?

4.5. Select a particular application domain, say manufacturing, and if possible specify certain processes that exist in that domain which are not easily representable by the generic processes described in this chapter. Recommend the additional details required to enhance the applicability of the generic processes for your particular scenarios.

4.6. If you were to design a special purpose simulation tool for traffic intersection simulation, what process modules would you design for your tool?

4.7. Describe a realistic scenario in which all of the queue and server selection situations (i.e., selection of a queue to join, a queue to serve, and a server) apply.

CHAPTER 5

EZSIM MODELING OBJECTS AND CAPABILITIES

5.1 INTRODUCTION

In this chapter most of the generic concepts described in Chapter 4 will be demonstrated using the EZSIM environment. The software will be described in more detail and several representative examples using EZSIM nodes and features will be discussed. The objective of this chapter is to familiarize the reader with the basic modeling techniques used in various application scenarios. Since the focus in this chapter is on modeling, the analysis of simulation outputs of the examples is not emphasized. Chapter 7 will elaborate on the analysis of simulation output for several application examples. Building and running the example models described in this chapter are both highly recommended. Sufficient explanation is given for construction of each of the example models. The data files corresponding to these examples are provided in the accompanying diskette. In the following sections each EZSIM node and the related capabilities are discussed.

5.2 SOURCE NODE

The SOURCE node creates entities. Every entity that the SOURCE node creates may have a name, which becomes an attribute for the entity (this attribute is called NAME). If additional attributes are desired for the entity, the ASSIGN node may be used as a node following the SOURCE node. The name attribute may be used for branching, for setting specific delay times, and for identifying the entity with a dedicated symbol in animation. If no name is specified for the entity it will be shown as a ? in animation. It is possible to control the number of creations at the SOURCE node. Also, creations may be started or stopped after certain times. It should be noted that stopping entity creation does not end the simulation. Simulation may continue until all entities leave the system, a TERMINATE node stops the simulation, or the simulation time reaches a prescribed length (specified at the Control menu).

As pointed out in Chapter 3, throughout EZSIM, various time elements such as time of first entity creation, time between creations, and delay times in DELAY and FACILITY nodes have a standard window that allows for specification of time as a constant, a statistical distribution, a user variable, or as an expression containing many elements. When the choice of statistical distribution is specified, a window listing several distribution names allows for the selection of the desired distribution. After the distribution is selected, the user is prompted with messages asking him or her to identify the values of each of the parameters of the distribution and the choice of the random number seed to be used for the specific process.

5.3 TERMINATE NODE

The TERMINATE node ends the path of entities. When an entity enters a TERMINATE node it is considered to be out of the network, and the computer memory allocated to its attributes is freed. Each TERMINATE node can end the entire simulation if the entity count specified in it is reached.

5.4 DELAY NODE

The DELAY node is used for creating a delay that corresponds to the traversal time of the entity from one node to another. The traversal time is selected from the standard time window and may be represented by a constant, various statistical distributions, an attribute value, a user variable value, or an expression. DELAY nodes may also be used for collecting several incoming branches and for creating several outgoing branches. The delay time in such cases may be set to zero.

Example 5.1. This example demonstrates a multiple incoming branch situation. Let us assume that in the freeway simulation example given in Chapter 3 there is an additional entrance connected to the middle of the freeway. Let us further assume that the interarrival time of cars entering the freeway through this entrance is exponentially distributed with a mean of 5 minutes. Travel time through the first and second segments of the

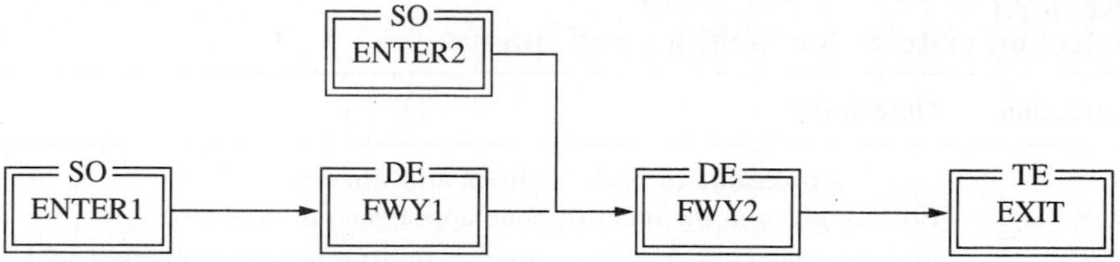

FIGURE 5.1
Demonstration of multiple incoming branches.

freeway (between the first entrance and the second entrance, and between the second entrance and the end of the freeway) are both distributed normally, each with a mean of 7.5 and a standard deviation of 1.5 minutes. We are interested in simulating the freeway system to observe the total number of cars in each segment of the freeway. Notice that only one freeway direction (freeway side) is being considered. The EZSIM network for the simulation model is shown in Fig. 5.1. SOURCE nodes named ENTER1 and ENTER2 represent each freeway entrance. You may specify different names (say, CAR1 and CAR2) and choose different animation symbols (say, 1 and 2) for cars arriving through the first and second entrances, respectively. DELAY nodes named FWY1 and FWY2 represent the travel times through the two segments of the freeway. Follow the procedure described in Chapter 3 for building the network and entering the nodal parameters. Run the model in the animation mode and observe the flow of entities in your model. Also observe the changes in the number of entities in each DELAY node as the model is being simulated.

5.5 FACILITY NODE

The FACILITY node acts as a server. Entities remain in the node for the duration of their service. When the node is occupied, the arriving entities have to wait until the node is free. FACILITY node must be preceded by a QUEUE node. The node allows several parallel servers. Parallel servers at a facility are assumed to be homogeneous; that is, they have the same characteristics. Provisions such as breakdown scheduling and simultaneous use of various resources are available at this node. Various selection criteria form multiple parallel queues before the facility may be chosen at the FACILITY node. The assembly option that joins several incoming entities is a selection criterion, for example. Table 5.1 lists the applicable criteria for facility and queue selections and their descriptions. This table can be seen while in the network construction mode by pressing Alt H when selection choices are listed at the FACILITY node. The entity leaving the facility node may be given a new name. This will allow for showing the unprocessed and processed parts, for example, with different symbols in the animation mode.

TABLE 5.1
Selection criteria for facilities and queues

Condition	Description
	Applicable to both facilities and queues
POR	Priority given in the order of node appearance in network.
CYC	Priority is given in a cylic manner — transfer to first available node after each selection.
RAN	Priority given randomly — assign equal probabilities to nodes.
LAV	Priority given to the node that has had the largest average number of entities in it to date.
SAV	Priority given to the node that has had the smallest average number of entities in it to date.
	Applicable to facilities
LRC	Priority given to the node with largest remaining unused capacity.
SRC	Priority given to the node with smallest remaining unused capacity.
LBT	Largest amount of usage (busy time) to date.
SBT	Smallest amount of usage (busy time) to date.
LIT	Select the node with longest idle time.
SIT	Select the node with smallest idle time.
	Applicable to queues
LWF	Priority given to the node that has had the longest waiting time.
SWF	Priority given to the node that has had the shortest waiting time.
LNQ	Priority given to the queue with largest number of entities in it.
SNQ	Priority given to the queue with smallest number of entities in it.
ASM	Select one entity from each queue and assemble them into a single entity.

5.6 QUEUE NODE

The QUEUE node represents buffers before FACILITY nodes and is always succeeded by a FACILITY node. A QUEUE may have capacity limitation, in which case the arriving entities may balk to a specified node, block the facility behind (if any), or terminate at the node. The latter case may happen when the blocking option is selected and there is no FACILITY node behind the QUEUE node. Whenever a limited queue capacity is specified, EZSIM automatically prompts the user for specification of balking or blocking options. When the balking option is chosen, EZSIM lists the names of all nodes in the network to which direct entity entrance is allowed. Balking traversal time may also be specified using the standard time window.

It is possible to specify a maximum acceptable entity waiting time after which the entity may balk to a specified node. Queues may have various priority disciplines. These may be first-in-first-out (FIFO), last-in-first-out (LIFO), or priority given to entities with highest or lowest value of a given attribute. Entities abandoning a queue (either because it is full or because their specified waiting time limit has been reached) move to their specified destination node through an invisible path in animation. Many selection criteria are available for entities choosing parallel queues to join and for facilities choosing parallel queues from which they receive entities. When a QUEUE node is connected to several facilities that follow it, several options for selection of a facility are available, as listed in Table 5.1.

Example 5.2. In this example, some of the capabilities of the QUEUE and FACILITY nodes are demonstrated. Assume that large parts arrive at a machining station for processing. The interarrival time of parts is 2 minutes. The buffer before the facility can accommodate a maximum of 5 parts. Arriving parts that find a full buffer are sent to a temporary storage area. The machining time is exponentially distributed with mean of 1.5 minutes. The machine is stopped every 30 minutes for a 5-minute adjustment during which parts may not be processed. The processing of each part requires 10 screws. 20,000 screws are initially available in the stock. We are interested in simulating the system for 1000 minutes of operation to find the queue and facility statistics, total time that the processed parts spend in the system, and the total number of unprocessed parts that are sent to the storage area.

Figure 5.2*a* shows the network diagram of the system. At the parameter specification stage for the QUEUE node a limited capacity of 5 is specified. The Balking option is chosen for entities that arrive at a full queue. The TERMINATE node named STORE represents the destination of these entities. The balking traversal time may be set at zero. All other fields for the QUEUE node are set at their default values. Since the path

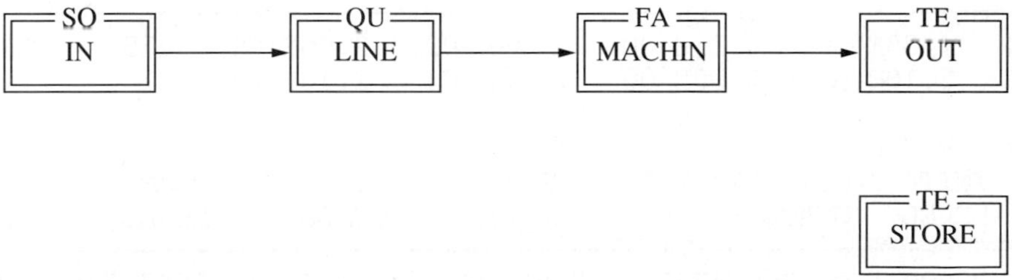

FIGURE 5.2*a*
Demonstration of QUEUE and FACILITY nodes.

of balking entities to their destination is transparent, the STORE node is not connected to any other node.

At the FACILITY node, a scheduled breakdown is specified with the indicated frequency and duration. Also, a need for utilization of a resource is specified. The resource is given the name SCREW and the question prompt asking the number of resources needed is answered by entering the number 10. No returned resources are specified. All other question prompts in the resource window are satisfied with default value selections. After the user specifies the parameters for all nodes and hits the Esc key, EZSIM finds out that a resource called SCREW is used in the model, but its initial level is not identified. Accordingly, it prompts the

```
                  ***  E Z S I M  STATISTICAL  REPORT  ***

Simulation Project: EX2
Analyst: BK
Date: 4/2/92
Disk file name: EX2.OUT

Current Time: 1000.000   Transient Period: 0.000

Q U E U E S:
NAME      MIN/MAX/LAST                MEAN      STD       MEAN      STD
          LENGTH                      LENGTH    LENGTH    DELAY     DELAY
-----------------------------------------------------------------------------
LINE       0/ 5/ 4                    1.53      1.58      3.04      3.75

F A C I L I T I E S:
NAME      NBR      MIN/MAX/LAST       MEAN      STD         MEAN    MEAN
          SRVRS    UTILIZATION        UTLZ      UTLZ        IDLE    BUSY
-----------------------------------------------------------------------------
MACHIN    1        0/ 1/ 1            0.79      0.41        0.44    1.66

V A R I A B L E S:
NAME      MEAN        STD         MIN         MAX         No.OBSRVD
-----------------------------------------------------------------------------
SYSTIME   4.85E+00    4.25E+00    1.89E-02    2.06E+01    475
UNPROCD   2.10E+01    0.00E+00    2.10E+01    2.10E+01    21

R E S O U R C E S:
NAME     INIT    MIN/MAX/LAST     MEAN       STD        MEAN        STD
         LEVEL   USAGE            USAGE      USAGE      LEVEL       LEVEL
-----------------------------------------------------------------------------
SCREW    20000   0/4760/4760      2352.28    1374.15    17647.72    1374.15
```

FIGURE 5.2*b*
Output for the model of Example 5.2.

user to initialize the resource level through a special initialization window. After the initialization stage, the statistics window appears. The statistics on the queue and the facility are automatically collected and reported by EZSIM. To obtain the statistics on the total system time of processed parts, the statistics name SYSTIME is defined in the Statistics menu for which the traversal time between the nodes IN and OUT is specified, in a manner similar to Example 5.1. To obtain the total number of unprocessed parts that are sent to the storage area a name, say UNPROCD, is entered in the first column of the statistics window. Next, the type of statistics (which in this case is entity count) is selected from the related window. Finally, in the list of nodes that EZSIM provides, STORE is selected, meaning that we want to obtain the entity count at that node. Figure 5.2*b* shows the simulation output for 1000 time units. To create this output, the Print Output option is chosen in the Output menu.

Example 5.3. This example shows a case of selection from multiple queues before a server. Assume that parts A and B arrive, each from a different route, at an assembly station. The interarrival time for both parts is uniformly distributed between 7 and 12 minutes. The assembly time is constant and lasts 4 minutes per each assembly operation. We would like to simulate the operation to find the statistics on the throughput of the facility (the number of parts assembled per unit time).

Figure 5.3 shows the network diagram of the model. Each SOURCE node creates a part type. Each part enters its own buffer before the server. At the parameter specification stage of the FACILITY node (once all questions are answered), a branch selection window pops up asking the user to identify the criterion for selection from the two queues before the facility. To see the list of the available criteria, press Alt H. The correct choice in this case is ASM, which takes one entity from each queue and assembles the two into one. To find the statistics on the system throughput, the statistics type selected in the statistics menu should be "interarrival at a node." Identifying PACK as the node for which the interarrival time statistics is desired provides a value that is the inverse of the throughput

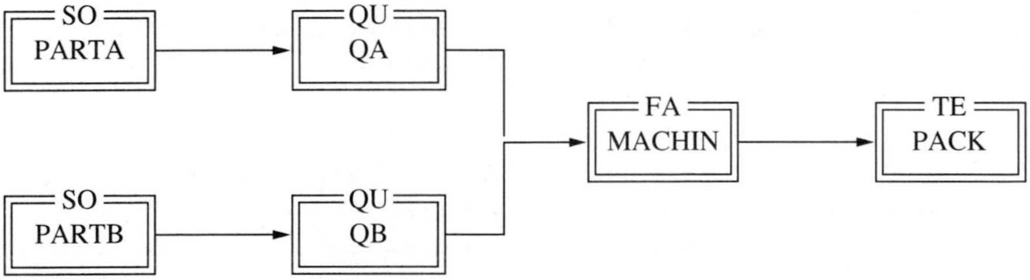

FIGURE 5.3
Selection from multiple queues.

of the facility. To view the assembly logic in operation, animation of this example is recommended.

Example 5.4. In this example a case of multiple facilities with a single queue is demonstrated. Let us assume that trucks arrive at a construction site to be loaded by one of three loaders. If there is more than one free loader available upon a truck arrival, a supervisor assigns the loader with the longest idle time to the arriving truck. Truck interarrival times are distributed exponentially with a mean of 1.2 minutes. Each loader has an exponentially distributed service time with means of 2, 4, and 5 minutes, respectively. We are interested in the queue statistics and the statistics on the total number of trucks served by each loader. We also want to know the total number of trucks that leave the system after 1000 minutes of operation.

Note that since these parallel facilities have different service characteristics, they should each be identified as a separate facility in the model; otherwise, a single FACILITY node with three parallel servers would have represented the system. The network for this example is shown in Fig. 5.4. At the parameter identification stage for the FACILITY node, a branch selection window will be shown. An option within this window is Selective Branching. Choosing this option will result in the appearance of yet another window that shows the list of facility selection options. The LIT (largest idle time) option is identified as the facility selection criterion. After completion of the network, the desired statistics, which are in the form of entity counts at the four nodes (LOADR1, LOADR2, LOADR3, and LEAVE) are identified. Watching this model in animation demonstrates the way the loaders are selected.

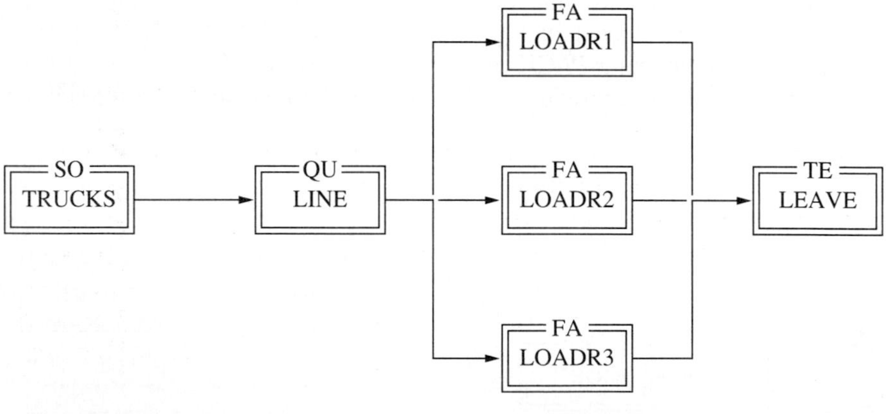

FIGURE 5.4
Illustration of selection from facilities.

Example 5.5. In this example, the use of probabilistic branching is demonstrated. Assume a simplified version of Example 5.2, where the queue has infinite capacity. Let us assume that 80 percent of the parts leaving the facility will have an acceptable quality, and the rest have to be sent back to the machining queue for rework. Figure 5.5 shows the network model for this situation. In this network, the FACILITY node is connected to the QUEUE node. (Notice that if this connection is made after the FACILITY node parameter identification is completed, then at the time of connection this node will change from a double-line border to a single-line border representation. This means that additional clarifications are required.)

When selecting the FACILITY node for parameter identification, a new window will appear for specifying the branching criterion. In this window, several types of criteria are provided. These are the Probabilistic, Conditional, Last Choice, and Always branching criteria. The Probabilistic branching option allows for allocation of mutually exclusive selection probability values to each emanating branch. The total value of these probabilities should be 1 (EZSIM rejects input that violates this rule by returning the cursor to the decimal point position of the last number violating the rule).

In this example, the Probabilistic option is selected for branches emanating from the FACILITY node. Values of 0.8 and 0.2 are specified for branches leading to the TERMINATE and QUEUE nodes, respectively. Watch this model in animation as reworked parts leave the facility and join the end of the queue.

Example 5.6. In this example the Conditional Branching option is demonstrated. This option allows for specifying various conditions applicable to each branch emanating from a node. Conditions may be based on values of (a) system variables (such as simulation time, number of entities in a given queue, or value of an available resource), (b) attributes of the entity that is about to travel through the branch, and (c) user variables. Complex expressions may be built for identifying conditions on branches.

Let us assume that a machine shop receives two types of parts for processing in a given day. The interarrival time for type-1 parts is

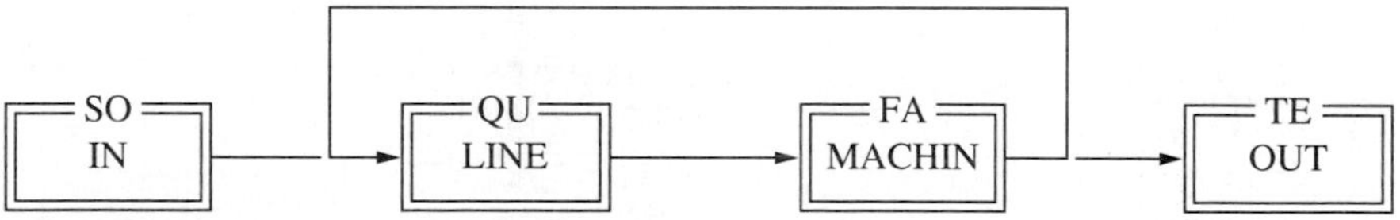

FIGURE 5.5
Illustration of probabilistic branching.

exponentially distributed with a mean of 4 minutes. Type-2 parts arrive every 5 minutes. The shop policy is to always process type-1 parts. If the queue before the machining station has less than 3 parts in it, then a type-2 part will be sent to the queue for processing. Those type-2 parts that arrive within the first two hours of operation and are not accommodated at the shop are sent to a subcontractor to be processed and returned before the end of the day. If a type-2 part arrives after the first two hours of operation and is not processed at the shop, it is sent to a storage area. We are interested in the number of parts that are processed in the shop, subcontracted, and sent to storage during one day of operation.

Figure 5.6*a* shows the network model for the above system. The source node PART1 generates the first part types and sends them to the queue named LINE. They then move to the machining station. The source node PART2 generates the second part type, which may go either to the queue for machining, to the subcontractor, or to the storage area. The conditions are specified for each branch using the Conditional Branching option. To specify the condition of the branch to the queue, the System Variable option is selected. The Length of Queue is then selected from the window of available options. Since there is only one queue in the model, EZSIM provides only one selection choice, which is LINE. As soon as this specific queue is identified, LEN(LINE) appears on the left-hand side of the condition expression that indicates the length of the queue named LINE. A window showing the relational operators is then presented. The $<$ is selected and number 3 is entered. Figure 5.6*b* shows the window for conditional branching at node PART2. Notice that the second expression uses an AND operator, which combines two expressions. TNOW (the simulation time) in this expression may be selected by first selecting the System Variable option and then choosing Simulation Time from the

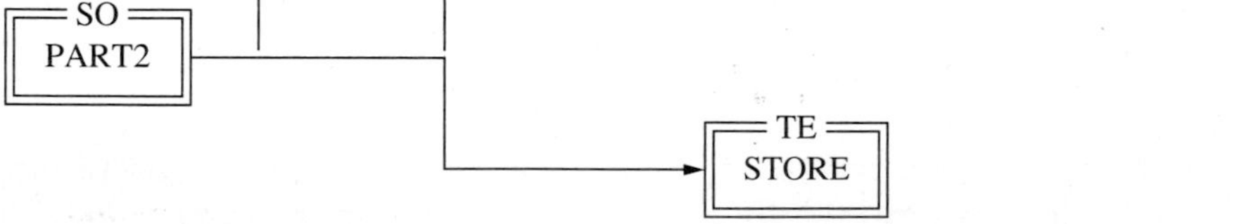

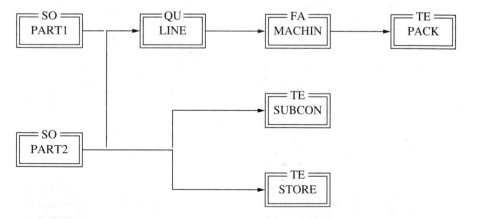

FIGURE 5.6*a*
Illustration of conditional branching.

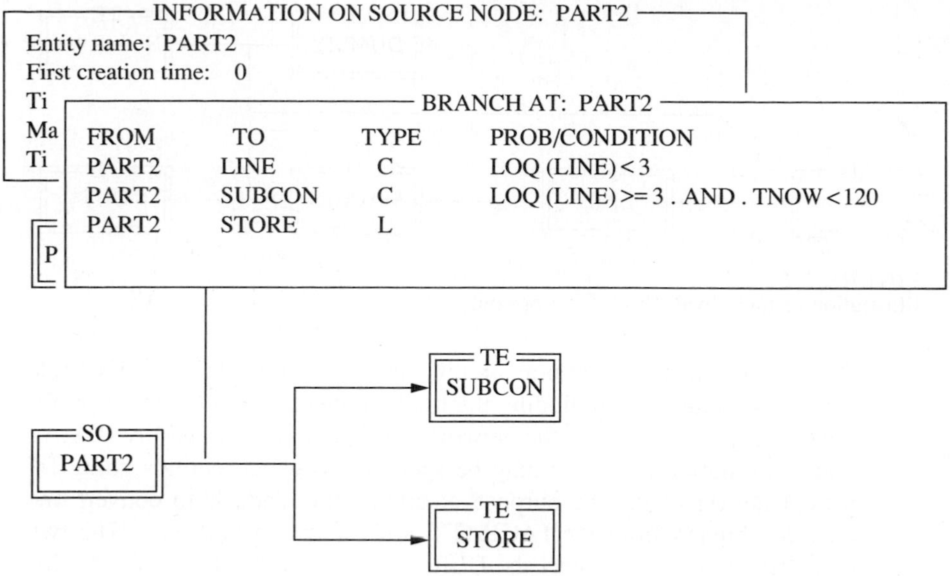

FIGURE 5.6*b*
The window for branching conditions.

corresponding window. The third condition uses Last Choice branching. The Last Choice branching option specified for a given branch results in the selection of the branch when no other branch can be taken.

The maximum number of branches that can be taken should be identified when all conditions are specified. In this case, this number is 1. If numbers higher than 1 are selected, entity multiplication may occur (an entity will be sent through each branch whose condition is satisfied). Of course, this will happen only if more than one condition is satisfied simultaneously. If none of the conditions are satisfied EZSIM destroys the entity. In the animation mode the user is warned about entity elimination with a message on the lower part of the screen.

In the Statistics menu, entity counts at nodes PACK, SUBCON, and STORE are specified. Running the simulation for the desired length of time provides the output statistics for these counts, as specified by the problem. It is recommended to first run this model in the animation mode to observe the passage of entities in various branches as the content of the queue and the value of simulation time change.

Example 5.7. This example demonstrates the use of the Always branching option. Specifying this option on a branch results in taking the branch regardless of the state of other branches. If other branches are taken simultaneously, the Always branching option may result in entity duplication.

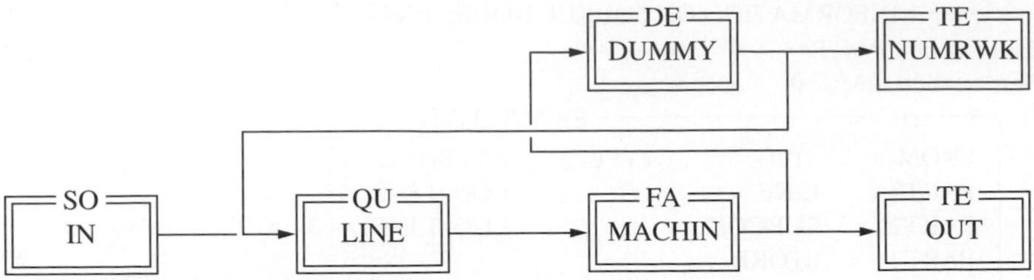

FIGURE 5.7
Illustration of the Always branching option.

To demonstrate the operation of this option, assume that, in Example 5.5, we are interested in finding the total number of parts that are sent for rework. Figure 5.7 shows the network for this model. A dummy DELAY node with duration of zero may be specified to which entities going for rework are directed. The purpose of adding this node is to convert one outgoing branch from the FACILITY node into two branches. The two emanating branches connect the DELAY node to nodes LINE and NUMRWK, respectively. The latter node receives a dummy duplicate entity for each rework transaction. The Always branching option is specified for each of the two branches emanating from the DELAY node. Entity count is specified as the choice of statistics for both OUT and NUMRWK nodes in the Statistics menu.

5.7 ASSIGN NODE

The ASSIGN node may be used to set new attributes or change attribute values of entities that pass through the node. The node can also be used to set the values of user-defined variables and to set new levels or increase or decrease the current levels of resources. Provisions for use of conditions and expressions are also made available at this node.

Example 5.8. This example shows the use of the ASSIGN node for setting attributes of entities. Assume that customers with either of the two types of service requirement, namely trimming or styling, arrive at a barbershop. The interarrival times of customers needing trimming and styling are distributed exponentially with means of 15 and 24 minutes, respectively. Trimming and styling times are also distributed exponentially with means of 10 and 20 minutes, respectively. There are two barbers at the shop. We would like to observe the queue statistics to decide on the number of seats to be placed in the waiting area.

Figure 5.8 shows the network for the model of the above system. The SOURCE nodes create the customer entities with each type of ser-

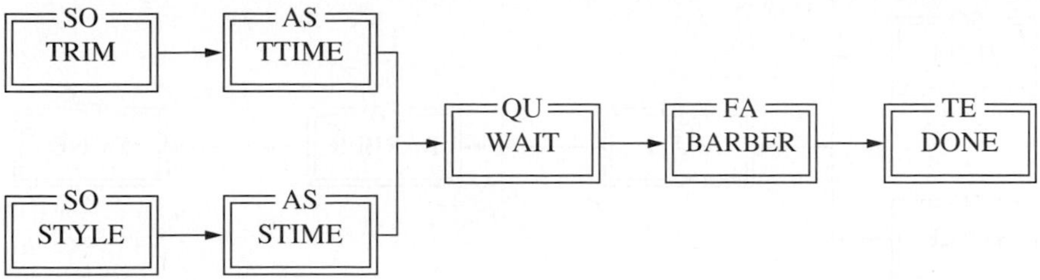

FIGURE 5.8
Use of ASSIGN node for setting attribute values.

vice requirements. The modeling approach is to store the corresponding service time requirement in an attribute of each entity immediately after it enters the system. This information is carried with the entity to the facility at which it is used as the service duration. At the ASSIGN node TTIME, the Attribute option is selected from the window of choices for assignment. The attribute is then given the name SRVTIME (an arbitrary name), and its value is specified as a sample from an exponential distribution with mean of 10 minutes. At the ASSIGN node STIME, the same option (Attribute) is selected and the attribute is given the same name (SRVTIME). The value of the attribute is specified as a sample from an exponential distribution with mean of 20.

At the FACILITY node, two parallel servers are specified. From the choices within the service duration window, Attribute is selected. As soon as this selection is made, a window appears with a list of all attribute names defined so far. Note that this list includes SRVTIME and NAME. The latter is a default attribute that is internally defined by EZSIM and carries the name of the entity. To specify the service duration, SRVTIME is selected. SRVTIME carries the service duration of the entity. Note that each entity carries its own service duration to the facility by means of its attribute.

Example 5.9. This example demonstrates an alternative approach to modeling the system described in the previous example. Rather than storing the required service time in an attribute of customer entities, a table that associates the entity "name" to the related service time may be built at the FACILITY node. In this case, when an entity arrives at the node its name attribute is observed, and the associated service time is retrieved from the table. Figure 5.9*a* shows the network for this model. At each SOURCE node, a name is given to the entities created. At the FACILITY node, the Entity Dependent Service Time option is selected for service duration, and the related service times are assigned to each entity name. Figure 5.9*b* shows the window in which the assignments have been made.

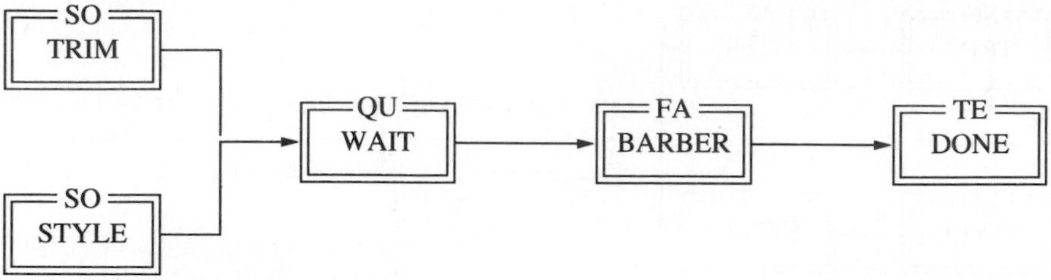

FIGURE 5.9*a*
Illustration of Entity Name Dependent Service Time.

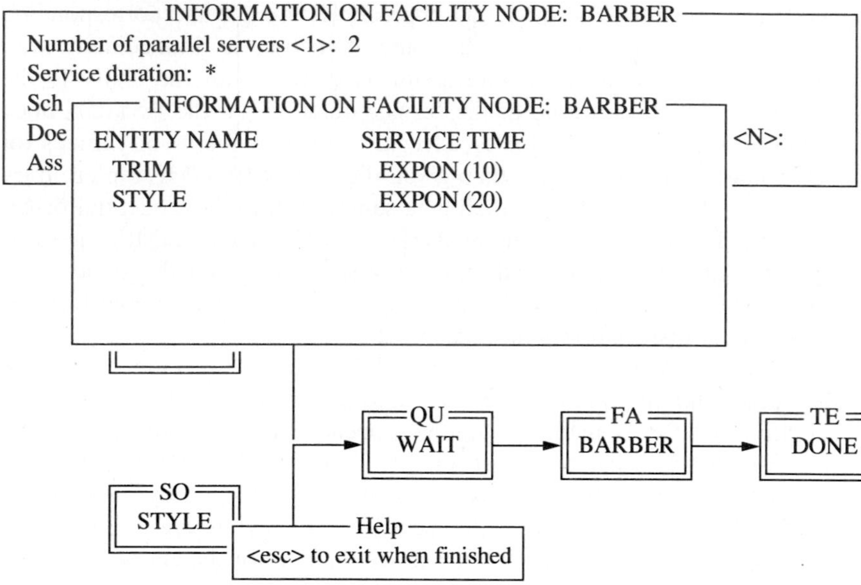

FIGURE 5.9*b*
Window for specifying service times.

Note that the capability to relate service durations to entity names at facilities is useful in manufacturing applications where machines process various parts, each of which has a different processing time requirement.

Example 5.10. This example shows the use of the ASSIGN node in manipulating a user variable. Assume that there are two tellers at a bank, each with his or her own customer waiting line. The first teller processes deposits, and the second one processes withdrawals. Interarrival times of customers have exponential distributions with means of 7 and 4 minutes for deposit and withdrawal transactions, respectively. The service time for deposit is uniform within a range of 2 to 4 minutes. Withdrawal time is

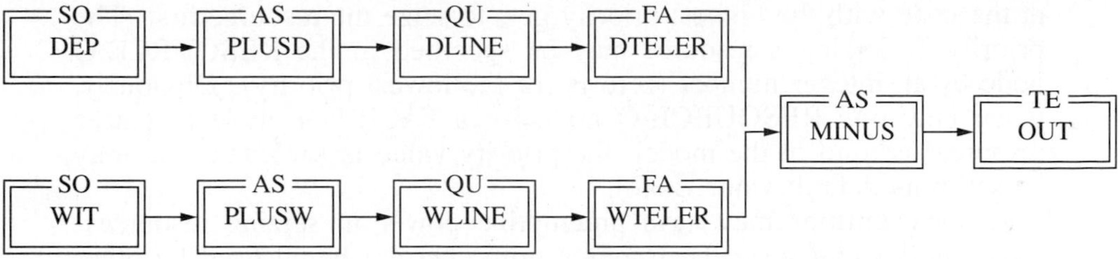

FIGURE 5.10
Use of ASSIGN node used for setting a user variable.

also uniform, with a range of 3 to 6 minutes. We are interested in the statistics on the total number of customers in the bank.

Figure 5.10 shows the network diagram for this example. The two ASSIGN nodes, named PLUSD and PLUSW in this example, increase a user variable, X, by one unit each time an entity goes through them. The ASSIGN node named MINUS decreases X by 1 unit each time an entity goes through it. This way, the value of X always represents the total number of entities that are in the system. To set the value of X in the ASSIGN node, the User Variable option of the node is selected. The variable is given the name X. The Expression option is then selected to set X to either $X + 1$ or $X - 1$. Upon departure from the graphics mode, the initial value of X is requested by EZSIM. This value is set at zero. At the Statistics menu, a variable name—say, NUMINSYS—may be given to the desired statistics. The User Variable option is then selected. EZSIM then provides a list of all variables defined in the model (the list contains only X in this case). Variable X is then selected. The statistics may be collected either as time based (which is more meaningful) or observation-based. In the latter case, a node should be identified as the point of observation. Each time an entity hits this node, the value of X is recorded as one statistical observation. Run this model in the animation mode, and while in animation press the V key to observe the value of X as it changes over time.

5.8 RESOURCE-Q NODE

The RESOURCE-Q node is a buffer in which entities wait to receive a specified number of one or more resource types. When the requested resources become available at their base stocks, the entity is released and the base stock levels of the seized resources are reduced by the prescribed numbers. Entities seizing resources may release them in whole or in part at specified points in the network. Priorities for seizing resources, times for resource retrieval, and return to base stock may also be specified at this node.

If various entities wait in different RESOURCE-Q nodes to seize a given resource, then when the resource becomes available the entity at the node with the highest priority gets to seize the resource first. The priority in seizing a resource may be specified in the RESOURCE-Q node by an integer number (zero is for the lowest priority). Obviously, when only one RESOURCE-Q node (or a FACILITY node that uses resource) is used in the model, the priority value is irrelevant and may be set at its default value (zero).

Some entities may have preemptive power in seizing resources. Such entities can seize the resource units that are being used by other entities if the resource base stock does not contain the required number of units of resource. The entities whose resources are taken away by preempting entities may resume using the resources once they are released by the preempting entity. In EZSIM the following resource preemption options may be specified:

1. Preemption of resources that have been seized the earliest
2. Preemption of resources that have been seized the latest
3. Preemption of all in-use resources
4. Preemption of all resources (i.e., in-use as well as base stock)

Options 1 and 2 apply to situations in which various in-use resource units have been seized at different points in time, and the preempting entity has the choice of taking its required number of resources from either those resource units that have been taken the earliest or those that have been taken the latest. Option 3 allows the preempting entity to seize all in-use resources, regardless of their number. An application example is when a power outage shuts down all operating machining stations in a factory. Option 4 allows the preempting entity to seize all in-use resources as well as all resource units in the base stock. An application example for this option is when a conveyor belt whose segments are represented as resources breaks down. In this case all of the segments of the belt will be unavailable whether or not they are in use.

Example 5.11. At an air terminal, cargo airplanes arrive at the rate of one per hour. Each plane waits until it loads 70 units of cargo and then takes off. Trucks, each bringing 10-unit loads, arrive at the terminal with an exponentially distributed interarrival time with mean of 9 minutes. There are initially 100 units of cargo at the terminal. We would like to simulate the system for 50 arriving planes to find the expected waiting time of each plane before takeoff.

Figure 5.11 shows the network diagram of the model. There are two disjoint network segments. The upper segment represents the movement of airplanes, and the lower one is for trucks. Each arriving plane goes to the RESOURCE-Q node and requests 70 units of the resource named CARGO (parameter specification for RESOURCE-Q node is similar to

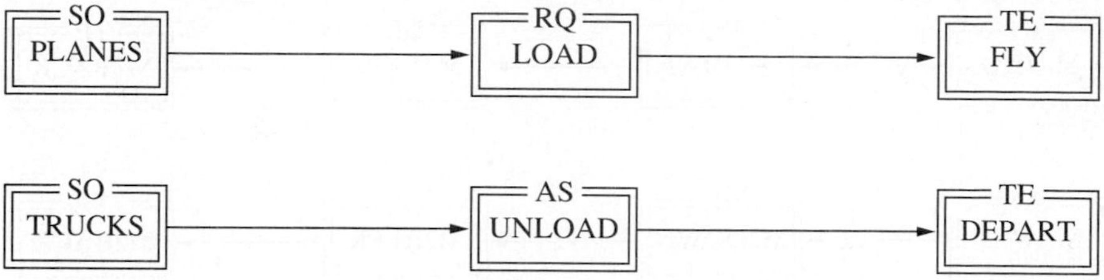

FIGURE 5.11
Illustration of RESOURCE-Q and ASSIGN nodes.

that of the resource section of FACILITY node). As soon as a sufficient number of cargo units becomes available, the plane takes off. The truck entities go to the ASSIGN node after their creation. At the assign node, they increase the level of the resource, CARGO, by 10 units, and continue on to their termination point. Notice that the base stock of the resource is the element that links the two seemingly unrelated network segments in this example. Run this model in the animation mode, and press V while in animation to observe the level of the resource as it changes over time.

Example 5.12. This example shows how a resource may be specified to play the role of mobile facilities. Assume that, in an automated textile manufacturing facility, an operator is in charge of monitoring several spinning machines that may stop because of material defects. This calls for the operator's involvement in clearing the problem. The operator is stationed in a control room where a monitor screen indicates the specific machine needing attention and the nature of the problem to be cleared. There are three types of machines situated in three different areas. The time between failures of groups of type-1 machines is Exp(40), and it takes Exp(2) to clear the problem. The times for group 2 and 3 machines are Exp(30)/Exp(3) and Exp(45)/Exp(2), respectively. Time units are in minutes. The operator's travel times from the control room to areas 1, 2, and 3 are 3, 2, and 4 minutes, respectively. After each repair, the operator should return to the control room. We would like to simulate this system to find the utilization level of the operator and each type of machine's average downtime resulting from operator unavailability.

Figure 5.12 shows the network model of the system. Notice here that each source node creates machine breakdowns for the corresponding groups of machines. Each machine requiring the operator's attention goes to a RESOURCE-Q node to seize the resource called MAN. The number requested is 1, and the times to retrieve and return the resource correspond to the operator's travel times from the control room to the corresponding area and back. After each RESOURCE-Q node, there is a DELAY node that represents the associated repair time. At each RESOURCE-Q node, it is specified that the seized resource should be released after the delay

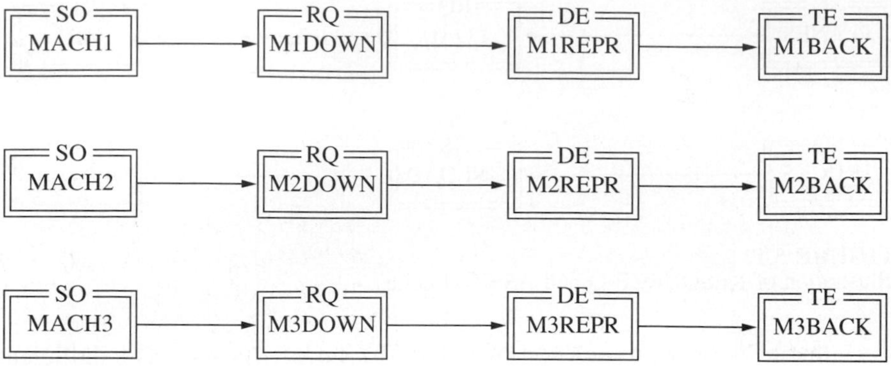

FIGURE 5.12
Illustration of resources used to represent mobile facilities.

node that follows the node. At the initialization stage, the initial number of resource, MAN, is set to 1. Note here that the queue statistics for each RESOURCE-Q node will be independently kept and printed by EZSIM, providing the statistics desired by the problem statement.

5.9 GATE-Q NODE

The GATE-Q node contains a buffer and a gate behind the buffer. When the gate is open, the entities in the buffer can leave the node; otherwise, they have to wait in the buffer until the gate is open. The initial status of a gate may be set either to open or closed. Gates may be opened and closed by SWITCH nodes that may be activated by some remote entities. When a gate is opened, entities in its buffer may all leave at once. To control the number of entities leaving at a gate opening time, a SWITCH closing the gate may be activated by the Nth entity that leaves the gate. The GATE-Q node also has a built-in timer control that can open and close the gate with the desired timing. The GATE node has its own buffer, and therefore should not be preceded by a QUEUE node. Each gate may be initialized to have either a closed or an open status at the beginning of the simulation.

5.10 SWITCH NODE

The SWITCH node can be used to open or close the gates from any remote point in the network. Any entity passing through the node at a desired time can activate the switch. The parameter identification window of the SWITCH node provides a list of all gates used in the model, and allows for specification of the gates that are opened or closed every time an entity passes through the SWITCH node.

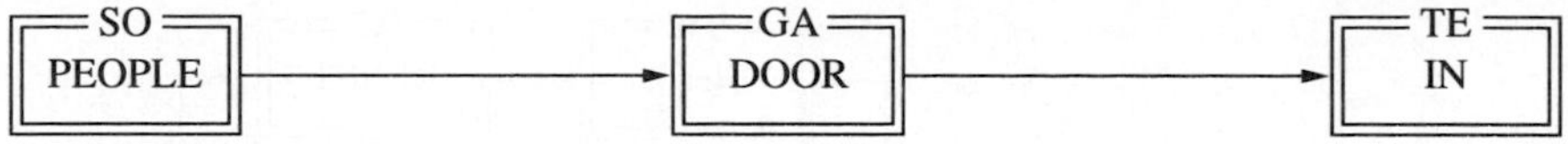

FIGURE 5.13*a*
Illustration of GATE node with timer.

Example 5.13. This example shows how a GATE node, with its self-timer activated, can be used to control the flow of entities. Assume that at an exposition a show starts every 30 minutes. After the start of each show, which takes about 15 minutes, the door to the showroom is closed for 20 minutes. It is then opened for 10 minutes. People arrive to see the show with a uniform interarrival time that ranges between 0.8 and 1.2 minutes. In order to design a waiting area with a shade, we are interested in the statistics of the queue where people are waiting for the show.

Figure 5.13*a* shows the network model of this simple system. The SOURCE node generates the arrivals of people to the GATE node, which represents the waiting area. The gate is set to be initially open and has a timer that alternately keeps it open for 10 minutes, closes it, keeps it closed for 20 minutes, and opens it again. Entities leaving the gate node are sent to a TERMINATE node. To observe the statistics on the queue of the waiting line, the queue statistics may be selected from the output menu.

Figure 5.13*b* shows the equivalent of the above model using external switches instead of the timer to open and close the gate. The model is made up of two disjoint networks. The upper network is for the flow of people entities. The gate here does not have a timer. The lower disjoint network is for the flow of a single imaginary entity that opens and closes the gate with the desired timings. Note that the SOURCE node should be specified to create one entity only. This single entity then circulates in the loop and opens and closes the gate as it goes through the SWITCH nodes called OPEN and CLOSE. The DELAY nodes specify the length of times during which the gate is to remain open and closed. Watch this model in animation. Note that when the gate opens, all entities leave the gate at the same simulation time.

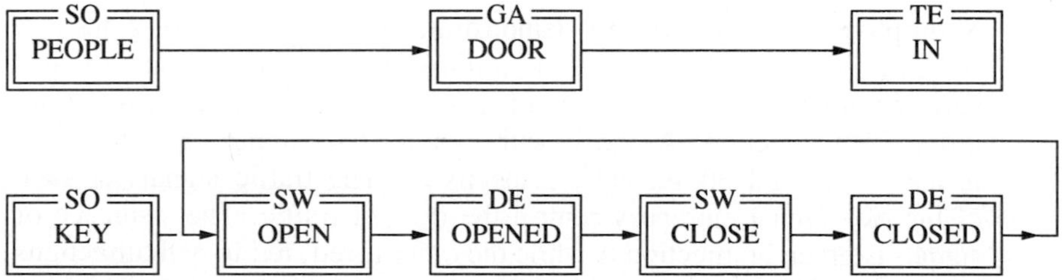

FIGURE 5.13*b*
Illustration of GATE node controlled by SWITCH nodes.

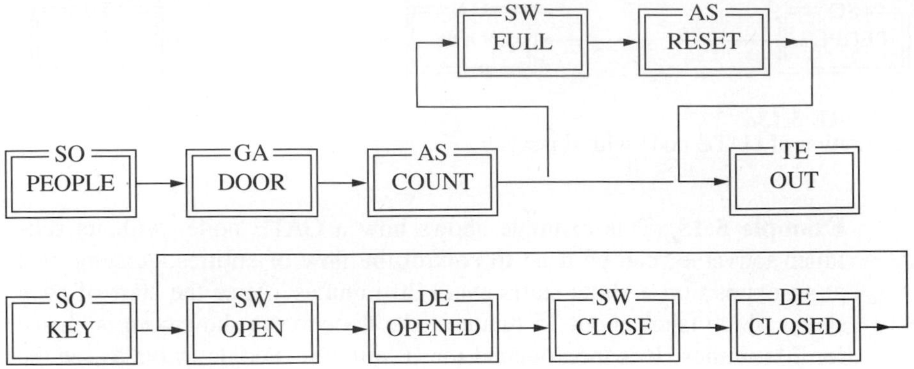

FIGURE 5.14
Illustration of controlled entity release from a gate.

Example 5.14. This example demonstrates how a controlled number of entities may be released from a gate. Assume in the previous example that the showroom has a maximium capacity of 45 people. At the start of each show, the door is opened, and a maximum of 45 people enter the showroom. We are interested in the statistics on the number of people who cannot enter the showroom at the beginning of the show.

The network in Fig. 5.14 shows the model for this example. As the entities leave the GATE, they enter an ASSIGN node at which a variable, say COUNTER, is incremented by one. The branch from this node to OUT is of the Always branching type. The branch from the ASSIGN node to the SWITCH node named FULL is conditional and is taken only when COUNTER becomes 45, indicating that 45 people have entered the showroom. At the switch node, the gate is closed, and at the subsequent ASSIGN node the value of COUNT is set to zero to prepare for counting the number of people entering for the next show. To collect the observation-based statistics on the number of people who cannot enter at showtime, the queue length at the gate (a system variable) may be observed each time the node FULL is hit by an entity. These requirements may be identified in the Statistics menu.

Example 5.15. This example demonstrates the application of gates and switches for determining a desirable traffic light timing. Assume that a narrow tunnel allows the passage of cars in one direction at a time. Cars arrive from directions A and B with exponential interarrival times with means of 20 and 30 seconds, respectively. The traffic signals at each of the two tunnel entrances control the flow of traffic. The sequence of signals is green in direction A while the other is red, red in both directions for cars to clear the tunnel (this takes 2 minutes), and green in direction B while the other is red. We would like to find the best traffic light timing that results in the smallest overall waiting time of all cars.

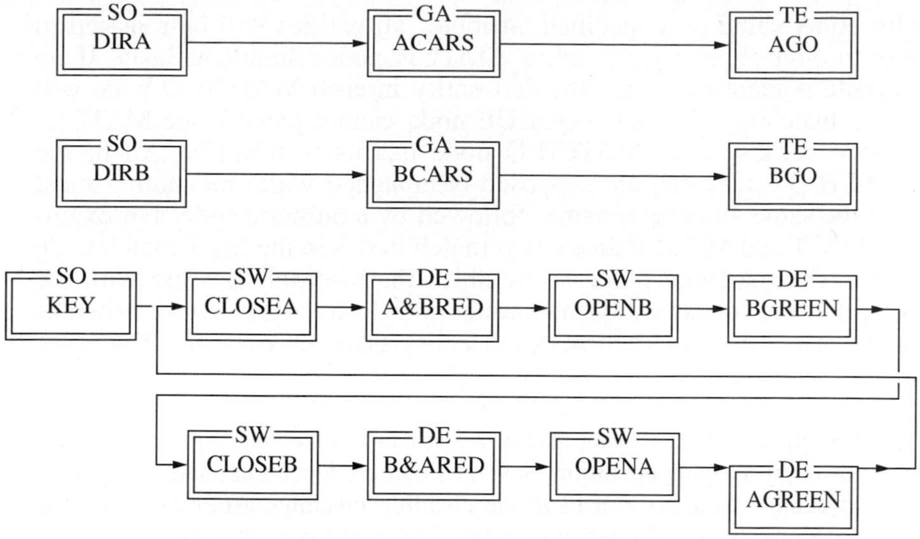

Figure 5.15
Use of SWITCH and GATE nodes for a traffic light simulation.

To find the best timing, we should systematically try various pairs of green light times in directions A and B, and, for each selected pair, we should simulate the system and observe the total average waiting times in both directions. This search may take several trials to reach a satisfactory solution and could use an available technique such as an adaptive pattern search method. Since the intention here is the demonstration of the capabilities of GATE and SWITCH nodes, we will try the model for only one pair of green light timings, which is 4 minutes in direction A and 3 minutes in direction B.

The network model for this example is shown in Fig. 5.15. The upper two SOURCE nodes create the car arrivals. Each GATE represents the queue of cars before the tunnel entrance. Both gates are initially closed. The SOURCE node named KEY creates one entity that circulates in the bottom disjoint network and activates the switches to repeat the sequence: open the gate in direction A, delay for 4 minutes, close the gate, delay for 2 minutes, open the gate in direction B, delay for 3 minutes, close the gate, and delay for 2 minutes. Various timings may be simulated for each direction. The queue statistics for the gates provide the desired information for each simulated scenario.

5.11 MATCH-Q NODE

The MATCH-Q node matches entities that arrive at the other associated MATCH nodes. When there is one entity at each of such nodes with equal value of a specified attribute, all entities will be released to pass through their corresponding MATCH nodes simultaneously. If no attribute is identified, then the first entity in each MATCH-Q node will be the matching candidate. QUEUE node cannot precede the MATCH-Q node because each MATCH-Q node has its own buffer. All of the MATCH-Q nodes that are supposed be matched with one another must have the same alphabetic name, followed by a numeric code. For example, MAT1 and MAT2 indicate two match nodes in the MAT match node group. The matching process resembles the assembly process with the exception that in assembly the entities join to form one entity, whereas in the match process each matched entity takes its own direction after being matched.

Example 5.16. In a manufacturing process, a cleaning operation is followed by a drying operation. Several small product components are placed in cubic containers that fit in the cleaning machine. After cleaning, the components are emptied into cylindrical containers that fit in the spin-drying machine. There are three cleaning machines and two drying machines. Each machine has one container of its type. After being released from the corresponding operation, each container type awaits the arrival of the other type for the transfer of its contents. If a cubic container leaving the cleaning machine does not find an available cylindrical container to empty into, it has to wait until one becomes available. Likewise, a cylindrical container when freed should await the arrival of a cubic container. The cleaning time per machine is uniformly distributed within 2 to 4 minutes. The drying time per machine is 2 minutes. The loading and unloading times of containers are negligible. We are interested in finding the average idle time of each kind of container. This is equivalent to the average idle time of the corresponding machine.

Figure 5.16 shows the network diagram of the system. The upper and lower disjoint networks in this figure represent the flow of the cubic and cylindrical containers, respectively. The SOURCE node named

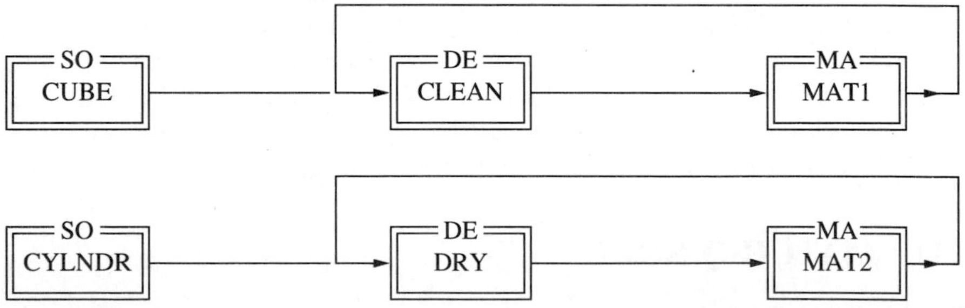

FIGURE 5.16
Illustration of application of the MATCH node.

CUBE initially generates three entities representing the cubic containers. The other SOURCE node generates two entities representing the cylindrical containers. Time between these creations is zero. Entities representing the cubic containers are delayed in the DELAY node named WASH and are then sent to MATCH-Q node to await the arrival of matching cylindrical containers at the other MATCH-Q node. When a cubic and a cylindrical container are matched, the transfer of the contents from one into another takes place, and each container takes its own route. The queue statistics for the MATCH-Q nodes provide the desired information. It is recommended to animate this model in order to observe the performance of the match process. Note that since the number of entities (containers) needing the facilities (machines) is equal to the number of facilities, machines are not represented as facilities in this example. If, however, specific statistics about machine utilization are desired or breakdown possibility exists, it will be necessary to use queues and facilities with parallel servers.

5.12 GROUP-Q NODE

The GROUP-Q node is used to either combine or batch the arriving entities in the form of a single outgoing entity. When ungrouping of the grouped entities at a later stage is not requested, the entities will be combined to form an entity whose attributes may be set by certain rules, which are provided in a selection window. When later ungrouping is requested, the incoming entities will be grouped into an outgoing batch in which each entity preserves its original attribute values. The batch itself may be given a name and be assigned attributes. The batched entities may be unbatched later using the UNGROUP node. Carriers may be defined with various capacities and other features to carry the grouped entity through the network. Carriers are entities, and when they are used, the attribute set of the grouped entities leaving the GROUP-Q node is the attribute set of the carrier transporting the group. When the carrier arrives at an UNGROUP node, it may release a portion or all of the batched entities.

Example 5.17. In a manufacturing activity, 3 units of component A and 2 units of component B are assembled to form a final product. The arrival rate of components A and B average 4 and 3 per minute, respectively. Interarrival times are exponentially distributed. The assembly operation takes 30 seconds. We are interested in finding the buffer size for storage of each component before the assembly station.

Figure 5.17 shows the network diagram of this example. The SOURCE nodes create the components with time between creations exponentially distributed with means of 20 and 30 seconds for components A and B. The GROUP-Q nodes have Grouping by Number as their

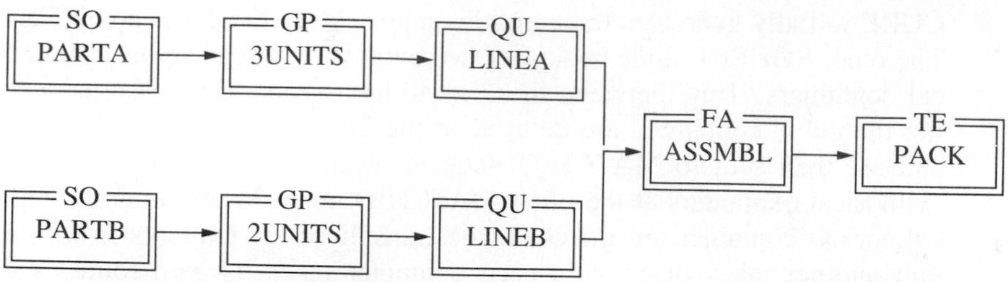

FIGURE 5.17
Illustration of the Assembly option in branch selection.

option; they group 3 units of component A into one and 2 units of component B into one, respectively. The grouped entities then go to QUEUE nodes before the FACILITY node. The selection option at the facility node is ASM. The queue statistics for the GROUP-Q and QUEUE nodes collectively provide the desired information. To observe the performance of the grouping process, it is recommended to run this model in the animation mode.

5.13 UNGROUP NODE

The UNGROUP node is used to release a portion or all of the batched entities arriving at the node, with or without a carrier. The possible number of outgoing branches of the UNGROUP node is either one or two. When there is only one outgoing branch, all batched entities are unbatched and sent along with the carrier (if used) through the outgoing branch. When there are two outgoing branches, a portion or all of the batched entities are unbatched and released through one branch, while the carrier and the remaining entities (if any) in the batch continue on through the other outgoing branch. No more than two outgoing branches are allowed at the UNGROUP node.

Example 5.18. Passengers arrive at a shuttle bus station in the long-term parking area of an airport with an exponentially distributed interarrival time which has a mean of 0.7 minutes. There are 5 shuttle buses operating between the parking area and the airport. Each bus can carry a maximum of 20 passengers. The trip time of each bus to the airport is uniformly distributed within a range of 7 to 11 minutes. The return time is also uniform, with a range of 5 to 9 minutes. We are interested in the queue statistics of passengers at the shuttle station in the parking area under either of the two departure rules for shuttle buses. Under the first operating rule, buses take whatever number of passengers are available at the station (up to their capacity) and leave immediately. Under the second rule, buses wait at the station until they are full to capacity, after which they leave for the airport.

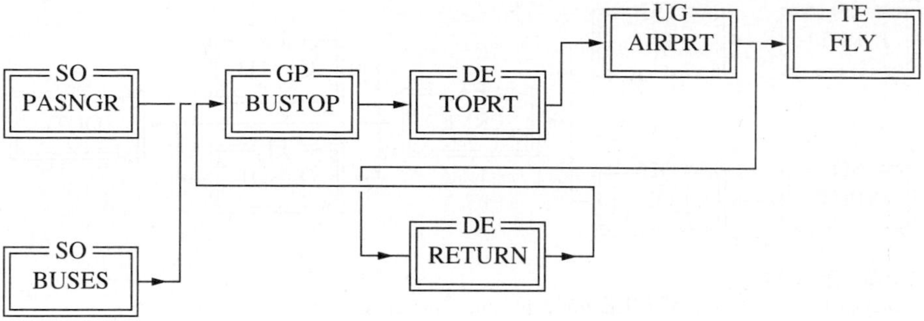

FIGURE 5.18
Illustration of application of GROUP and UNGROUP nodes.

Figure 5.18 shows the network diagram of this system. The SOURCE node named PASNGR creates the passenger entities that are sent to the GROUP-Q node. At this node, a carrier is defined with the name BUS with a capacity of 20. The choice of grouping rule in the first scenario is Available, and in the second scenario it is Capacity. EZSIM provides a detailed description of carrier departure criteria in a help window. The SOURCE node named BUSES initially creates five shuttle buses with the given name of BUS (with zero time between creations). The grouped entities go to the DELAY node that represents the travel time to the airport. At the UNGROUP node called AIRPRT, the ungrouping rule is ALL, meaning that all passengers get off the bus. The passenger route is to the TERMINATE node while the carrier route takes the BUS entities back to the GROUP node through a DELAY node that represents the return time. The queue statistics provide the desired information. Animation of this model provides insight into the operation of GROUP-Q and UNGROUP nodes.

5.14 FILE NODE

The FILE node is used to manipulate the contents of queues in a model. The operations at this node include transferring, copying, and deleting entities along with their attributes. Position in queues (rank) for the subject entity may be specified as either last, first, or *N*th. All or a portion of entities (specified by a given attribute) in a queue may be manipulated simultaneously by the FILE node. Transfer and copy operations may be performed on the same file; that is, entities may be transferred or copied from a given queue onto the same queue. File operations may be executed as an event by sending any entity to the FILE node at the desired time.

Example 5.19. There are two parallel queues for a facility. Each time a departure takes place, the last entity in the longer line leaves the line to

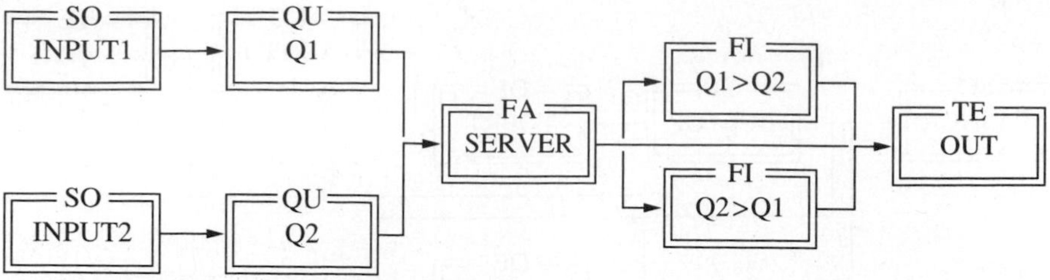

FIGURE 5.19
Illustration of the use of FILE node for moving entities between queues.

join the other one, if the shorter line has at least two less entities in it. The interarrival time to each line is exponentially distributed with a mean of 10 minutes. Service time at the facility is uniformly distributed within a range of 3 to 5 minutes. Queue selection at the facility is random. We are interested in the queue statistics.

Figure 5.19 shows the network diagram of this example. Entities leaving the facility take one of the three possible branches. Conditional branching is specified for the upper and lower branches. The upper branch goes to the FILE node that takes the last entity in QUE1 and places it in QUE2, if LOF(QUE1)>LOF(QUE2)+1. The lower branch goes to another FILE node, which takes the last entity in QUE2 and places it in QUE1, if LOF(QUE2)>LOF(QUE1)+1. The Expression option may be chosen to create the expressions for the above conditions. The middle branch goes directly to the TERMINATE node, and the Last Choice branching option is specified for it.

5.15 SUMMARY

This chapter provides a broad overview of the EZSIM software features in relation to the common processes in discrete systems. The capabilities of individual nodes in representing common processes have been discussed and demonstrated using several modeling examples. Also covered are various features of EZSIM that address modeling needs related to branch selection, attribute and user variable manipulation, and a variety of approaches to statistics generation. Following the individual examples given in this chapter should enable the user to use EZSIM and other simulation tools, especially process-oriented simulation languages, for a variety of simulation modeling applications. An extensive coverage of realistic simulation application examples will be provided in Chapter 8.

5.16 EXERCISES

5.1. There is a barbershop in which only one barber works at any given time (i.e., a single-server system). The interarrival time of customers to this barbershop is exponentially distributed with a mean of 12 minutes. The length of time required to get a haircut is normally distributed with a mean of 15 minutes and a standard deviation of 3. It is assumed that there is no limit to the number of customers who can be waiting for a haircut and that once a customer arrives at the barbershop he will not leave until his hair has been cut. Initially there are no customers waiting and the barber is idle. Perform a simulation of the barber shop for an 8-hour period. The goal of this simulation is to estimate the average time spent in the barbershop by each customer.

5.2. In a factory, parts are produced at the rate of one every 5 minutes. An inspection station checks each part. Eighty-five percent of the units are accepted and are sent to the shipping department. The rejected sets go to a repair station where there are two operators working in parallel. The repaired parts are then sent to the inspection station. The inspection time is uniformly distributed between 3 and 5 minutes. The repair time for each operator is exponentially distributed with a mean of 7 minutes. Simulate this system for 1000 time units to determine the buffer sizes needed in the inspection and repair areas. Would you recommend an additional inspection operator and/or repairman? Find the statistics on the total time that each part spends in the system.

5.3. In Exercise 5.2, assume that those parts which need repair for the second time are scrapped. Model the system and simulate the model until 3 scrap occurrences are observed (you may use a TERMINATE node with entity count of 3 for the scrapped entities).

5.4. At an air terminal, cargo airplanes with capacities of either 50 or 70 cargo units arrive at the rate of one every 60 minutes. Each plane waits until it is loaded to capacity, and then it takes off. Fifty-five percent of the arriving planes have a 50-unit capacity. Cargo arrives in batches of 10-unit loads with an exponential interarrival time with a mean of 9 minutes. Modeling the unit loads as resources, run the simulation model of this system for 100 arriving planes to determine the average waiting time of each plane before takeoff.

5.5. Forty percent of the customers arriving at a barbershop require styling; the rest require trimming. The styling time is Exp(15) and the trimming time is Exp(8). There is one barber at the shop. The interarrival time of customers is Exp(12). Simulate this system for 1000 time units, (a) assuming the barber as a facility and (b) assuming the barber as a resource. Compare your results.

5.6. Interarrival times of passengers leaving the ticket counter at a bus stop can be represented by a normal distribution with a mean of 3 minutes and a standard deviation of 0.5 minutes. Once they have purchased their tickets, the customers wait in the bus stop's waiting room until 10 customers have accumulated. If a bus is available at that time, the customers board it, which takes 5 minutes. Simulate the processing of

10 busloads of customers through the bus stop, and compute the average length of time a customer has to wait in the waiting room. Assume that the interarrival time of buses is exponentially distributed with a mean of 15 minutes.

5.7. Consider the processing of customers at a bank with two tellers. The service time of each teller is uniformly distributed between 6 and 12 minutes. Customers that arrive at the bank when both servers are busy wait in a single waiting line. Initially, there are two customers in line. If there are five customers in line when a customer arrives at the bank, the incoming customer will leave the bank without being waited on. A customer will also leave the bank without being waited on if he or she has been waiting in line for 5 minutes. The bank is to be analyzed until 100 customers have been waited on. The time between arrivals is prescribed as samples from an exponential distribution with a mean of 10 minutes. The first customer is scheduled to arrive 5 minutes after the simulation begins. Determine how many customers are lost during the simulated time period.

5.8. Consider a situation involving two types of jobs that require processing by the same worker. The job types are assumed to form a single queue before the worker. The first job entity is scheduled to arrive every 8 minutes, and only 100 of them are to be created. These jobs have an exponentially distributed service time with a mean duration of 7 minutes. For the other type of job, the time between arrivals is 12 minutes, and 50 of these jobs are to be created. The estimated service time for each of these jobs is exponentially distributed with a mean duration of 5 minutes. Both types of jobs are routed to a single queue. Entities (jobs) at this queue are ranked on small values of their actual service time. The actual service time of the server is specified as the estimated service time plus a sample from a normal distribution with a mean of 3 and a standard deviation of 0.4 minutes. Thus, the actual processing time is equal to the estimated processing time plus an error term that is assumed to be normally distributed. This model might be used to represent a job shop in which jobs are performed in the order of the smallest estimated service time. Estimate the average time that each job spends in the shop.

5.9. Consider a company with a maintenance shop that involves two operations in series. When maintenance is required on a machine and four machines are waiting for the first operation, the maintenance operations are subcontracted to an external vendor. The queue for the second operation has a maximum length of two. Therefore, if two machines are waiting in this queue, any new machines will be blocked along with the first operation. Maintenance requirements are generated every 0.5 time units and are routed to the first operation, which has a triangularly distributed service time with a mode of 0.4 and minimum and maximum values of 0.2 and 0.8, respectively. The second operation has a service time that is uniformly distributed between 0.5 and 1. Simulate this process until 10 machines have been sent to the external vendor for maintenance.

5.10. Consider the simulation of a situation involving an inspector and an adjustor. Seventy percent of the items inspected are routed directly to packing; 30 percent of the items require adjustment. Following adjustment, the items are returned for reinspection. The inspection time is a function of the number of items waiting for inspection and the number waiting for adjustment. If 5 or fewer items are waiting to be inspected, the service time of the next entity is 6 minutes. If less than 9 units are waiting to be inspected and 2 or fewer units are waiting to be adjusted, the service time of the next entity is 8 minutes. If neither of these conditions apply, the item will not be processed (i.e., it will be terminated). When 20 such entities materialize, the simulation run is to be completed. The run can also be completed when 300 items have been sent to packing. The interarrival time of items to be processed is 10 minutes, and the adjustment time is exponentially distributed with a mean of 10 minutes. Model and simulate this system.

5.11. Simulate 10 different timing configurations for the traffic light described in Example 5.15. Run each scenario for 1000 seconds and find the best timing for which the overall average waiting time of cars is minimized.

5.12. Cars arrive at a traffic signal according to a normal distribution. The mean of the distribution is 12 seconds and the standard deviation is 3 seconds. The light changes from green, its initial status, to red every 1 minute, and vice versa. When the light turns green, the cars waiting at the light take 4 seconds each to pass through the intersection before the light turns red. Simulate the passage of 100 cars through this intersection, and calculate the average number of cars that wait at the traffic light. Assume a single lane of traffic in each direction.

5.13. For the problem described in Example 5.18, first try various grouping rules for the carrier to determine the passenger loading policy which results in the least average passenger waiting time. Next, assume that the parking lot manager can utilize shuttle vans which can carry a maximum of 7 passengers. Using simulation, find the number of required vans that will result in approximately the same (within 10 percent) average customer waiting time. Choose a simulation time of 8 hours for each modeling scenario.

5.14. Customers arrive at a supermarket every 1.5 minutes. The first thing they do upon entering the supermarket is obtain a shopping cart, five of which are available for customer use. If there are no carts available when a customer arrives, assume that they will wait for the next cart. After obtaining a shopping cart the customers proceed through the supermarket to do their shopping, which takes approximately 15 minutes. After having completed their shopping, the customers form a single line in front of 2 checkers. Assume that the length of time required for a checker to ring up a customer's groceries, and for the customer to pay for those groceries, is 3 minutes. As soon as the customer is done paying for the groceries, the shopping cart is available for the next customer to use. It then takes the customer 1 minute to exit the store. Simulate the serving of 50 customers by the checkers and their subsequent exiting from the store. We are interested in the statistics regarding the available carts and the customer waiting time for a cart.

5.15. Memorandums arrive at a copy machine, which is run by a single operator, with a uniform interarrival time distributed between 1 and 3 minutes. The operator makes one copy of the memorandum. The copying time is exponentially distributed with an average of 0.5 minutes. The original document is then sent to the file room to be filed, and the copy is sent to the department that the memorandum was written to. Simulate the copying and routing process for one hour of operation. All memos copied in the first 20 minutes of the simulation are sent to the engineering department; all memos copied in the next 25 minutes are sent to the manufacturing department; all subsequent memos are sent to the business office. We are interested in knowing the copier utilization statistics and the total number of copies sent to the engineering and manufacturing departments.

5.16. Boxes of paper are delivered to a supply room at a rate of one box every two minutes. An additional five minutes are required to store each of the boxes. Because of the size of the storage room, the boxes are removed on a last-in-first-out (LIFO) basis. Removal of the boxes from the storage area is performed by two persons who work for 20-minute intervals. After having worked for 20 minutes, both of the workers take 40-minute breaks. Carrying each box from the storage area takes approximately 0.6 minutes. Simulate the removal of 200 boxes from the supply room to find the queue and server utilization statistics.

5.17. Two bulldozers dig the earth to prepare for a building foundation at a construction site. Each dumps its load once every 0.8 minutes on an area with a limited size where small and large trucks are loaded by loaders. 10 bulldozer loads make one load for the small trucks. 18 bulldozer loads make one load for the large trucks. The loading area has a capacity of 40 bulldozer loads. As soon as this capacity is reached the bulldozers stop operating. They resume their operation when there is enough dumping area for at least 15 bulldozer loads. There is one loader, which takes 2.5 and 4 minutes to load each small and large truck, respectively. There are two large trucks and three small trucks in operation. Trucks dump their loads outside the city limits and return to the construction site. Their round trip time is normally distributed with a mean of 44 minutes and a standard deviation of 8 minutes. Simulate this operation for 1000 minutes to determine the total number of bulldozer loads that are carried and the total length of time when bulldozers are idle.

5.18. Suppose that you are the construction manager in charge of the operation described in Exercise 5.17. Can you reduce the resources used (bulldozers, loaders, and trucks) and still maintain approximately the same performance (number of bulldozer loads carried in 1000 minutes)? How?

CHAPTER 6

ANALYSIS OF SIMULATION INPUTS AND CREATION OF THEIR EFFECTS

6.1 INTRODUCTION

Real life is rarely deterministic. Many of the outside influences to systems under study (such as entity arrivals) and internal system component behaviors (such as duration of service time) follow nondeterministic, or random, patterns. The arrival of cars at an intersection and the duration of airplane maintenance operations in the earlier examples of this book are representative instances. To build a representative simulation model of the system under study it is necessary to re-create the random effects that are present in the system. In this chapter we present the methods for analysis of the input data, some numerical methods for random number generation, and a method for generation of random variates.

6.2 RANDOM EFFECTS IN SIMULATION

One possible method of re-creating the random effects that are present in the system is to use the stream of data collected from the real world (field data) and make the model subject to exactly the same data patterns. The traffic intersection simulation discussed in Chapter 3 used this approach since the model cars were released to the intersection in accordance with the same arrival time data collected at the real intersection. There are, however, several problems with using field data. First, field data is generally limited in number; the number of data points that can be gathered is often limited due to the time and cost associated with the data collection process. This limits the possible length of simulation to the length of the data collection period. Second, field data is available only for systems that are currently operational. As indicated previously, an important role of simulation studies is in the design of nonexisting future systems. Obviously, there would be no field data for some external influences or internal component behavior of most hypothetical systems. Third, it is not possible to easily perform a sensitivity analysis using field data (e.g., finding the effect of a 10 percent increase in the standard deviation of service times of a facility). Fourth, since data is generally not in computer-readable forms, entering large amounts of data into the computer is too time-consuming.

Another drawback of using field data as the source of randomness is that once entered it occupies valuable computer memory, which could be used otherwise to build larger models and to generate more useful statistics about the simulated system performance. It should be noted, however, that field data, when available, is sometimes a useful means for checking the validity of simulation models. Ideally, the model subjected to the field data should perform exactly like the system it simulates, and thus its output statistics should closely correspond to the statistics collected on the actual system performance.

Because of the above limitations inherent to field data, it is necessary to devise a mechanism to generate random data artificially according to the analyst's desired specifications. These characteristics in many instances are those of the collected field data. The procedure has the following steps:

Step 1. Collect a sufficient amount of field data to serve as a reliable sample of the actual statistical population.

Step 2. Perform a statistical analysis of the sample in order to identify the nature (probability distribution and its parameters) of the statistical population from which the sample is taken.

Step 3. Use or devise a mechanism that is able to create an unlimited number of random variates that are represented by the population identified in Step 2.

6.3 IDENTIFYING THE DISTRIBUTION OF FIELD DATA

In order to generate random effects with a behavior similar to that of the field data, one must first identify a model of behavior for the field data and then use that model for generation of the corresponding random effects. This model is called a *probability distribution function* (pdf); it provides the probability of occurrence of each given value or a given range of values for the random variable.

If each of the commonly observed classes of random phenomena is classified according to the characteristics of its pdf, the number of classes will not be very large. In other words, a small number of pdf structures can represent most commonly observed random phenomena. Inspired by this fact, mathematicians have devised general pdf structures for each of these commonly observed random effects. These structures are called *theoretical distribution functions*. Uniform, triangular, normal, exponential, binomial, Poisson, Erlang, Weibull, beta, and gamma are some of the popular theoretical probability distribution functions. Appendix B presents the structure of these distribution functions and the types of situations in which they usually apply.

Having predefined functional relations in pdf form greatly facilitates the process of generating random variates that are representative of real random phenomena. Whenever possible, it is desirable to use an available theoretical pdf that fits the field data. Before selecting an available theoretical pdf it is important to identify which function will provide the best representation of the parent population of the data, using a method such as the statistical goodness-of-fit test.

One method for identifying candidate probability distribution functions is to use the visual approach, using frequency histograms. To create a histogram of the field data, the observed range of the data is divided into a number of cells of equal size. The data is then recorded by placing an observation mark within the cell in which the observed quantity belongs. The number of marks on each cell indicates the frequency of occurrence of random variable values within the upper and lower bounds of each cell.

The relative frequencies of occurrence can be found by dividing the number of observations in each cell by the total number of observations. The result is an estimate of the actual probability of an observation falling within a given cell. This is why frequency histograms resemble the shapes of the corresponding probability distribution functions. Selection of the number of histogram cells, which affects the cell width and the shape of the histogram, is important. Having too few or too many cells results in histograms that do not reveal any useful information. There is no specific rule or set formula for selection of the number of cells for a histogram; the decision is generally intuitive. However, it is known that setting the number of cells approximately equal to the square root of the number of data points often works well (see Hines and Montgomery, 1980).

Once a frequency histogram for the field data is established, the shapes of the available theoretical probability distribution functions can be visually compared with the shape of the histogram, and the candidates for goodness-of-fit tests can be selected. The following example shows how a histogram may be constructed for a given data set.

Example 6.1. Let us assume that 60 data points representing the time in seconds between arrivals of cars from a certain direction at an intersection are as follows:

12	07	26	06	18	15	44	28	09	44	16	19	37	29	08
10	09	18	35	17	20	31	08	24	15	18	30	11	28	68
07	19	04	26	25	37	46	09	18	14	07	34	26	09	49
09	16	32	07	04	06	23	08	36	19	05	21	09	03	22

Using a cell width of 10 seconds, the above data may be arranged in such a way that each observation could indicate an occurrence within its corresponding cell. This arrangement is shown in Table 6.1. Note that this table is presented only to show the procedure for building a histogram. In a normal application, it is not necessary to write the value of each observation on the histogram. A typical histogram shows only the frequencies for each cell in the form of vertical bars, as shown in Fig. 6.1, which demonstrates the histogram for the above data. Several spreadsheet and database management programs offer capabilities for building histograms for user data. Appendix C presents a simple BASIC program that may be used to create histograms.

TABLE 6.1
Arrangement of data in corresponding cells

Cell	Frequency	Data
[0–10)	19	07 06 09 08 09 08 07 04 09 07 09 09 07 04 06 08 05 09 03
[10–20)	16	12 18 15 16 19 10 18 17 15 18 11 19 18 14 16 19
[20–30)	12	26 28 29 20 24 28 26 25 26 23 21 22
[30–40)	8	37 35 31 30 37 34 32 36
[40–50)	4	44 40 46 49
[50–60)	0	
[60–70)	1	68

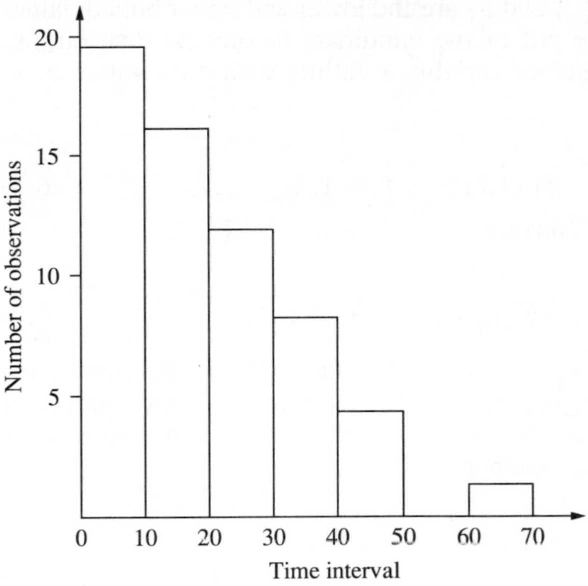

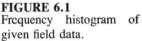

FIGURE 6.1
Frequency histogram of given field data.

6.3.1 Goodness-of-Fit Tests

The goodness-of-fit tests are used to test the hypothesis that a given number of data points are independent samples from a particular probability distribution. It should be noted that failure to reject a hypothesis does not correspond to accepting it; therefore, if a candidate distribution is found to have a good fit with given field data, these tests at best reject the hypothesis that the candidate distribution is not the actual distribution of the population from which the data is drawn. Furthermore, these tests are not very reliable for a small number of data points; in such situations, they are capable of detecting only gross differences. On the other hand, for a large number of data points these tests are usually sensitive to small departures from the candidate distribution; therefore, they may recommend rejecting the hypothesis of goodness of fit even if the candidate distribution has a close fit. In the following sections the two most popular goodness-of-fit tests and the corresponding illustrative numerical examples are presented.

6.3.1.1 CHI-SQUARE TEST. One of the most common tests for goodness of fit is the chi-square test. First, a histogram of the field data is built. The upper and lower bound pair of each cell of the histogram are then taken, and using the candidate pdf, the probability of the random variable falling within each pair of upper and lower bound values is found. For example, if a_{i-1} and a_i are the lower and upper bound values of cell i, and $f(x)$ is the pdf of the candidate theoretical distribution, then the probability of random variable x falling within the range a_i to a_{i+1} is

$$p_i = \int_{a_{i-1}}^{a_i} f(x)\,dx \qquad i = 1, 2, \ldots, N \tag{6.1}$$

or

$$p_i = F_x(a_i) - F_x(a_{i-1}) \qquad i = 1, 2, \ldots, N \tag{6.2}$$

where N is the number of histogram cells and $F_x(X)$ is the cumulative distribution function evaluated at $x = X$. Assuming that the number of observations in cell i of the histogram is O_i, then the test statistics are computed in the following manner:

$$\chi^2 = \sum_{i=1}^{N}(O_i - n\,p_i)^2/n\,p_i \tag{6.3}$$

where n is the total number of observations and $n\,p_i$ is the theoretical frequency for cell i. If the value of the above test statistics does not exceed the value found from the chi-square table for $N - k - 1$ degrees of freedom at a given significance level, then the hypothesis that the data comes from the candidate distribution may not be rejected. The parameter k is the number of parameters of the candidate distribution being tested for goodness of fit to the field data. These parameters are estimated using the field data. Note that the chi-square test requires $n\,p_i > 5$ for all i. If some cells have too few observations, they must be combined with other adjacent cells to meet this requirement. N is the final number of valid cells.

Note that since the chi-square test for goodness of fit depends on the histogram of the data (which is generally configured intuitively), it may not yield reliable results when a histogram does not capture the true grouping of data elements. Nevertheless, this test is very popular and in most practical instances it may generate reliable results. There are, however, other tests (such as the Kolmogorov–Smirnov test) that do not need a histogram. These alternative tests, however, are more suitable for continuous distributions and may not perform satisfactorily for discrete distributions.

Example 6.2. This example demonstrates the use of a goodness-of-fit test using the chi-square method. Let us apply this method to the data presented in Example 6.1. Since the shape of the histogram for that data set resembles that of the exponential distribution, we choose this

TABLE 6.2
Computations for chi-square test

<table>
<tr><th>Cell</th><th>Observed frequency (O_i)</th><th>Theoretical frequency (np_i)</th><th>$\frac{(O_i - np_i)^2}{np_i}$</th></tr>
<tr><td>[0–10)</td><td>19</td><td>23.52</td><td>0.869</td></tr>
<tr><td>[10–20)</td><td>16</td><td>14.31</td><td>0.200</td></tr>
<tr><td>[20–30)</td><td>12</td><td>8.69</td><td>1.261</td></tr>
<tr><td>[30–40)</td><td>8</td><td>5.29</td><td>1.388</td></tr>
<tr><td>[40–50)</td><td>4</td><td>3.21</td><td rowspan="2">1.236</td></tr>
<tr><td>[50–70)</td><td>1</td><td>4.97</td></tr>
<tr><td>Totals</td><td>60</td><td>60</td><td>4.954</td></tr>
</table>

distribution as the candidate to be tested for its goodness of fit to the above field data. The exponential distribution has only one parameter, the mean, which in this case is 20.1 seconds.

Solution. The pdf of the candidate exponential distribution is thus as follows:

$$f(x) = (1/20.1)e^{-x/20.1} \qquad x > 0$$

The corresponding cumulative distribution function (cdf) of the above distribution is

$$F(x) = 1 - e^{-x/20.1} \qquad x > 0$$

Using the related equations we obtain the results shown in Table 6.2. Since in the chi-square test the expected number of observations in any one cell should not be too small (not less than 5), in Table 6.2 cell 5 is combined with cells 6 and beyond, because the number of observations in each of these cells is relatively small. Since there is one distribution parameter, the mean, that is estimated using the available data, the number of degrees of freedom for the chi-square is $5 - 1 - 1 = 3$. Assuming a significance level of 5 percent, the corresponding test value as read from the chi-square table given in Appendix D, Table D.1 is 7.81. Since this quantity exceeds our computed value of 4.954, we may conclude that the selected exponential distribution has a reasonably good fit to our field data.

6.3.1.2 KOLMOGOROV–SMIRNOV TEST. Another goodness-of-fit test that is frequently used is the Kolmogorov–Smirnov (KS) test. An advantage of this test over the chi-square test is that it does not require histogramming the field data (which, because of its subjective basis, may result in the loss of some pertinent information). Another advantage of the KS test is that it performs well even for a small number of data points.

The major disadvantage of the KS test is that it applies only to continuous distributions. Also, the original form of the KS test required that all the parameters of the candidate distribution under the test be known (i.e., the parameters could not be estimated using the field data). Given that the actual parameters of the distribution of the field data are rarely known, this seriously limited the applicability of the original KS test. More recently, a new form of the KS test has been developed which allows for the estimation of the parameters using the field data, but this recent test more favorably applies only to normal, exponential, and Weibull distributions. Note that the KS test has often been applied to other continuous distributions (and even to discrete distributions) using parameters that are estimated on the basis of the field data. Although this practice may result in finding reasonably good fits, the user should be aware that it will produce a conservative test in which the chance of rejecting a satisfactory candidate distribution will become greater than desired (see Connover, 1980).

The KS test is conducted by developing an empirical cumulative probability distribution function based on the field data and comparing it with the cumulative probability distribution function of the candidate theoretical distribution. If $X_1, X_2, \ldots, X_n$ are the given field data ranked in an increasing order, then the empirical distribution function is defined as

$$F_n(x) = \frac{\text{Number of } X_i \leq x}{n}$$

Thus, $F_n(x)$ is a step function such that $F_n(X_i) = i/n$ for $i = 1, 2, \ldots, n$.

The test is then based on the largest absolute deviation between the empirical and the theoretical cdf for every given value of x. This deviation is compared to the tabulated KS critical values (see Table D.2) to determine if the deviation can be attributed to random effects and thereby the candidate distribution can be accepted as having a good fit to the field data. More specifically, the test has the following steps:

Step 1. Rank the field data in an increasing order

Step 2. Using the theoretical cdf, $\hat{F}(x)$, compute

$$D^+ = \max_{1 \leq i \leq N} \left[\frac{i}{N} - \hat{F}(x_i)\right]$$

$$D^- = \max_{1 \leq i \leq N} \left[\hat{F}(x_i) - \frac{i-1}{N}\right]$$

Step 3. Let $D = \max(D^+, D^-)$

Step 4. Find the critical value from the KS table for the given significance level and the sample size N.

Step 5. If $D \leq$ the critical value, accept the candidate distribution as having a good fit to the field data; otherwise, reject it.

Example 6.3. In this example we use the KS test to examine under a significance level $\alpha = 0.05$ if a given set of data represents random numbers (i.e., is distributed uniformly between zero and 1). Suppose that five data points 0.53, 0.35, 0.03, 0.94, 0.22 are given.

Solution. For the uniform distribution the cdf is

$$\hat{F}(x) = \frac{1}{b-a}x \qquad a \leq x \leq b$$

For this particular case $a = 0$ and $b = 1$. Therefore $\hat{F}(x) = x$. We now arrange the cdf values in an increasing order and perform the related computations as shown in Table 6.3.

According to the computations, $D = \max(0.27, 0.14) = 0.27$. The KS critical value from the corresponding table provided in Appendix D, Table D.2 for a sample size of 5 and a significance level of 0.05 is 0.565. Since D is less than the KS critical value, the hypothesis that distribution of the given data is uniform is not rejected.

Example 6.4. In this example we use the KS test to observe if a given field data has an exponential distribution. Suppose the following 10 data points representing the interarrival times of customers to a bank teller are collected (times are in minutes): 3.10, 0.20, 12.10, 1.40, 0.05, 7.00, 10.90, 13.70, 5.30, 9.10. The data points are collected within a 63-minute time period. According to theory, if the distribution of interarrival

TABLE 6.3
Computations for test of randomness using the KS test

i	$\hat{F}(x_i)$	i/n	$i/n - \hat{F}(x_i)$	$\hat{F}(x_i) - \frac{i-1}{n}$
1	0.03	0.20	0.17	0.03
2	0.22	0.40	0.18	0.02
3	0.35	0.60	0.25	−0.05
4	0.53	0.80	0.27	−0.07
5	0.94	1.00	0.06	0.14
			$D^+ = 0.27$	$D^- = 0.14$

TABLE 6.4
Computations for testing the goodness of fit of an exponential distribution using the KS test

i	$x_i = \sum_{j=1}^{i} t_i$	$\hat{F}(x_i)$	i/n	$i/n - \hat{F}(x_i)$	$\hat{F}(x_i) - \frac{i-1}{n}$
1	3.10	0.049	0.1	0.051	0.049
2	3.30	0.052	0.2	0.148	−0.048
3	15.40	0.244	0.3	0.056	0.044
4	16.80	0.267	0.4	0.133	−0.033
5	16.85	0.267	0.5	0.233	−0.133
6	23.85	0.379	0.6	0.221	−0.121
7	34.75	0.552	0.7	0.148	−0.048
8	48.45	0.769	0.8	0.031	0.069
9	53.75	0.853	0.9	0.047	0.053
10	62.85	0.998	1.0	0.002	0.098
				$D^+ = 0.233$	$D^- = 0.098$

times within T time units is exponential, the arrival times are uniformly distributed within 0 and T. To find the arrival times we simply add the interarrival times. Thus, $t_1, t_1 + t_2, t_1 + t_2 + t_3, \ldots, t_1 + t_2 + \cdots + t_{10}$ are the arrival times of the first through tenth customers. Dividing these times by the length of the data collection period (63 in this case), a normalized set of data which is distributed between zero and one will result. We may then proceed by applying the procedure used in the previous example to test the fit for the uniform distribution.

Table 6.4 summarizes the related computations. According to this table, $D = 0.233$. Choosing a significance level of 0.05 and a sample size of 10, the critical value from the KS table is 0.409. Since the calculated value of the test statistic is smaller than the tabulated value, there is no reason not to accept that the given tabulated data is distributed according to a uniform distribution with parameters 0 and 1. Equivalently, we conclude that the interarrival times are exponentially distributed.

6.3.1.3 CONSIDERATIONS WHEN USING GOODNESS-OF-FIT TESTS. The question that naturally arises is when to use the chi-square test and when to use the KS test. Generally, for large numbers of data points (greater than 30) and discrete distributions, the chi-square test is more appropriate. For smaller sample sizes and continuous distributions, the KS test is recommended. It should be noted, however, that the KS test has also been used with reasonable success for discrete distributions.

The reader should be cautioned that according to some experts (see, for example, Schmeiser, 1992) goodness-of-fit tests do not consider the context in which the input data is to be used (e.g., the quality of the model, the importance of the particular input data to the simulation experiment, and the user's desired precision in the results). Sometimes too much data and the excessive test power may cause a good candidate distribution to be rejected. Too little data and insufficient test power gives the user a false sense of security in selecting a distribution which in fact is a bad choice.

A new curve-fitting approach (see Flanigan and Wilson, 1993) based on Bézier curves, which are usually used in computer-aided design (CAD), is recommended. This approach may perform better than the classical goodness-of-fit test recommended by most simulation and statistics textbooks. The new approach models univariate continuous probability distributions and provides an inherently graphical distribution whose degrees of freedom, or number of parameters, can be increased as needed. Distribution fitting using this approach combines all available information (i.e., subjective, visual, and empirical) in the formulation of the distribution.

6.3.2 Selection of Distribution in the Absence of Field Data

In many circumstances, especially when nonexisting systems are studied, field data may not exist. Also, when too many stochastic activities are involved in the system being studied, in the interest of analysis time a rough estimate of the distributions of these activities may suffice. An expert in the field may be consulted to provide for each activity the values for three parameters a, b, and c, which are the minimum, maximum, and most likely estimates of the values of the related random variable, respectively. The most likely value is the distribution mode. The triangular distribution (or preferably the beta distribution) may then be used to represent the process. The advantage of using the beta distribution is that a desirable shape may be obtained for it by selecting proper values for its parameters. If an estimate for the mean, μ, of the distribution is also provided, the following relationships may be used to obtain the proper parameter values for the beta distribution:

$$\hat{\alpha}_1 = \frac{(\mu - a)(2c - a - b)}{(c - \mu)(b - a)}$$

$$\hat{\alpha}_2 = \frac{(b - \mu)\hat{\alpha}_1}{\mu - a}$$

Experience has shown that most distributions that represent the time to perform a task are skewed to the right (see Law and Kelton, 1991). Note that when μ is greater than c, the distribution function is skewed to the right; otherwise, it is skewed to the left.

6.4 RANDOM NUMBER GENERATION

The significance of random numbers was established in Chapter 3. Some methods of creating certain random outcomes using physical devices (coin, die, and disk) were also discussed in that chapter. The internal random number generator of the software systems used in Chapters 3 and 5 allowed experimentation with several probability distribution types. In this section we review the details of some numerical techniques that may be used to create random numbers. These random numbers may then be used to generate random variates.

Since we discussed the use of physical devices for random number generation, let us first present the limitations of using these devices, which provide a reasonable justification for the use of numerical techniques:

1. Physical devices, unless they are highly elaborate and expensive, cannot generate truly random numbers. For example, it is almost impossible to make a perfectly balanced wheel.
2. Generation of random numbers using physical devices is a slow process which is difficult to integrate with computers that run simulation models at much higher speeds.
3. The streams of random numbers generated by physical devices are nonrepeatable. The "repeatability property" is desirable considering the necessity that various modeling scenarios be subjected to the same set of random effects.

Software methodologies have been developed for generating random numbers via numerical techniques. This process is called *pseudorandom number generation*. The word *pseudorandom* suggests that the numbers generated via numerical methods are interlinked through numerical relations and are really not independent of one another. In good random number generating algorithms, this interlinkage is made in such a way that the numbers generated do not reveal their dependencies to the statistical techniques used for checking the quality of pseudorandom numbers. The desirable properties of pseudorandom number generators are as follows:

1. They should generate uniformly distributed numbers between the continuous range of zero and one.

2. The numbers they generate should be as independent as possible of one another; ideally, no autocorrelation should exist between them.
3. The generator must be fast and should not require excessive computer memory.
4. The numbers generated should have a long cycle before their sequence repeats. The length of a cycle is the number of random numbers in a sequence before the repetition of the next similar sequence.
5. It should be possible to regenerate a stream of random numbers to repeat similar random patterns to which various modeling scenarios may be subjected. The ability to regenerate a known stream is also useful for debugging simulation programs and validating simulation models.
6. It should be possible to generate multiple streams of random numbers which are independent of each other. This property allows for the allocation of dedicated streams to certain modules within the model, so that when the model configuration changes under various scenarios the performance of those modules remains unaltered. Consider for example a situation in which the effect of adding a server to a single server queuing system is to be studied (such as the case of the freight company example in Chapter 3). If the stochastic arrival and service processes use the same stream of random numbers, the use of an additional server would alter the original allocation of random numbers to the arrival process, since the same numbers in the stream must be used by three processes instead of two (one arrival and two service processes). This results in an arrival pattern in the double-server scenario which is different than the one in the single-server scenario. Thus, the two scenarios would be compared under different arrival patterns, which is undesirable. Preferably, a unique random number stream should be dedicated to the arrival process in both scenarios so that the same arrival pattern takes place in the two models. The only change imposed on the model in the new scenario would be the addition of the new server.

The importance of the above properties will be highlighted in the course of reviewing the simulation examples that will be given in the following chapters.

6.4.1 Pseudorandom Number Generation Techniques

In this section some representative pseudorandom number generation techniques are introduced. One of the simplest methods for pseudorandom number generation is the midsquare method. Although this method generally has very poor performance, the simplicity of its algorithm makes it suitable for demonstration of the pseudorandom number generation concept. To obtain n-digit random numbers, the midsquare procedure is as follows:

0. Select an n-digit integer, called the *seed*.

1. Find the square of the number. If the number of digits of the result is less than $2n$, append leading zeros to the left of the number to make it $2n$ digits long.

2. Take the middle n digits of the number found in step 1.

3. Place a decimal point before the first digit of the number found in step 2. The resulting fractional number is a random number.

4. Feed the number found in step 2 into step 1 and repeat the process.

The following numerical example demonstrates the foregoing simple algorithm for the case of 4-digit random numbers:

```
S0=Seed=5625
S1=(5625)^2=31640625        r1=0.6406
S2=(6406)^2=41036836        r2=0.0368
S3=(0368)^2=00135425        r3=0.1354
S4=(1354)^2=01833316        r4=0.8333
S5=(8333)^2=69438889        r5=0.4388
⋮
```

The midsquare method, despite its simplicity, is not usually used because of its many weaknesses. For example, the method can quickly degenerate depending on the choice of the seed number. This may be demonstrated by choosing a seed value of 5500:

```
S0=Seed=5500
S1=(5500)^2=30250000        r1=0.2500
S2=(2500)^2=06250000        r2=0.2500
⋮
```

Note in the above that all subsequent values of random numbers will be the same! Another major drawback of the midsquare method is that if a generated random number is close to zero (having several zeros after the decimal point), all subsequent numbers will also be very small.

A more widely used method for pseudorandom number generation is the linear congruential method. This method uses the following recursively evaluated equation with a random number generated at each iteration:

$$Z_i = (aZ_{i-1} + c)\bmod m \qquad r_i = Z_i/m$$

where Z_0 is the seed. The first part of the above equation includes a modular division operation, which indicates that Z_i is set equal to the remainder of the division of the quantity in parentheses by m. Parameters a, c, and m, as well as the seed, are nonnegative integers and must satisfy $0 < m$, $a < m$, $c < m$, and $Z_0 < m$.

The following numerical example demonstrates the operation of the congruential method:

```
Seed=1, a=6, c=1, m=25

Z1=(6x1+1) mod 25          Z1=7          r1=7/25 = 0.28
Z2=(6x7+1) mod 25          Z2=18         r2=18/25= 0.72
Z3=(6x18+1)mod 25          Z3=9          r3=9/25 = 0.36
```

The program listed in Fig. 6.2 is written in BASIC and generates N pseudorandom numbers using the congruential method. When $c = 0$ the algorithm is called the *multiplicative congruential method*; otherwise, it is called the *mixed congruential method*. The choice of parameter values in the linear congruential method significantly affects the quality of the random numbers generated by this method. For instance, if the value of parameter a in the above example is changed from 6 to 5, the first number generated will be 0.4399, and all numbers generated after this will be the same (0.2399)! Therefore, the congruential method may degenerate. When its parameters are properly selected, however, it is more successful in passing the available tests of randomness.

```
10 INPUT ''How many random numbers to generate?'';N
20 REM Initialize parameters
30 Z=7  REM This is the seed
40 A=6:C=1:M=25
50 REM Generate and print N random numbers:
60 FOR I=1 to N
70 Z=(A*Z+C) mod (M)
80 R=Z/M
90 PRINT R
99 NEXT I
```

FIGURE 6.2
BASIC program listing of the congruential random number generator.

The performance of all numerical methods depends on the word size of the computer executing the related operations, since the possible sizes of the largest and smallest numbers are influenced by the number of bits that comprise the computer word size (usually 16 or 32). Accordingly, the parameters of the pseudorandom number generators are usually tuned to the choice of the computer executing the related algorithm. In the congruent method, the length of the random number cycle is always smaller than parameter m. Therefore, a relatively large value for m is desirable. A value of $2^k - 1$, where k is the word size of the computer, works well, as does $a = 2^k + 5$. As long as parameter c is prime relative to m, the result is satisfactory. This method is not very sensitive to the choice of Z_0; however, in the case of the multiplicative congruent method, a value of 0 should not be used for Z_0 because a degenerate sequence will be produced.

It is important to note that random number generation is an involved field that encompasses a considerable amount of prior and current research work. Caution should be exercised in selecting an available method for a given computer hardware platform. Only the methods that are tested and recommended by field experts should be used. Users who do not have sufficient knowledge of the domain are strongly advised against using self-invented generators! It should also be noted that since virtually all programming language compilers and simulation software tools are equipped with pretested random number generation routines, users rarely find it necessary to create their own random number generation programs.

6.4.2 Methods for Testing Random Numbers

In this section, a brief overview of some of the test methods available for checking various aspects of the quality of pseudorandom numbers is presented. If an expert report recommending a particular generator is unavailable, analysts are advised to use these test methods when conducting a simulation project.

The two most important properties expected of random numbers are uniformity and independence. The uniformity test can be performed using the available tests for goodness of fit. For example, a statistically sufficient number of random numbers may be used to check the distribution of the numbers against the theoretical uniform distribution using either the chi-square or the KS method and a significance level, say 0.05. Example 6.3 demonstrated the use of the KS test for random numbers. This test is called the *frequency test*.

Numbers may be uniformly distributed and still not be independent of one another. For example, a sequence of monotonically increasing numbers within the range of zero to one is uniformly distributed if the incremental quantity is constant for all (e.g., $0, 0.1, 0.2, \ldots, 0.9$). The *runs up and down test* is usually the major test used to check for dependency. This test detects if a statistically unacceptable pattern of increase and/or decrease exists among the adjacent numbers in a stream. Section 7.9.1 presents the details of this test.

The *gap test* is used to ensure that reoccurrence of each particular digit in the stream happens with a random interval. The KS test is then used to compare these intervals with the expected length of gaps.

The *autocorrelation test* checks the correlation among the random numbers and compares them with the desirable correlation of zero.

The *poker test* groups the numbers together as a poker hand and compares each hand to the expected hand using the chi-square test.

True random numbers should not possess any conceivable pattern. Obviously, one may conceive of a large number of possible patterns between numbers, and a special test may be devised to detect each particular pattern. It is neither advisable nor practical to perform all of these tests to check the reliability of a random number generator. One should compare the expense of performing these tests against the expense resulting from the imperfection of the generator in actual simulation projects. For more details regarding random number tests, see Banks and Carson (1984) and Law and Kelton (1991).

6.5 RANDOM VARIATE GENERATION

The major motivation behind generating random numbers is to use them for creation of random variates with various characteristics. After determining the theoretical distribution that fits the field data, the next step is to devise a method to generate random variates with the selected distribution using a random number generator. The use of random numbers for generation of uniformly distributed random variates was demonstrated in section 3.4.1. In this section we will demonstrate how random numbers may be used to generate other types of random variates.

A commonly used method is the *inverse transform technique*. This method takes advantage of the fact that the cumulative distribution function (cdf) for all statistical distributions ranges between zero and one for various values of the random variable. Also, the values of the cdf which correspond to the random variables are always distributed uniformly, regardless of the distribution of the random variable (this does not mean that all cdfs have a uniform shape). Using these properties, a random number is generated; then, using the cdf, the value of the random variate that corresponds to the random number value may be found (see

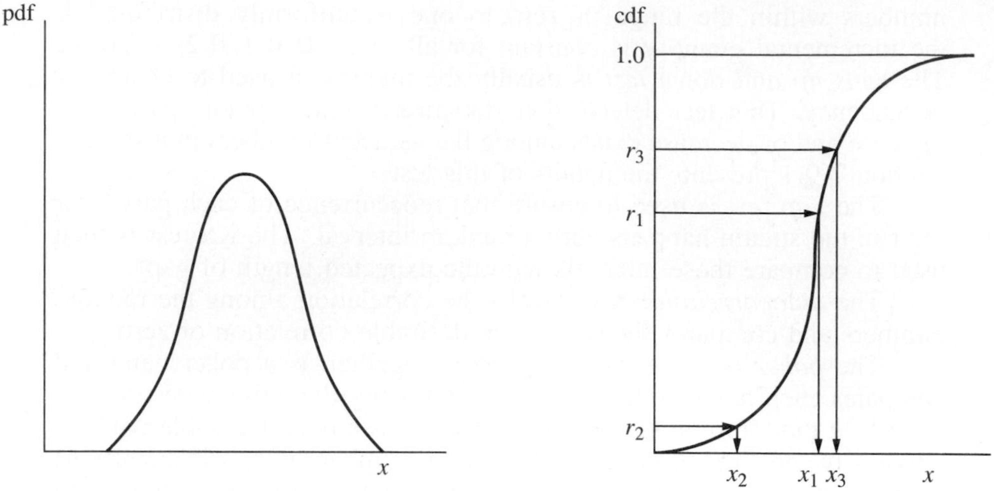

FIGURE 6.3
Demonstration of the inverse transform method.

Fig. 6.3). Denoting the random number by r and the cdf of the random variate x by $F(x)$, the following equation holds:

$$F(x) = r$$

or

$$x = F^{-1}(r)$$

Therefore, having the cumulative distribution function, one can use random numbers to generate the desired random variates by means of the inverse transform technique, provided that the inverse of the cdf is obtainable. The following examples clarify the procedure.

Example 6.5. Uniform distribution. Earlier in the book we intuitively built the following formula to generate uniformly distributed random variates using random numbers:

$$x = a + (b - a)r$$

where a and b are the lower and upper bounds of the distribution. The pdf of uniform distribution is defined as

$$f(x) = 1/(b - a) \qquad a \leq x \leq b$$

The cdf of x is found by integrating the pdf over the range of a to x:

$$F(x) = \int_a^x \frac{1}{b - a} dt = \left. \frac{t}{b - a} \right|_a^x$$

or

$$F(x) = (x - a)/(b - a) \qquad a \leq x \leq b$$

Replacing $F(x)$ by random number r yields the following equation:

$$r = (x - a)/(b - a)$$

which yields

$$x = a + (b - a)r$$

This is the same result that was intuitively found.

Example 6.6. Exponential distribution. The exponential distribution is commonly used to represent interarrival times of entities. The pdf for this distribution is defined as follows:

$$f(x) = \mu e^{-\mu x}$$

where $x \geq 0$ and the parameter μ is the mean number of occurrences per time unit (rate of occurrence). The mean interarrival time is $1/\mu$.

The cdf of x is given by

$$F(x) = \int_0^x \mu e^{-\mu t} dt = -e^{-\mu t}\Big|_0^x = 1 - e^{-\mu x}$$

Setting r equal to the above cdf value and taking natural logarithms of both sides yields

$$-\mu x = \ln(1 - r)$$

From the above equation the random variate may be expressed in terms of the random number according to the following equation:

$$x = -\frac{1}{\mu}\ln(1 - r)$$

Therefore, the above equation may be used to generate a random variate x_i (which may represent an interarrival time) for each random number r_i.

Since $1 - r$ in the above equation is also a random number, we may substitute r in place of $1 - r$ in the equation. This results in the following relationship:

$$x = -\frac{1}{\mu}\ln(r)$$

It should be noted that the above simplification may result in some undesirable effects related to the dependency of the random variates (see Schmeiser, 1992).

Example 6.7. Erlang distribution. An Erlang distributed random variable may be generated on the basis of the method provided for sampling from exponential distribution. By definition, an m-Erlang distribution with parameter μ is the result of summation (convolution) of m identical exponential distributions each with parameter μ. Therefore, given the exponential distribution for each random variable X_i,

$$f(X_i) = \mu e^{-\mu X_i}$$

The Erlang random variable Y is defined as

$$Y = X_1 + X_2 + \cdots + X_m$$

From the previous example we have

$$X_i = -\frac{1}{\mu}\ln(r_i) \qquad i = 1, 2, \ldots, m$$

Thus we have

$$Y = -\frac{1}{\mu}[\ln(r_1) + \ln(r_2) + \cdots + \ln(r_m)]$$

or

$$Y = -\frac{1}{\mu}\ln(r_1 \cdot r_2 \cdot \quad \cdots \quad \cdot r_m)$$

In the above equations $r_1, r_2, \ldots, r_m$ are random numbers. Therefore, to generate a single sample from an m-Erlang distribution using this method, m random numbers are required.

Example 6.8. Poisson distribution. The Poisson distribution is a discrete distribution which represents the number of occurrences of an event (e.g., arrival) within one time unit. The probability mass function (pmf) for the Poisson distribution is as follows:

$$P(n) = P(N = n) = \frac{e^{-\lambda}\lambda^n}{n!} \qquad n = 0, 1, 2, \ldots$$

where λ is the mean number of occurrences per time unit. The interval between the occurrences of successive events that have a Poisson distribution is exponentially distributed (in fact, the Poisson distribution is derived from the exponential distribution). If interarrival times before the nth arrival are denoted by $X_1, X_2, \ldots, X_n$, the following equation indicates that there are exactly $N = n$ arrivals during one time unit:

$$X_1 + X_2 + \cdots + X_n \leq 1 \leq X_1 + X_2 + \cdots + X_{n+1}$$

We can now proceed by generating exponential times until an arrival [say, the $(n + 1)$th] occurs after one time unit. We then set $N = n$. Using the left-hand side of the above relationship and the equation for generation of the exponential random variate, we have

$$\left(-\frac{1}{\lambda}\right)\ln(r_1) + \left(-\frac{1}{\lambda}\right)\ln(r_2) + \cdots + \left(-\frac{1}{\lambda}\right)\ln(r_n) \leq 1$$

or

$$-\frac{1}{\lambda}\left[\ln(r_1) + \ln(r_2) + \cdots + \ln(r_n)\right] \leq 1$$

or

$$-\frac{1}{\lambda}\ln(r_1 \cdot r_2 \cdot \ \cdots \ \cdot r_3) \leq 1$$

or

$$r_1 \cdot r_2 \cdot \ \cdots \ \cdot r_3 \leq e^{-\lambda}$$

This means that we can continue to generate random numbers until their product exceeds $e^{-\lambda}$. At this point the number of random numbers generated, n, is the sample derived from the Poisson distribution. To generate Poisson random variates within t time units we simply substitute $e^{-\lambda t}$ instead of $e^{-\lambda}$ in the above relationship.

Example 6.9 Normal distribution. The normal distribution is a very popular probability distribution function. Because of its complicated structure, the normal distribution function does not have an inverse representation. Consequently, the inverse transform technique may not be directly applied for sampling from the normal distribution. A method that uses the inverse transform technique to generate approximate normal variates takes advantage of the central limit theorem, which asserts that the sum of n identically distributed and independent random variables $Y_1, Y_2, \ldots, Y_n$ with mean μ and variance σ^2 is approximately normally distributed with mean $n\mu$ and variance $n\sigma^2$. If we take n random numbers to represent the above random variables, then, since random numbers have a uniform distribution ranging between 0 and 1 with $\mu = 0.5$ and $\sigma^2 = \frac{1}{12}$ (refer to the characteristics of the uniform distribution in Appendix B), the variable X defined as

$$X = \sum_{i=1}^{n} r_i$$

is approximately normally distributed with mean $0.5n$ and variance $n/12$. It follows that variable Z defined as

$$Z = \frac{\sum_{i=1}^{n} r_i - 0.5n}{\sqrt{n/12}}$$

is approximately normally distributed with mean zero and variance 1 (i.e., Z is the normalized form of X).

The approximation in this method improves as n increases. But the larger n is, the more time is required per sample generation. A value for n which is large enough to provide a reasonable accuracy and simplify the computation is 12. This yields the following equation:

$$Z = \sum_{i=1}^{12} r_i - 6$$

Now, variable Z may be used to generate approximately normal variates with mean μ and standard deviation σ using the following equation:

$$X = \mu + Z\sigma$$

or

$$X = \mu + \left(\sum_{i=1}^{12} r_i - 6 \right) \sigma$$

Hence, to generate each normally distributed sample using this method we must generate 12 random numbers to be used in the above equation.

6.5.1 Generation of Random Variates with Arbitrary Distributions

There may be situations in which field data has an arbitrary distribution which does not make an acceptable fit with any theoretical distribution. The inverse transform method may be applied in such situations, provided that the probability distribution function is first determined. Let us consider examples for continuous and discrete random variables.

Example 6.10. Arbitrary continuous distribution. Assume that a variable has the following pdf:

$$f(x) = 0.25 \qquad \text{for } 0 \le x < 1$$

$$f(x) = 0.75 \qquad \text{for } 1 \le x \le 2$$

Since the pdf for the above distribution has two different functions over two ranges of the random variate values, the cdf has a different form for each of the given ranges (see Fig. 6.4):

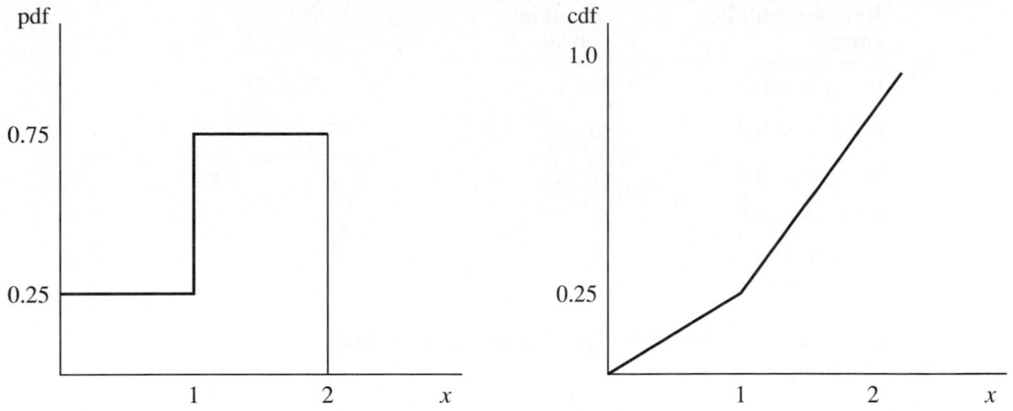

FIGURE 6.4
Graphs of the pdf and cdf of the given random variable.

$$r = F(x) = \int_0^x 0.25\,dt = 0.25x \qquad \text{for } 0 \le x < 1$$

$$r = F(x) = 0.25 + \int_1^x 0.75\,dt = 0.75x - 0.5 \qquad \text{for } 1 \le x \le 2$$

From the above equations the following may be concluded:

$$x = 4r \qquad \text{for } 0 \le r < 0.25$$

$$x = (2/3)(2r + 1) \qquad \text{for } .25 \le r \le 1.0$$

Therefore, to generate a random variate, a random number is first generated and its value is checked against the above ranges. The corresponding equation is then used to compute the value of the random variate.

Example 6.11. Arbitrary discrete distribution. As an example of an arbitrary distribution for a discrete random variable, assume that the field data has been shown to take values 5, 6, 7, 8, and 9 with the following probability distribution:

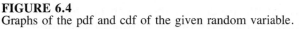

x =	**5**	**6**	**7**	**8**	**9**
$p(x) =$	.1	.2	.3	.3	.1
$P(x) =$	.1	.3	.6	.9	1

where $p(x)$ and $P(x)$ are the pdf and the cdf of the random variable, respectively. Given a random number value and using the above cdf, the following table yields the corresponding random variate values:

Random number range	Random variate
$0 \leq r < 0.1$	5
$0.1 \leq r < 0.3$	6
$0.3 \leq r < 0.6$	7
$0.6 \leq r < 0.9$	8
$0.9 \leq r \leq 1.0$	9

6.5.2 Other Methods for Random Variate Generation

The inverse transform method discussed in this chapter is directly applicable only to situations in which the inverse of the given cdf exists in a closed form. As demonstrated in the case of the normal distribution, special approaches may be used for some distributions for which the inverse is not obtainable.

One simple approach for using the inverse transform technique for distributions with an undefined inverse is to compile the results of numerical evaluation of a standardized form of the cdf of the distribution for the practical range of random variable values (e.g., the standard normal distribution table for $-3 \leq Z \leq +3$). The inverse of the function can then be found through a search and extrapolation procedure in the compiled list of values. This method is relatively fast; however, it generates approximate results. The precision of this method depends on the number of data points in the compiled list. Nonetheless, as the list of the compiled points becomes longer, memory allocation and search time increase, resulting in slower computer execution of the procedure.

There are other methods for random variate generation which do not require the inverse of the probability distribution functions. One such method is called the *acceptance–rejection* method. This method involves the use of a user-devised proxy function (not a probability distribution function) which covers (majorizes) the original distribution (i.e., for every value of x the value of the proxy function is larger than that of the original pdf). The proxy function is then normalized by being divided by the area under its curve (its integral within the range of all possible x values). This process converts the proxy function to a probability distribution function (i.e., its integral for various x values would range between 0 and 1).

Let us refer to the majorizing (proxy) function by $g(x)$, to its normalized form by $h(x)$, and to the original distribution pdf by $f(x)$. The majorizing function is devised such that it has a simpler structure than the original distribution function and, therefore, random variates from $h(x)$ can be readily obtainable using the inverse transform method. The acceptance–rejection procedure uses two random numbers in the following steps of its iterations:

Step 1. Devise a majorizing function, $g(x)$, which covers $f(x)$.

Step 2. Generate random number r_1, and find $Y = h^{-1}(r_1)$.

Step 3. Generate random number r_2. If $r_2 \leq f(Y)/g(Y)$, then accept $X = Y$ as a sample from $f(x)$; otherwise, discard Y and go back to Step 2 to generate another Y.

Example 6.12. This example demonstrates the application of the acceptance–rejection method for generating random deviates from a probability distribution for which the inverse is not defined. Suppose that we wish to generate random variates from a normal distribution whose mean and standard deviation are 10 and 2, respectively. As shown in Fig. 6.5, the maximum value of this normal pdf is at the mean and can be found by plugging $x = 10$ in the following function, which represents the normal pdf:

$$f(x) = \frac{1}{\sigma\sqrt{2\pi}} e^{-(x-\mu)^2/2\sigma^2} \qquad -\infty \leq x \leq +\infty$$

The evaluated value of $f(10)$ for the given mean and standard deviation values is 0.199. Since the normal function is open-ended from each side, we first need to truncate it. Let us choose the range [4, 16], which contains

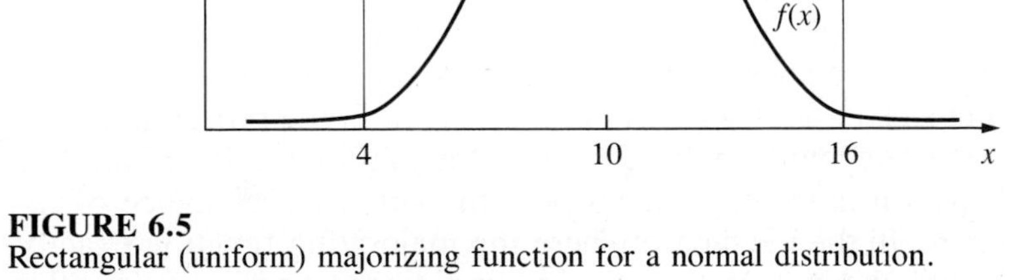

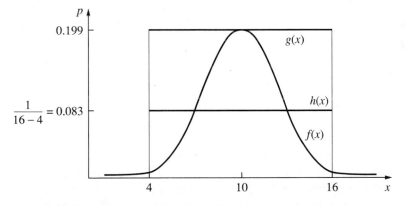

FIGURE 6.5
Rectangular (uniform) majorizing function for a normal distribution.

the mean in the middle and is as wide as six times the size of the standard deviation. Since more than 99.97 percent of the area under the normal pdf curve falls between $\mu - 3\sigma$ and $\mu + 3\sigma$, this truncation is not likely to result in serious loss of accuracy.

As shown in Fig. 6.5, one possible proxy function (and not necessarily the best one for efficiency purposes) which majorizes the above truncated pdf is a simple rectangle with a width of 12 and a height of 0.199. This majorizing function may thus be written as

$$g(y) = 0.199 \qquad 4 \leq y \leq 16$$

We can now find the normalized form of the majorizing function through dividing $g(y)$ by the area of the rectangle:

$$h(y) = \frac{0.199}{(0.199)(16-4)} = \frac{1}{16-4} \qquad 4 \leq y \leq 16$$

It may be noticed that $h(x)$ has the familiar uniform distribution which ranges between 4 and 16. Let us now proceed through the steps of the algorithm:

Step 1. Our devised majorizing function is $g(x) = 0.199$, for $4 \leq x \leq 16$.

Step 2. We generate a random number, r_1, and evaluate $Y = 4 + (16 - 4)r_1$. Suppose $r_1 = 0.37$. From the equation we obtain $Y = 8.44$.

Step 3. We generate a second random number, r_2. Suppose $r_2 = 0.68$. We must now evaluate $f(Y)/g(Y)$ for $Y = 8.44$. From the normal pdf we obtain $f(8.44) = 0.147$. Since r_2 is smaller than $0.147/0.199 = 0.74$, we accept $Y = 8.44$ as being a random variate from the given normal distribution and go back to Step 2 to generate more samples, if desired. Note that if r_2 were larger than 0.74, we would have rejected Y and gone back to Step 2 for another trial.

The above example shows that a considerable number of trials in the acceptance–rejection method may not result in acceptable samples and hence the corresponding computation time would be wasted. The efficiency of an acceptance–rejection model is measured by the proportion of acceptances in the total iterations involved in the procedure. This efficiency mainly depends on the choice of the majorizing function. The closer this function is to the original pdf, the smaller the chance of rejection becomes. In the ideal case where the majorizing function exactly overlaps $f(x)$ we have $f(x)/g(x) = 1$, and since every r_2 is smaller

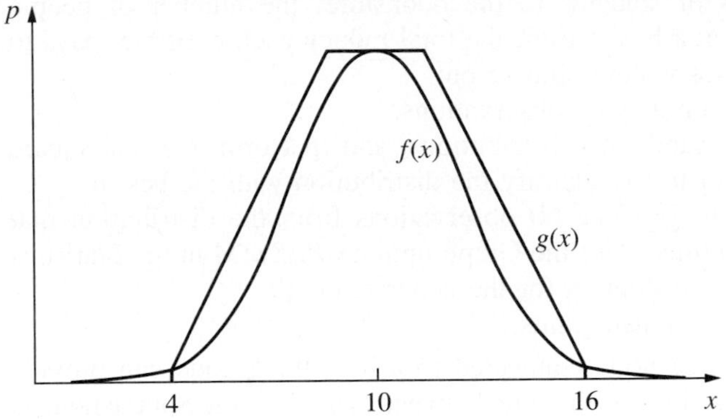

FIGURE 6.6
An arbitrary majorizing function for a normal distribution.

than 1, every Y will be accepted. (This, however, is not practical, since in this case $g(x)$ would not have an inverse; otherwise, there would be no need to use this procedure.) Consequently, it becomes obvious from Fig. 6.5 that the choice of the rectangular majorizing function is not expected to result in an efficient model for normal distribution. There are many better alternative proxy functions in this case. For example, an arbitrary function such as the one shown in Fig. 6.6 would result in superior computational efficiencies, if used as the majorizing function of the given normal distribution.

6.6 SUMMARY

This chapter has presented some methods for analyzing the simulation input data and fitting a statistical distribution to the data. Methods for using computers to generate random numbers and random variates that represent the input data or the stochastic behavior of some of the internal system components have also been introduced in this chapter. It should be pointed out that almost all simulation software tools are equipped with modules that generate random numbers as well as a variety of random variate types. Some of these tools, including EZSIM, even provide some facilities for generating random variates from arbitrary distributions defined by the user.

6.7 EXERCISES

6.1. Collect 60 data points on a stochastic process. These could be the times between the passage of cars on the street next to your room, the interarrival times of students to the bookstore, the number of people getting off a bus at a bus station, the total money each customer pays to a cashier in a grocery store, and so on.

(*a*) Plot a histogram of your observations.

(*b*) Select three candidate distributions and perform the chi-square goodness-of-fit test to identify the distribution with the best fit.

(*c*) Use EZSIM to generate 60 observations from the distribution that you have identified. Use the Graph option of EZSIM in the Statistics menu to plot a histogram for the generated data.

(*d*) Compare the two histograms.

Hint: A Source node connected to a Terminate node can provide interarrival observations. If the time between arrivals at the Source node is specified to have the statistical distribution of your choice, then collecting the traversal interarrival time of entities at the Terminate node provides the desired information. Observations that are not of time nature (e.g., amount of money paid) may be similarly created.

6.2. Starting with a seed of 5983 produce 10 four-digit random numbers using the midsquare method.

6.3. Using the linear congruential method with parameters $Z0 = 97$, $a = 62541$, $c = 19$, and $m = 1000$, generate 10 three-digit random numbers.

6.4. Test the numbers generated in the Exercise 6.3 for randomness using the KS test. Use a significance level of 0.05.

6.5. Generate 10 random numbers using the random number generation function of the computer language of your choice (BASIC, FORTRAN, C, etc.). Test the numbers for randomness using the KS test and a significance level of 0.05.

6.6. Write a program that generates 100 random numbers using the linear congruential method with the parameter values given in Exercise 6.3. Plot a histogram for these numbers. Modify the program to generate 100 random numbers using the random number generator provided by your programming environment [e.g., RND(.) in BASIC] and plot the corresponding histogram.

6.7. Using the KS test and a significance level of 0.05, test the following 20 data points, which represent the interarrival times of airplanes to an airport, for possible fit to the exponential distribution: 29, 02, 06, 33, 18, 23, 07, 01, 37, 47, 14, 03, 21, 31, 07, 05, 09, 11, 15, 24.

6.8. Write a computer program in the language of your choice to generate N samples, and plot the corresponding histograms for each of the following:

(*a*) Exponential distribution

(*b*) Poisson distribution

(*c*) m-Erlang distribution

(*d*) Normal distributions

The user should be able to easily specify N, the parameters of the distributions, and the specifications of the histogram (i.e., number of cells, cell width, and the lower bound of the first cell). Use the random number generation function of the compiler of your programming language.

6.9. Generate 50 random variates with exponential distribution and a mean of 10 using the program that you develop for part (*a*) of Exercise 6.8.
(*a*) Plot an appropriate histogram for 50 such generations.
(*b*) Using the chi-square test and a significance level of 0.05, test the data against exponential distribution.

6.10 Generate 10 random variates using the program that you developed for part (*d*) of Exercise 6.8. Use the KS test and a confidence level of 0.05 to test the generated samples against a normal distribution.

6.11 The probability density function of x is defined as

$$f(x) = (1/16)x \qquad \text{for } 0 \leq x \leq 4$$

$$f(x) = 1/4 \qquad \text{for } 4 < x \leq 6$$

(*a*) Draw this pdf.
(*b*) Using the inverse transform method, develop the necessary relationships to generate random variate x given random number r.
(*c*) Write a routine in any language that generates the above random variate. Assume that the function RND(1) which generates random numbers is available.

6.12 Using the inverse transform technique, develop the equations for creating triangularly distributed random variates.

6.13 Assume that the interarrival time and service time in a single-server queuing system are exponentially distributed. Modify the BASIC or C programs provided in Appendix A to model this system. Assume that the average interarrival and service times are 10 and 8 minutes, respectively.
(a) Run your program for 10,000 entities served.
(b) Compare your results with those of an equivalent EZSIM program.
(c) Compare the above results with the results that you obtain using the following analytical formula for the system:

$$\rho = \lambda/\mu$$

$$W_q = \frac{\lambda}{\mu(\mu - \lambda)}$$

$$L_q = \lambda W_q$$

The above equations provide average server utilization, average waiting time of entities in the queue, and the average number of entities in the queue, respectively. λ and μ are mean arrival and service rates (average number per time unit, not average time intervals), respectively.

6.14 Write a program that generates 1000 normally distributed random variates with mean of 50 and standard deviation of 12 using an acceptance–rejection model which utilizes a rectangular majorizing function. Use the random number generator of your programming language and make your program plot a histogram of the samples generated. Make provisions for calculating and printing the total number of rejections and the percentage of rejections.

6.15 Perform Exercise 6.14 using the majorizing function shown in Fig. 6.6. Compare the results of the two exercises.

6.8 REFERENCES AND FURTHER READING

Banks, J. and J. Carson: *Discrete-Event System Simulation*, Prentice-Hall, 1984.

Connover, W.: *Practical Nonparametric Statistics*, 22nd edition, John Wiley & Sons, New York, 1980.

Flanigan Wagner, M., and J. Wilson: "Using Univariate Bezier Distributions to Model Simulation Input Processes," *Proceedings of Winter Simulation Conference*, Los Angeles, Dec. 1993.

Hines, W., and D. Montgomery: *Probability and Statistics in Engineering and Management Science*, 2nd edition, John Wiley & Sons, New York, 1980.

Hogg, R., and A. Craig: *Introduction to Mathematical Statistics*, 4th edition, Macmillan, New York, 1978.

Law, A., and W. Kelton: *Simulation Modeling and Analysis*, 2nd edition, McGraw-Hill, 1991.

Pritsker, A.: *Introduction to Simulation and SLAM II*, John Wiley & Sons, 1986.

Schmeiser, B.: "Modern Simulation Environments: Statistical Issues," *Proceedings of 1st Industrial Engineering Research Conference*, pp. 139–143, May 1992.

Shannon, R.: *Systems Simulation—The Art and Science*, Prentice-Hall, 1975.

Taha, H.: *Simulation Modeling and SIMNET*, Prentice-Hall, 1988.

Watson, H., and J. Blackstone: *Computer Simulation*, 2nd edition, John Wiley & Sons, New York, 1989.

CHAPTER 7

ANALYSIS OF SIMULATION OUTPUT

7.1 INTRODUCTION

Most systems studied by simulation have stochastic behavior in their inputs (e.g., entity arrival processes) and in some of their internal components (e.g., service times at facilities). The relationship between various internal system components may also be subject to chance (e.g., probabilistic routing of entities from one process to another). Simulation models convert the stochastic influences that they receive in the form of inputs and internal processes into statistical data which constitutes their output. From the output analysis point of view, simulation is simply another statistical sampling and analysis method, with its own peculiarities. The objective in this chapter is to provide the rudimentary concepts related to the role of statistical analysis in simulation and to demonstrate the applications of these concepts by means of some numerical examples. As there are numerous research articles and books available on the topic of statistical analysis in simulation, the focus of this chapter is to acquaint the reader with a broad spectrum of related issues that a simulation analyst is likely to encounter in his or her simulation practices. Major issues related to verification, validation, experimentation, and documentation in the context of simulation analysis will also be discussed in this chapter.

7.2 THE IMPORTANCE OF SIMULATION OUTPUT ANALYSIS

Simulation studies may be performed for one or both of the following purposes:

1. To determine the characteristics (mean, variance, minimum, maximum, etc.) of certain variables for given input conditions, parameter values, and model configurations to analyze and understand the behavior of an existing system or to predict the behavior of a future system at the system design stage.
2. To compare the characteristics (mean, variance, minimum, maximum, etc.) of certain variables under various input conditions, parameter values, and model configurations. Manipulation of these factors and comparison of their effects for each simulated scenario can result in finding the condition under which the system performs satisfactorily. The ultimate intent of the analyst may be either to improve the performance of an existing system or to design a future system.

Most systems studied with simulation are stochastic. Simulation by nature is, therefore, a statistical sampling, estimation, and analysis process. As is the case in any statistical study, the sampling choice and size affect the quality of the estimates for the parameters of the population from which samples are taken. Conversely, to decide on the sample size, some knowledge about these parameters is necessary.

Systematic analysis of simulation output is an essential component of any successful simulation study. It is for these results, after all, that simulation models are built. Many simulation users, however, tend to ignore the importance of appropriately analyzing the output of their simulation programs. In fact, many simulation users draw conclusions on their simulation studies based on the output of only a single simulation run with an arbitrary simulation run length.

To illustrate how simulation results for a given model can vary under different experimental settings, consider Table 7.1, which for a single-server queuing system shows the average server utilization and average waiting time in the system for various lengths of simulation runs. In the model that generated these results the times between the arrivals to the system are distributed uniformly between 5 and 35 time units. Service times are distributed uniformly between 14 and 21 time units.

Table 7.2 shows the same statistics for various runs, each using a different random number stream for interarrival times (i.e., a different seed is used for the random variate generator). The run length for all runs in this case is 5000 time units. The differences between the results

TABLE 7.1
Simulation results for various lengths of run

Run length	Average server utilization	Average waiting time
500	0.91	28.6
1,000	0.95	36.3
5,000	0.91	32.3
10,000	0.87	29.6
20,000	0.86	27.3

in Tables 7.1 and 7.2 would have been even more dramatic if interarrival times, service times, or both were distributed according to an open-ended distribution, such as an exponential distribution (see Exercise 7.1).

The tendency for users not to pay sufficient attention to output analysis may have several causes. First, many users become amused with the process of model building and computer program construction, especially if animation is used. Consequently, these users usually prefer to spend their time performing the types of activities that are interesting to them, rather than becoming involved in the analysis of a bunch of numbers which they may view as a boring task. Second, an objective analysis of the output may require a well-planned experimental design, several simulation runs, and a thorough statistical analysis of the simulation results. These activities are usually time-consuming and unattractive to an analyst who may have limited time or funds to meet the project deadline. Third, because of some inherent dependencies among variables in simulation models, statistical analysis of simulation results is not always straightforward and may require knowledge of advanced statistics.

The special characteristics of sample data derived from simulation in many instances impose certain restrictions on the use of the traditional

TABLE 7.2
Simulation results under various random number streams

Run number	Average server utilization	Average waiting time
1	0.91	32.3
2	0.88	26.1
3	0.91	31.6
4	0.89	26.8
5	0.90	32.0

statistical methods in the analysis of simulation results. Also, the choice of analysis of systems for their transient state or steady state affects the procedure for output analysis. Consequently, it is imperative to first discuss various types of simulation applications with respect to output analysis.

7.3 TYPES OF SIMULATION WITH RESPECT TO OUTPUT ANALYSIS

Certain peculiarities of simulation experiments should be taken into account upon experiment setting and output analysis. An important distinction that should be made when setting up a simulation experiment is between terminating and nonterminating systems. In general, as far as the simulation output analysis is concerned, systems may be classified according to the structure shown in Fig. 7.1. The following sections discuss the related issues in more detail.

7.3.1 Nonterminating Systems

Nonterminating systems are those that do not have an end to their operation during a practical time horizon. A communications network, the emergency room of a hospital, a computer center, and a traffic system are examples of nonterminating systems. Additionally, some seemingly terminating systems are actually not so. A factory, for example, that

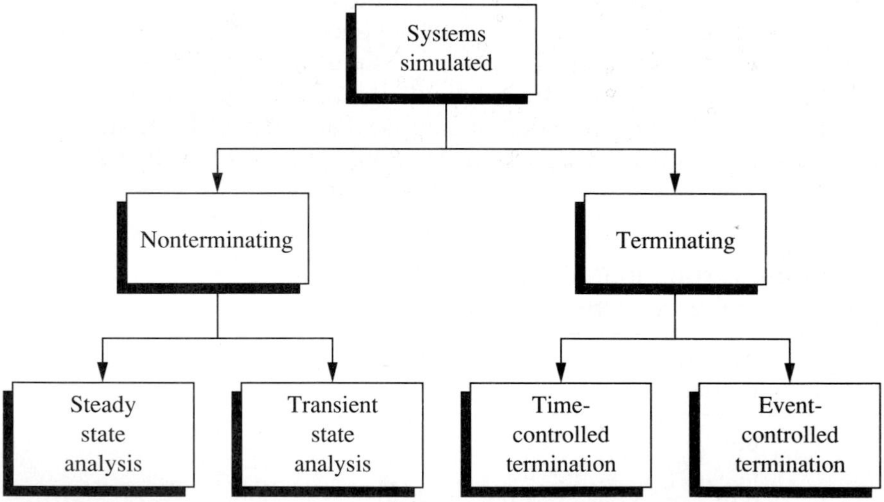

FIGURE 7.1
Classification of simulation studies with respect to output analysis.

closes at the end of each day and opens the next day may be correctly viewed as a nonterminating system as far as the flow of parts and positions of various inventories are concerned, since the starting condition for each day is the ending condition of the previous day.

Most nonterminating systems eventually reach a steady state. The steady state behavior is typically of interest in most studies of nonterminating systems. Figure 7.2 shows typical phases of transient and steady state behavior of a variable in a nonterminating system.

Nonterminating systems do not always have a steady state. For example, there may always be a flow of traffic in a section of a street, but this flow may change in different times of the day. Furthermore, a nonterminating system may not have a single steady state; instead, it may cycle through many phases, with each phase having its own transient and steady state conditions. For example, the traffic flow in a street during the morning rush hour may start from 6:00 A.M. and gradually increase until it reaches a steady state. At about 9:30 A.M. the traffic may slow down and reach another steady state that lasts until noon. Figure 7.3 depicts various cycles that a typical traffic system may routinely go through.

In the simulation study of nonterminating systems, we may be interested in the system behavior either during the transient or during the steady state conditions. To study the transient behavior of a nonterminating system for a period of time (which may or may not extend to the end of the transient period) we treat the system as if it is a terminating one whose end is controlled by time (i.e., the system operation ends

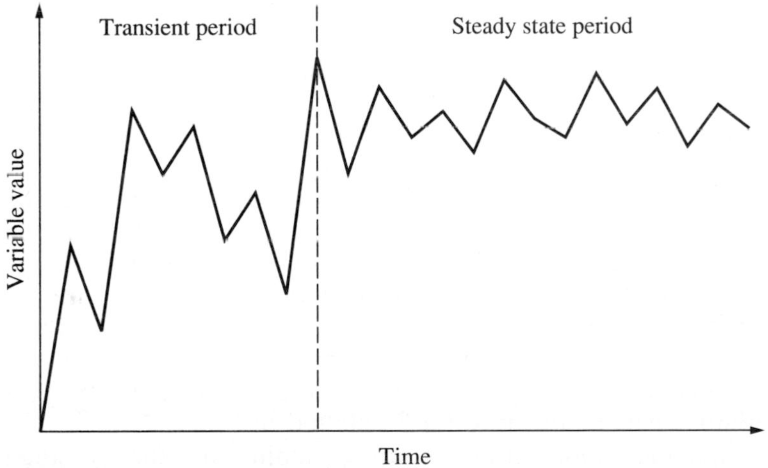

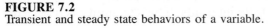

FIGURE 7.2
Transient and steady state behaviors of a variable.

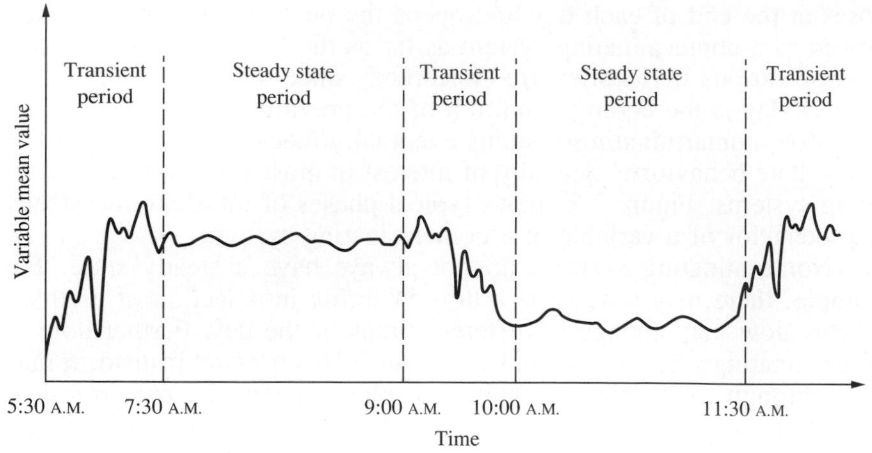

FIGURE 7.3
Various state cycles in a traffic system.

after a certain time). The discussion related to terminating systems is presented later.

When studying the steady state behavior of a system, it is essential to know when the transient state ends and the steady state begins. The data collected during the transient state represents a different statistical population than that of the steady state. If the transient state data is mixed with the steady state data in the computation of statistical estimates, unreliable estimates of steady state behavior will result. The detection of the end of the transient period can signal the proper time to start collecting statistics for steady state analysis.

Several statistical and experimental methods (such as the test of runs) have been suggested for detecting the end of the transient period. The easiest and perhaps the most reliable method is visual inspection of the pattern of changes in the variable of interest, an example of which is shown in Fig. 7.2. An even better approach is to inspect the plot of the cumulative average of the variable, rather than the variable itself. Note that individual observations that fall at extreme ends of the variation range may distract the visual inspection and conceal signs of an approaching steady state. The cumulative average of the individual observations often has a smoother pattern of change and can better reveal the underlying changes in the system behavior. Figure 7.4 shows the plot of the cumulative average of the variable plotted in Fig. 7.2.

Many simulation tools provide plot capability for the specified variables. EZSIM provides plots of cumulative averages of the specified variables at the specified time intervals. This capability is useful for

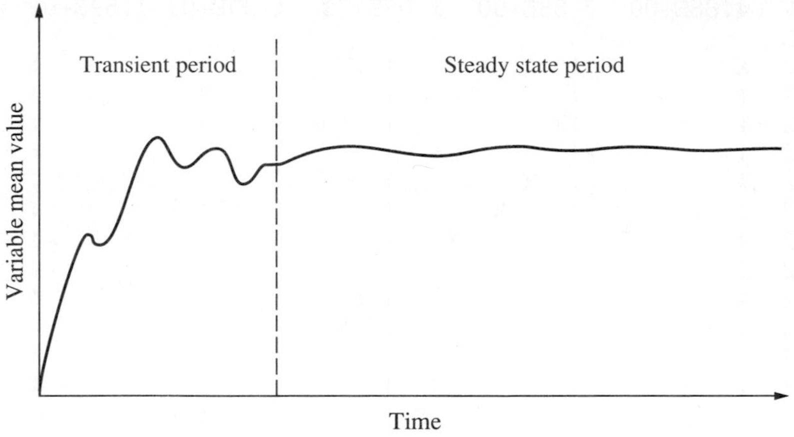

FIGURE 7.4
Behavior of the cumulative average of a variable.

detecting the end of transient periods in simulation studies. Figure 7.5 shows a typical EZSIM plot of the cumulative average of a variable.

After detecting the length of the transient period in the pilot run, a new run may be made in which statistics are cleared at the end of the transient period. Some simulation tools, including EZSIM, allow for specification of the length of the transient period after which statistics are to be collected. When this is done, statistics are not collected during the specified transient period; collection starts from the beginning of the steady state period, thus excluding the effects of the transient period. Note that in this case simulation proceeds as usual during the transient period, and the system state is updated, even though no statistic is collected.

If the objective of the analysis is to study the steady state performance of the system, then the faster the steady state is reached, the less computer time is wasted for the model executions that take place during the transient period. The initial condition of the system usually has a significant impact on the length of the transient period. Figure 7.6 shows the plot of the average number of entities in a queue with various initial numbers in the queue at the beginning of the simulation. Note that some runs converge to the steady state considerably faster than others.

If the model variables could be initialized with their average values during the steady state condition (e.g., queues initialized with their steady state average lengths), the steady state may be reached in the very early stages of simulation. This, however, requires the analyst to have a prior knowledge of the steady state behavior of the system, that is, the knowledge for which the system is simulated. An approach for properly initializing the model is to observe the actual performance of

```
XLEVEL=X  4.88E+00  7.88E+00  1.09E+01  1.39E+01 1.69E+01
 T I M E
1.00E+02  X          !           !           !          !
2.00E+02  !          ! X         !           !          !
3.00E+02  !          !X          !           !          !
4.00E+02  !         X!           !           !          !
5.00E+02  !          !  X        !           !          !
6.00E+02  !          !    X      !           !          !
7.00E+02  !          !       X   !           !          !
8.00E+02  !          !           X           !          !
9.00E+02  !          !           ! X         !          !
1.00E+03  !          !           !  X        !          !
1.10E+03  !          !          X!           !          !
1.20E+03  !          !        X  !           !          !
1.30E+03  !          !         X !           !          !
1.40E+03  !          !           ! X         !          !
1.50E+03  !          !           ! X         !          !
1.60E+03  !          !           !   X       !          !
1.70E+03  !          !           !      X    !          !
1.80E+03  !          !           !        X  !          !
1.90E+03  !          !           !          X!          !
2.00E+03  !          !           !           !X         !
2.10E+03  !          !           !           ! X        !
2.20E+03  !          !           !           !  X       !
2.30E+03  !          !           !           !   X      !
2.40E+03  !          !           !           !   X      !
2.50E+03  !          !           !           !  X       !
2.60E+03  !          !           !           !X         !
2.70E+03  !          !           !           X          !
2.80E+03  !          !           !          X!          !
2.90E+03  !          !           !          X!          !
3.00E+03  !          !           !           X          !
```

FIGURE 7.5
An EZSIM plot of the cumulative average of a variable.

the system in the past and initialize the system variables at the average observed levels. This can apply only to existing systems, however, and not to nonexisting systems for which historical data is unavailable. For the case of nonexisting systems, domain experts may be consulted to provide estimates for the expected characteristics of the steady state condition of the system being designed.

It should be noted at this point that for certain systems, such as terminating systems with a single observation per replication (to be discussed later in this chapter), the issues of transient and steady states are irrelevant.

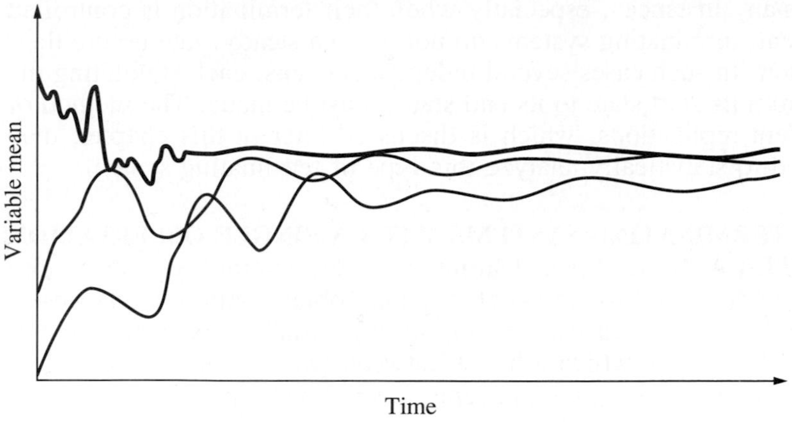

FIGURE 7.6
Transient periods under various initial conditions.

7.3.2 Terminating Systems

Terminating systems are those that typically start from a no-action or empty state and end with either of these conditions. The termination of such systems occurs either after a certain time lapse or at the occurrence time of a certain event. A bank that starts in the morning with an empty state and closes after eight hours and a retail outlet's inventory planning for an item over a planning horizon of three months are examples of terminating systems whose ends of operation are controlled by time. A construction company that has won the bidding for building a hospital, a shipyard that has received an order for building five oil carriers, failure of a complicated device, and a battle between two armies are examples of terminating systems whose operation end is signaled by an event. In these latter examples the end of system operation may not be known in advance.

Terminating systems may or may not reach their steady state (if they have any at all) before their operation ends. If they actually do have steady state behavior, then in certain situations they may be treated as nonterminating systems. For example, if we are interested in finding the number of seats to be placed in the waiting area of a barbershop, we would be interested in knowing if the system has a steady state behavior to base our decision on that system state. It is likely for the barbershop to actually have a steady state behavior which starts long before the closing time of the shop. In that case we may extend the simulation time beyond the natural end of the system operation to collect sufficient samples for our estimation purposes.

In many instances, especially when their termination is controlled by an event, terminating systems do not reach a steady state before their termination. In such cases several independent runs, each simulating the system from its start state to its end state, must be made. The method of independent replications, which is discussed later in this chapter, may be applied to statistically analyze this type of terminating system.

7.3.2.1 TERMINATING SYSTEMS WITH A SINGLE OBSERVATION PER REPLICATION. When dealing with terminating systems whose termination is caused by an event, we may obtain only a single observation of a certain (and usually important) variable for each run. For example, the time at which a battle between two armies ends is only a single observation point (not an average) which is found after simulating the entire combat process from start to end. Obviously, more than one sample is needed to arrive at reliable conclusions regarding a statistical population; therefore, several independent runs, each using a different allocation of random numbers, must be used. (This method may be interpreted as a special case of the independent replications method in which only one sample from each replication is used.)

In the simulation of terminating systems with a single observation per replication, usually a single run which incorporates multiple replications is used instead of performing several individual runs. To implement multiple replications in a single run, the entity causing the termination event at the end of one replication is sent back to the beginning point of the model to start a new replication. In this manner, the task of performing multiple independent runs is avoided. Notice in this method that each replication immediately follows the previous one, and since the random number stream is not initialized at the beginning of each replication (i.e., simulation execution is not terminated), each replication receives random numbers from different section(s) of the original stream(s) of random numbers (which is desirable).

Using the above method, a single run can provide the mean and standard deviation statistics for the variable of interest. Since there is no dependency (other than that of random numbers), observations are expected to be independent of one another and identically distributed. Because of these desirable properties, the traditional statistical methods of confidence interval construction, sample size determination, and hypothesis testing are more robust when applied to this class of simulation problems. Following are some representative examples.

Example 7.1. Let us start the examples in this section with the classical thief of Baghdad problem, which is used in many probability textbooks: The thief of Baghdad is trapped in a dungeon that has three doors. Two of the doors are each connected to a tunnel. The first tunnel returns the thief to the dungeon after three months, and the other returns him after five months. The third door leads the thief to freedom. If the thief has an equal probability of choosing each of the three doors every time he returns to the dungeon, what is the expected duration of time before he meets freedom?

As shown in Fig. 7.7, the problem may be represented with an entity flow diagram. At the SOURCE node one entity representing the thief is created. The DELAY nodes START, DUNGON, and FREED are dummy nodes which have durations of zero and are used for branching purposes only. DELAY nodes SHORT and LONG represent the tunnels. When the entity arrives at the FREED node, one simulation replication is completed. The entity is duplicated at this node, and a copy of it is sent back to the START node to start another replication. The TERMINATE node is used to specify the total number of replications before the end of the simulation. In the Statistical menu the statistics on the traversal time between the DUNGON and FREED nodes, which is the time before freedom, is specified.

Notice that in this example, instead of sending a copy of the freed entity back to the START node, we could have alternatively created several entities with an arbitrarily selected time between creations (say 1 time unit) at the SOURCE node. The traversal time statistics of these entities between the SOURCE node and the TERMINATE node would then give the desired statistics. Under this condition, however, several entities (thieves) would flow in the tunnels simultaneously. In this particular example this situation does not create a problem, but in many others, including the project management example given below, the system should be in an empty state each time the returning entity starts the system; otherwise, corruptions in the estimate of the statistics will occur.

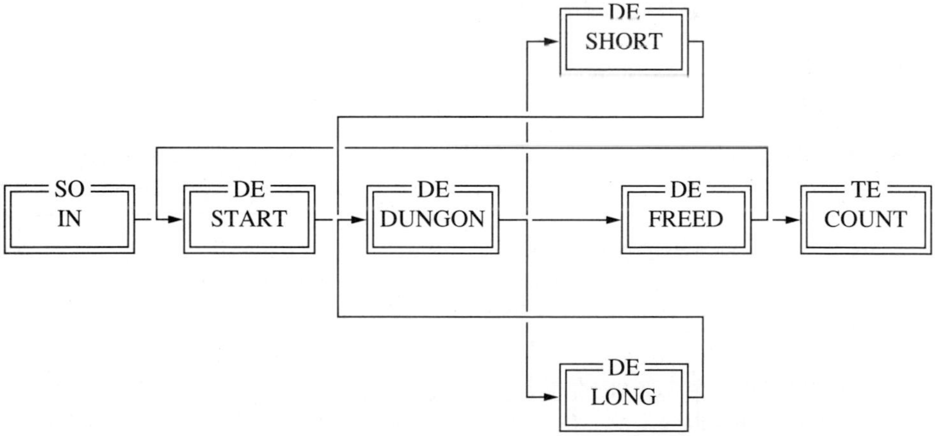

FIGURE 7.7
EZSIM representation of the thief of Baghdad problem.

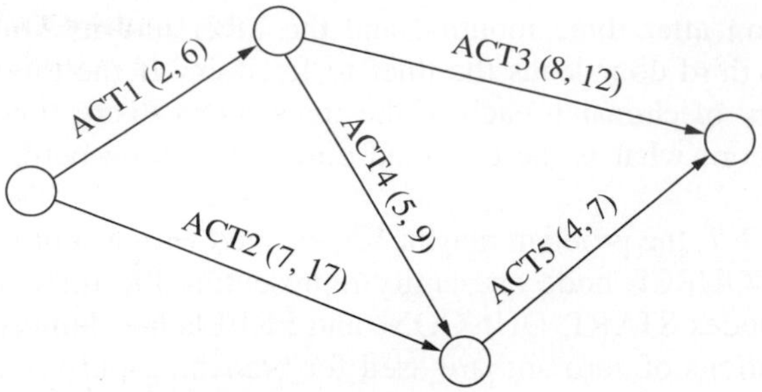

FIGURE 7.8
A sample project management network.

Example 7.2. A project that involves several activities with precedence relationships (such as a construction project) is represented in classical CPM (critical path method) as a network, with branches representing activities and nodes representing the events that signal the start and end of activities. Each time an activity ends, all activities that immediately succeed it may start; an activity may not start until all of its preceding activities end. Figure 7.8 shows a sample project management network. Durations of activities are usually random variables for which the nature of distribution is known (due to past experience). In this example it is assumed that the activity durations are uniformly distributed with the indicated parameter values. One of the major objectives of using an activity network is to estimate the overall project duration for contract bidding and resource management purposes.

Obviously, a project may not be considered as being nonterminating, because it is supposed to have a definite end. Furthermore, the issues of transient and steady state are irrelevant in this type of problem. An approach that may be used to model a project management network for simulation is to assume that each activity is a delay process. Figure 7.9

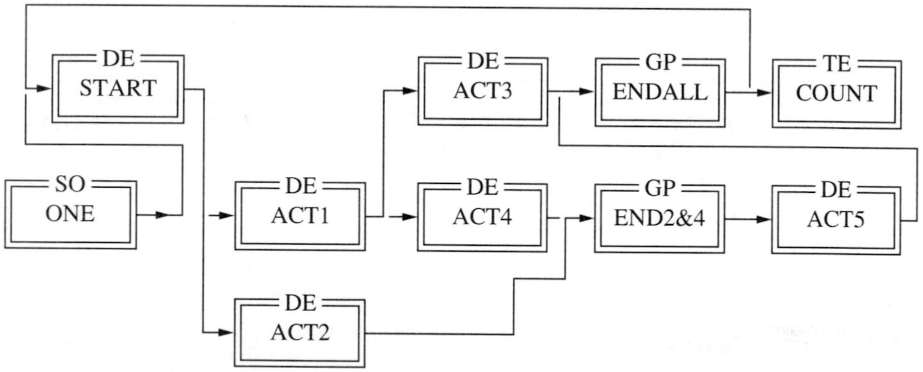

FIGURE 7.9
EZSIM representation of the project management network.

shows an EZSIM model for the above project network. At the SOURCE node one entity is created. The DELAY node START is a dummy node with duration of zero. An entity flowing through an activity represents the progress being made in performing that activity. At nodes where there are more than one emanating branch the Always branching option is specified for each branch to start each emanating activity simultaneously. At nodes where there are more than one incoming branch, grouping of incoming entities into one entity can signal the completion of all leading activities. This indicates that activities emanating from the node may not start until all activities leading to the node are completed. In this example, two entities are needed to satisfy each GROUP-Q node in the network (i.e., there are two incoming branches for each GROUP-Q node). Therefore, at the GROUP-Q node the number accumulated according to the grouping criterion (which is "by number") is 2. Default values are chosen for other grouping options.

Note that after an entity enters the START node, it traverses through the network and multiplies at nodes that have more than one emanating branch, until eventually all entities combine to form one entity that hits the ENDALL node. A traversal time between the beginning and end nodes of the network represents a single observation of the project duration (i.e., only one sample point). To perform multiple replications within the same run the entity leaving the ENDALL node is then sent back to the START node to generate another observation. A copy of the entity is sent to the TERMINATE node, which stops the simulation after the desired number of replications is achieved. Collecting statistics on the traversal time between the START and ENDALL nodes provides the desired statistics on the project duration. The desired number of replications, say 100, may be specified as the termination count in the TERMINATE node. Running this model in the animation mode illustrates the modeling logic.

Example 7.3. This example demonstrates the use of simulation in system reliability studies. Consider a device in which four major components are used. Components A and B are in series and components C and D work in parallel. The system fails when either component A or B fails, or when both components C and D fail. Given that the years of operation before failure for each component is exponentially distributed with a mean of 10 years, we are interested in simulating the device to find the expected length of time before its failure. Figure 7.10 shows the system diagram for this example. The EZSIM network for this system is shown in Fig. 7.11.

In this model one entity is created at the SOURCE node. This creation signals the beginning of the system operation. START is a dummy DELAY node at which the Always branching choice is specified. This results in the creation of one entity for each of the delay nodes associated with the components at the beginning of each replication. Each of the entities then takes one of the DELAY nodes called COMP-A to COMP-D.

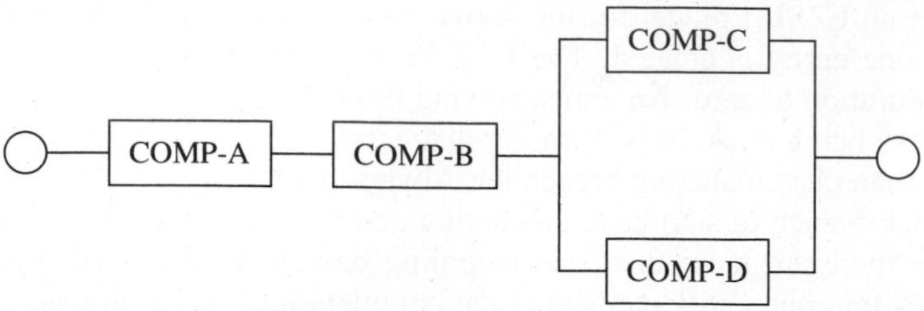

FIGURE 7.10
System diagram for the reliability problem.

The delay time in these nodes is exponentially distributed with a mean of 10. When one of these entities reaches the dummy DELAY node called FAIL, the system fails. Notice that the entities leaving the COMP-C and COMP-D nodes are grouped into one entity. The grouped entity may not be created unless two entities are received at the node, meaning that both components C and D must fail before they cause system failure. After the FAIL node the entity causing the system failure is sent to the TERMINATE node in which the total number of replications is specified.

A flag represented by a user variable (say, X) with an initial value of zero is specified. The entity causing the system failure sets this flag to 1 as it passes through the ASSIGN node FLAG1 on its way to the TERMINATE node. This flag is used on the branches emanating from the FAIL node to direct the remaining entities to the upper route of the network (i.e., conditional branching is used at the FAIL node; if $X = 1$, entities

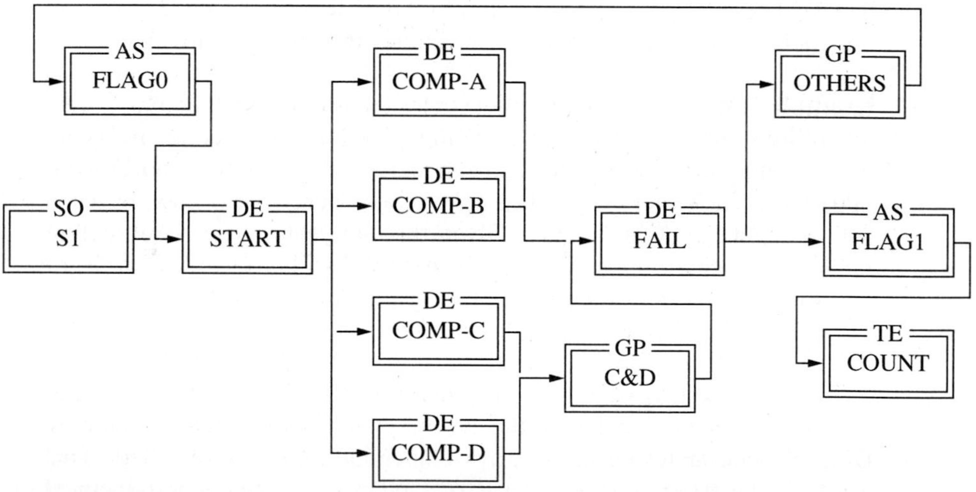

FIGURE 7.11
EZSIM representation of the reliability problem.

are sent to the node called OTHERS). The GROUP-Q node named OTHERS ensures that only one entity representing the last failure event goes back to start a new replication. This node groups the last two entities (the ones that did not cause system failure) into one entity. The grouped entity first resets the flag back to zero at the ASSIGN node named FLAG0, and then starts a new cycle. Collecting the entity traversal time statistics between nodes START and COUNT provides the statistics on the life of the device. Watching the simulation of this model in the animation mode illustrates and clarifies the modeling approach.

7.4 PROBLEM OF DATA DEPENDENCY OR AUTOCORRELATION

Note that in Examples 7.1 through 7.3, which concerned terminating systems with a single observation per replication, each observation in the simulation is independent of other observations. The only possible cause of dependency in the sample data in these cases results from the inherent dependencies among the random numbers. However, as discussed in Chapter 6, random number generators that pass the statistical tests for randomness provide numbers that may be considered independent. It should be pointed out that the sample independency phenomenon is not limited to the terminating systems with a single observation per replication. There are many other terminating and nonterminating system simulations in which statistical samples are independent of one another (see Example 7.4). The independency of sample data in all of these situations allows the direct use of the classical statistical analysis approach.

In most instances data collected from simulation models lacks the independency property. The data dependency problem is usually due to queues. In other words, if there is a queuing process in a system, the samples are generally autocorrelated. Queues, however, are not the only cause of sample dependency. For example, conditional branching of entities usually causes a dependency among sample data since it makes the movement of some entities subject to the status of other entities in the system.

To clarify how queues create entity interdependency, consider for example that if the time in a queue for the ith entity is long, this time will most likely be long for the $(i-1)$th, and $(i+1)$th entities as well. This is true because the long waiting time is due to congestion in the system, and a congested system is not likely to instantly change its state to an uncongested system. Therefore, consecutive entities that arrive during the congestion period will all have long waiting times. These positively correlated dependencies create undesirable effects that lead to imprecise estimations, if the traditional statistical approaches are directly used. It should also be pointed out that not all queues create data dependency. For instance, in the project management and device reliability examples, the GROUP-Q nodes do not create undesirable dependency effects, since the early entities, which have to wait for the last arriving entity to the queue, are simply ignored (i.e., they are combined with the last arriving entity).

It is clear from the above discussion that regardless of the independency of the input data, the simulation output data may lack the independency property. To alleviate the problem of data dependency several approaches are suggested. For example, one approach is based on selectively collecting statistics on samples that are far enough apart that they are not subject to the effects of similar system states (e.g., randomly select 1 out of every 10 entities to collect the waiting time in the queue). Each of the suggested techniques has certain strengths and drawbacks. For example, the method of collecting statistics on only a portion of entities results in a smaller sample size for a given run length. In this chapter two major methods of dealing with the problem of data dependency—independent replications and batch means—are presented.

In the following sections the traditional statistical confidence interval methods for the estimation of mean, proportion, difference between means, sample size determination, and tests of hypothesis are first presented for the types of simulation application in which samples are independent and hence the use of the traditional statistical methods is justified. We will then present the popular methods used for situations in which there is a dependency between the statistical samples derived from simulation.

7.5. CONFIDENCE INTERVALS

The accuracy of a statistical estimate is always expressed over an interval rather than by a fixed value. These intervals are only estimates; they may or may not contain the true value of the parameter which is being estimated. The degree of confidence in the interval actually containing the parameter is represented by a probability value, generally referred to as the *confidence level*, and the interval itself is called the *confidence interval*. For a given confidence level, a smaller confidence interval is considered to be better than a larger one. Likewise, for a given interval, a high confidence level indicates a better estimate. Consequently, using a given sample data, an analyst could conceivably create many confidence intervals with varying confidence levels (i.e., the higher the confidence level, the wider the corresponding confidence interval). In practice, however, analysts a priori accept a confidence level and adhere to it.

The choice of sample size also affects the confidence interval and level. Generally, for a given confidence level, the larger the sample size, the smaller the confidence interval. In other words, knowledge of the confidence interval and confidence level for the parameter to be estimated is needed to specify the sample size. The sample size in the context of simulation affects the run length and the total number of events and entities that are handled in the model.

7.5.1 Estimation of Population Mean

Mean values of variables in simulation models are often important measures, and in most studies they are the focus of analysis. Average waiting time in a queue, average time spent in the system, average length of a queue, and average time between departures are some examples of frequently used mean estimates.

Generally, the two most important values reported in a simulation output for a variable are the mean and the standard deviation. If the individual observations of a random variable are denoted by $X_1, X_2, X_3, \ldots, X_n$ for n observations (e.g., waiting times in the system for n entities), the sample mean and variance are calculated using the following equations:

$$\overline{x} = (x_1 + x_2 + \cdots + x_n)/n \tag{7.1}$$

$$S^2 = \frac{1}{n-1}\sum_{i=1}^{n}(x_i - \overline{x})^2 \tag{7.2}$$

Assuming that the random variables are independent of one another and are identically distributed, the distribution of the mean of the above variables, $\overline{x}$, is normal (for sufficiently large sample size, e.g., above 30). This holds because of the central limit theorem, which states that for sufficiently large sample sizes (regardless of the distribution of the individual samples), the distribution of the sample mean is normal if the individual samples are independent of one another and are identically distributed. Hence, random variable Z is distributed normally with a mean of zero and a standard deviation of one (standard normal distribution), where Z is defined as

$$Z = \frac{\overline{x} - \mu}{\sigma_{\overline{x}}} \tag{7.3}$$

and

$$\sigma_{\overline{x}} = \frac{\sigma}{\sqrt{n}} \tag{7.4}$$

where μ is the population mean that we are attempting to estimate, and σ is the standard deviation of the population of the random variables. Accordingly, with a probability of $1 - \alpha$, the interval between the two symmetric values given by the standard normal distribution table for the corresponding α value will contain Z. The confidence interval relationship may thus be written as

$$P\{-Z_{\alpha/2} \leq Z \leq Z_{\alpha/2}\} = 1 - \alpha \tag{7.5}$$

Substituting the value of Z in the above expression and some rearrangements, the following expression for the confidence interval of the population mean may be derived:

$$P\left\{\overline{x} - \frac{Z_{\alpha/2}\sigma}{\sqrt{n}} \leq \mu \leq \overline{x} + \frac{Z_{\alpha/2}\sigma}{\sqrt{n}}\right\} = 1 - \alpha \tag{7.6}$$

Since the population standard deviation is generally not known, and since the sample standard deviation, S, serves as a reasonable estimate for σ, S will henceforth be used in place of σ.

It should be mentioned that typically no less than 30 independent samples should be used in order to justify the use of the normal distribution for confidence interval computation and other related applications that use the above statistical relationships. When the sample size is smaller than 30, the student t distribution should be used. The use of the latter distribution is demonstrated later in the discussion of the independent replications method. Since in single simulation runs the sample size is generally much larger than 30, the use of the normal distribution is justified.

Example 7.4. Besides providing a numerical example for the confidence interval, this example also serves to demonstrate that data independency is not limited to terminating systems. Suppose that a gas station is located on the side of a freeway which connects two cities A and B. The station manager places telephone orders for gasoline from both cities. The time intervals between placing orders to city A are uniformly distributed between 5 and 9 hours. Ordering intervals for city B are uniformly distributed between 10 and 14 hours. The trip time of gasoline tankers from city A to the gas station is normally distributed with a mean of 7 hours and a standard deviation of 0.5 hour. The trip time from city B is normally distributed with a mean of 12 and standard deviation of 2 hours. Assume that immediately after an order is placed, a tanker is sent to the gas station. Through simulating the system for 100 tanker arrivals to the station, we would like to find a 95 percent confidence interval for the mean time between the arrival of tankers to the gas station. Remember that 30 or more independent observations are sufficient to justify the use of normal distribution. Therefore, the sample size of 100 is sufficient for this justification.

Figure 7.12 shows the EZSIM network of this simple model. The SOURCE nodes named FROMA and FROMB generate entities representing tankers sent from cities A and B, respectively. The time between

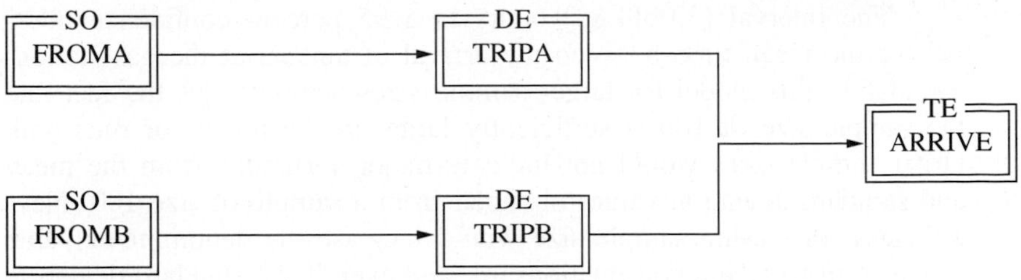

FIGURE 7.12
EZSIM representation of the gas station model.

creations corresponds to the time between placing the orders. The DELAY nodes named TRIPA and TRIPB represent the tanker trip times to the gas station from cities A and B, respectively. In the Statistics menu the interarrival times of entities at node ARRIVE are specified with the statistics name of TBETWEEN. Note that in order to collect 100 interarrival times at the ARRIVE node, 101 entity counts should be specified for the corresponding TERMINATE node. Figure 7.13 shows the EZSIM output of this model.

To read the value of Z from the standard normal table given in Appendix D, Table D.3, the quantity $1 - \alpha/2$ should be found, where in this case $\alpha = 1 - 0.95 = 0.05$. Therefore, a Z value of 1.96 which corresponds to $1 - 0.05/2 = 0.975$ may be read from the normal table for this example. Thus, using the values from the EZSIM output, we have

$$4.37 - \frac{(1.96)(2.39)}{\sqrt{100}} \leq \mu \leq 4.37 + \frac{(1.96)(2.39)}{\sqrt{100}}$$
$$3.09 \leq \mu \leq 4.83$$

```
***    E Z S I M  STATISTICAL  REPORT   ***

Simulation Project: GAS STATION
Analyst: BK
Date: 7/25/93
Disk file name: EX7-4.OUT

Current Time: 443.94    Transient Period: 0.00


V A R I A B L E S:
NAME       MEAN        STD         MIN         MAX         No. OBSRVD
----------------------------------------------------------------------
TBETWEEN   4.37E+00    2.39E+00    6.79E-02    9.31E+00    100
```

FIGURE 7.13
Simulation output of the gas station model.

The interval [3.09,4.83] hours is a 95 percent confidence interval for the mean time between the arrival of tankers at the gas station. Simulating this model for larger sample sizes demonstrates the fact that the sample size of 100 is sufficiently large, as the results of runs with larger sample sizes would not indicate major variations from the mean and standard deviation values obtained from a sample of size 100. Nevertheless, increasing sample size would increase the denominator value in the confidence interval relationship, and even if the standard deviation remains unchanged, a smaller confidence interval will result. This simply indicates that for a given confidence level, larger sample sizes provide tighter estimates of the population parameters.

Note in this example that it may be argued that in the long run the mean arrival rate of tankers at the station is the sum of the mean departure rates of the tankers from the two cities, and is independent of the tanker trip times (i.e., eventually every tanker that leaves the two cities reaches the gas station, regardless of the trip time). This is definitely a legitimate argument.The following relationship provides the actual mean time between arrivals of tankers at the station:

$$\text{Mean arrival rate at gas station} = [(5 + 9)/2]^{-1} + [(10 + 14)/2]^{-1}$$
$$= 0.226 \text{ Tankers per hour}$$

Thus,

$$\text{Mean interarrival time at gas station} = 1/0.226 = 4.42 \text{ Hours}$$

Note that the simulation estimate for the above mean based on 100 samples has been 4.37 hours, which supports the above argument.

In the context of this example the impact of trip times is actually not on the mean but on the standard deviation of the interarrival times. The standard deviation in turn affects the width of the confidence interval. To observe the impact of trip times on the standard deviation of interarrival times, various parameters for trip durations may be tried.

Example 7.5. As an application of a terminating system, let us suppose that the management in charge of the project described in Example 7.2 is interested in finding a 95 percent confidence interval on the project duration based on 500 simulated project completions.

In the TERMINATE node ENDALL a termination count of 500 should be specified. The model may be run in the batch mode with a very large run length (say, 100,000 time units) to make sure that the control of the end of simulation is done by the TERMINATE node. Figure 7.14 shows the simulation results. The queue statistics in this output are indications of the activity slack times, but are irrelevant in this case. Using the confidence interval relationship, the relatively tight interval of [18.2, 18.6] days is found for the project duration.

```
***   E Z S I M  STATISTICAL  REPORT   ***

Simulation Project: PROJECT MANAGEMENT
Analyst: BK
Date: 12/10/92
Disk file name: EX7-2.OUT

Current Time: 9190.29    Transient Period: 0.00

Q U E U E S:
NAME      MIN/MAX/LAST            MEAN     STD      MEAN     STD
          LENGTH                  LENGTH   LENGTH   DELAY    DELAY
-------------------------------------------------------------------
END2&4    0/ 2/ 0                 0.15     0.36     1.42     1.98
ENDALL    0/ 2/ 0                 0.24     0.43     2.23     2.88

V A R I A B L E S:
NAME        MEAN       STD        MIN        MAX       No. OBSRVD
-------------------------------------------------------------------
DURATION    1.84E+01   2.32E+00   1.28E+01   2.35E+01  500
```

FIGURE 7.14
Simulation output of the project management model.

Note that many useful statistical inferences may be made using the data provided by a simulation of this project. For example, a project manager may want to know the probability of finishing the project before a given date to assess the potential for making a profit versus the risk of taking a loss, prior to participating in the project bidding event. Using the mean and standard deviation values and the standard normal table, the probabilities of completing before a certain date may be found.

For example, let us assume that the project client requires the project to be completed in 15 days, or a large penalty will be charged to the contractor. The Z value associated with the probability of completing the project in 15 days is $(15 - 18.4)/2.32 = -1.47$. (Note that in this equation, the denominator is the standard deviation of individual random variables representing the project duration and not the mean of the random variables; therefore, division by $\sqrt{n}$ is not performed.)

Using the standard normal table, the probability associated with $Z = +1.47$ is found to be approximately 0.93; therefore, the corresponding probability value for $Z = -1.47$ is $1 - 0.93 = 0.07$, meaning that there is only a 7 percent chance of completing the project in 15 days or less. Consequently, on the basis of this simulation analysis, the contractor will probably decide that it is too risky to bid on this project.

7.5.2 Estimation of Proportion

Proportion is another figure of interest in most simulation studies. Fractional values such as percentage of parts that do not pass the quality control test, percentage of cars that turn left at an intersection, and lost sales as a percentage of total number of demand transactions are some examples of the proportion figure.

A proportion figure could represent the percentage of one of the possible types of outcomes in a number of trials. Referring to the outcomes as success and failure with probabilities of p and $1 - p$, respectively, a random variable which takes the values of 0 (failure) or 1 (success) may be defined. This binary random variable has a Bernoulli distribution for which the mean is p and the variance is $p(1 - p)$. For a sample of size n the ratio of the number of successes to the total number of trials, $\overline{p}$, is the estimator for p and is normally distributed if n is larger than 10, $np > 5$, and $n(1 - p) > 5$. The following equation may then be written:

$$P\left\{-Z_{\alpha/2} \le \frac{\overline{p} - p}{\sqrt{p(1 - p)/n}} \le Z_{\alpha/2}\right\} = 1 - \alpha \tag{7.7}$$

Unlike the case of the confidence interval for the mean, the confidence interval for p may not be found by a simple rearrangement of the above equation. If we substitute $\overline{p}$ for p in the denominator (a good approximation when n is large), we obtain the following confidence interval:

$$p\left\{\overline{p} - Z_{\alpha/2}\sqrt{\overline{p}(1 - \overline{p})/n} \le p \le \overline{p} + Z_{\alpha/2}\sqrt{\overline{p}(1 - \overline{p})/n}\right\} = 1 - \alpha \tag{7.8}$$

Since the sample size in a simulation is usually large enough to meet the normal approximation requirement used for arriving at the above equation, the only limiting factors in the use of the equation are $np > 5$ and $n(1 - p) > 5$ conditions. If these conditions are not met in a certain experiment, then a more precise confidence interval may be built on the mean number of successes (instead of the proportion of successes) using binomial distribution. The proportion figure may then be estimated by dividing the estimate for the total number of successes by the total number of trials. Many textbooks in statistics present the method for constructing confidence intervals for binomially distributed random variables.

Example 7.6. Suppose that the project activity network given in Example 7.2 represents the activities required to make one unit of a product. Let us also assume that the manufacturer is interested in finding a 95 percent confidence interval for the proportion of products that take 15 days or less to manufacture, based on the simulation of production of 500 product units.

The EZSIM network given for Example 7.2 may be slightly modified as shown in Figure 7.15 to provide the statistics on the number of parts produced in 15 days or less. The modeling approach is to send a copy of the entities whose traversal time is 15 days or less to the TERMINATE node named EARLY, and then to collect entity count statistics at this TERMINATE node. To accomplish this, an ASSIGN node named INIT is used to set the value of a user variable, T, equal to the current simulation time (a system variable) every time a replication is started. A copy of the entity leaving ENDALL is sent to EARLY, if the entity traversal time from START to ENDALL exceeds 15 days. This is done by specifying a condition on the branch which connects ENDALL to EARLY. The condition is specified using the Expression choice of the conditional branching options. The condition appears as 15 ≥ TNOW − T in the corresponding EZSIM window. The maximum possible number of branches that can emanate from ENDALL is specified to be three. In the Statistics menu the entity count statistic at EARLY is requested and is given the name FASTPROD. The output result of simulation of this model is shown in Figure 7.16.

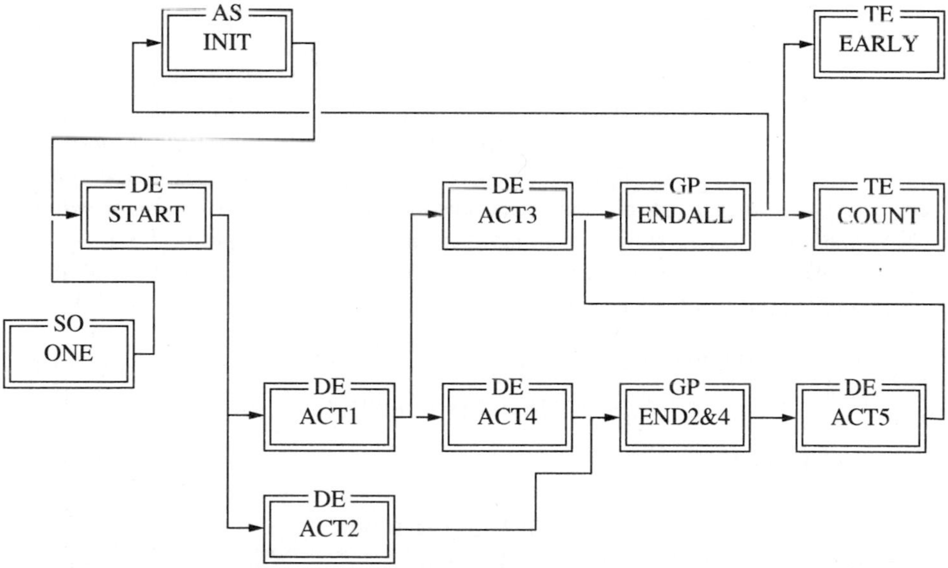

FIGURE 7.15
EZSIM representation of the production model.

```
***   E Z S I M   STATISTICAL   REPORT   ***

Simulation Project: PRODUCTION PLANNING
Analyst: BK
Date: 6/6/1993
Disk file name: EX7-6.OUT

Current Time: 9190.29    Transient Period: 0.00

Q U E U E S:
NAME      MIN/MAX/LAST          MEAN      STD       MEAN      STD
          LENGTH                LENGTH    LENGTH    DELAY     DELAY
--------------------------------------------------------------------
END2&4    0/ 2/ 0               0.15      0.36      1.42      1.98
ENDALL    0/ 2/ 0               0.24      0.43      2.23      2.88

V A R I A B L E S:
NAME        MEAN        STD         MIN         MAX         No. OBSRVD
--------------------------------------------------------------------
DURATION    1.84E+01    2.32E+00    1.28E+01    2.35E+01    500
FASTPROD    3.80E+01    0.00E+00    3.80E+01    3.80E+01    38
```

FIGURE 7.16
Simulation output for the production model.

Note that, based on this simulation, 38 out of a total of 500 units produced took 15 days or less to make. In other words, the proportion of the fast units is $38/500 = 0.076$, or 7.6 percent of total production. (Also note that this proportion is reasonably close to the 7 percent value that we obtained from the standard normal table in Example 7.5.)

Using the corresponding confidence interval relationship and the above proportion value we have

$$0.076 - 1.96\sqrt{0.076(1 - 0.076)/500} \leq p$$
$$\leq 0.076 + 1.96\sqrt{0.076(1 - 0.076)/500}$$

which yields

$$0.053 \leq p \leq 0.099$$

This interval may be interpreted as follows: We are 95 percent confident that the unit production time for approximately 5 to 10 percent of products is 15 days or less.

7.5.3 Estimation of Difference between Means

It is often helpful to simulate a system under two different scenarios (e.g., a single-server system and a two-server system) and compare the means of some measure of effectiveness (e.g., customer waiting time) under each scenario. If the difference between the two means is significant, the alternative indicating the more desirable measure of performance is recommended.

According to classical statistics theory, if two independent random variables are normally distributed, then the random variable representing their sum or their difference is also distributed normally. Assuming that the means of the samples in each simulated scenario (say, $\overline{x}$ and $\overline{y}$) are normally distributed (i.e., assuming that the central limit theorem applies), then the difference between means will also be normally distributed with the following population mean and population variance:

$$E[\overline{x} - \overline{y}] = \mu_x - \mu_y \tag{7.9}$$

$$\text{Var}[\overline{x} - \overline{y}] = S_{\overline{x}}^2 + S_{\overline{y}}^2 = \frac{S_x^2}{n_x} + \frac{S_y^2}{n_y} \tag{7.10}$$

The assumption of normality leads to the following equation:

$$P\left\{-Z_{\alpha/2} \le \frac{(\overline{x} - \overline{y}) - (\mu_x - \mu_y)}{\sqrt{S_x^2/n_x + S_y^2/n_y}} \le Z_{\alpha/2}\right\} = 1 - \alpha \tag{7.11}$$

A simple transformation of the above equation results in the following confidence interval:

$$P\left\{(\overline{x} - \overline{y}) - Z_{\alpha/2}\sqrt{\frac{S_x^2}{n_x} + \frac{S_y^2}{n_y}} \le \mu_x - \mu_y \le (\overline{x} - \overline{y}) + Z_{\alpha/2}\sqrt{\frac{S_x^2}{n_x} + \frac{S_y^2}{n_y}}\right\} = 1 - \alpha \tag{7.12}$$

Example 7.7. Suppose that in the device reliability problem presented in Example 7.3, the design engineer is considering an alternative design in which component D (which serves as a standby unit for component C) is eliminated, but a new component C is used which has an exponentially distributed operating life with a mean of 30 years (three times the average life of the other components). Although the new component costs more than the total cost of the two components that it replaces, the simplified design reduces assembly and other related production costs. The engineer is, however, concerned about the device's reliability under the new design configuration. Consequently, the designer is interested in simulating the device under both design scenarios to find a 95 percent confidence interval for the difference between the expected life of the device under each design configuration on the basis of 100 simulation replications for each scenario.

Figure 7.17 shows the EZSIM network for the new design configuration. Given the discussion regarding Example 7.3, the new model is self-explanatory. The simulation outputs for the original and new models are given in Fig. 7.18.

If we denote the life of the device under the original and new design scenarios by random variables x and y, respectively, then the following information is provided by the two simulation outputs:

$$\overline{x} = 4.9 \qquad S_x = 3.47 \qquad \overline{y} = 4.04 \qquad S_y = 4.18$$

Note that the sample size for both scenarios is 100. Plugging these values into the equation for the confidence interval on the difference between the means yields the following confidence interval:

$$-0.21 \leq \mu_x - \mu_y \leq 1.93$$

This interval indicates that the original design is likely to have a better reliability. The procedure for more objectively comparing two alternative scenarios will be discussed further in this chapter under the section for hypothesis testing.

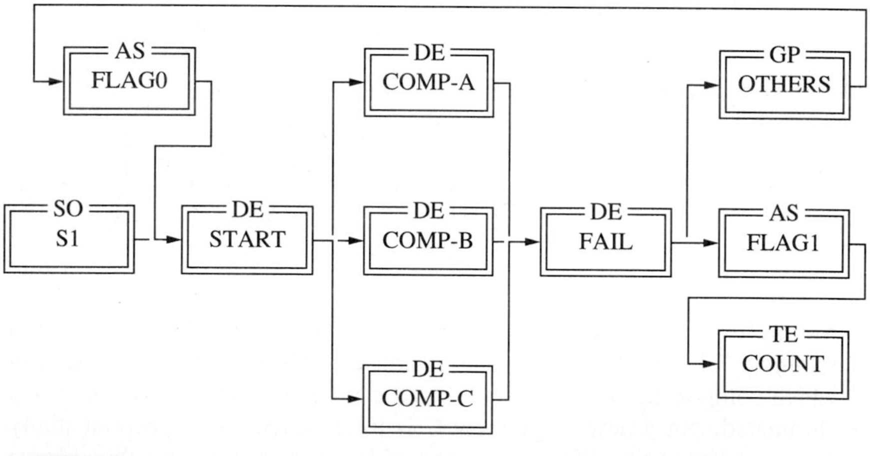

FIGURE 7.17
EZSIM representation of the new configuration of the device reliability model.

```
***   E Z S I M  STATISTICAL  REPORT   ***

Simulation Project: DEVICE RELIABILITY
Analyst: B.K.
Date: 7/15/93
Disk file name: EX7-3.OUT

Current Time: 2095.47    Transient Period: 0.00

Q U E U E S:
NAME      MIN/MAX/LAST       MEAN       STD        MEAN       STD
          LENGTH             LENGTH     LENGTH     DELAY      DELAY
-------------------------------------------------------------------
C&D       0/ 2/ 1            0.44       0.50       4.68       8.24
OTHERS    0/ 2/ 0            0.51       0.50       5.44       8.80

V A R I A B L E S:
NAME   MEAN        STD         MIN         MAX         No. OBSRVD
-------------------------------------------------------------------
LIFE   4.09E+00    3.47E+00    1.27E-01    1.96E+01    100

Simulation Project: DEVICE REL. - NEW DESIG
Disk file name: EX7-7.OUT

Current Time: 3335.81   Transient Period: 0.00

Q U E U E S:
NAME      MIN/MAX/LAST      MEAN       STD        MEAN       STD
          LENGTH            LENGTH     LENGTH     DELAY      DELAY
-------------------------------------------------------------------
OTHERS    0/ 2/ 0           0.64       0.48       10.75      17.88


V A R I A B L E S:
NAME   MEAN        STD         MIN         MAX         No. OBSRVD
-------------------------------------------------------------------
LIFE   4.04E+00    4.18E+00    4.05E-02    2.24E+01    100
```

FIGURE 7.18
Simulation outputs of the models for the original and new device configurations.

7.6 SELECTION OF SAMPLE SIZE

In the above calculations of confidence intervals we liberally assumed that the sample size was given. An important issue in the real world is the determination of sample size, which identifies the length of the simulation run. According to the statistics theory known as the *law of large numbers,* independent random variables have an important convergence property which for very large sample sizes enables them to provide nearly exact estimates of their population parameters. Given the power of today's computers, it is now possible to run simulation programs for very large run lengths in reasonable periods of time. But the answer to the question of what sample size is "large enough" depends on the inherent variations in the random variables. Therefore, in order to determine an appropriate stopping rule for a given simulation experiment the sample variance must first be estimated.

Obviously, the choice of sample size depends on the degree of precision that we expect from the output. As shown previously, this degree of precision is represented by the confidence interval; that is, the characteristics of a desired confidence interval can lead to the determination of the proper sample size. More specifically, given an actual interval (e.g., [15,20]) and a confidence level (e.g., 0.95), the sample size, n, may be derived from the confidence interval equations in terms of these known values and some population parameters.

Paradoxically, to find the population parameters needed for sample size computation, the model must first be simulated. In other words, to find the sample size (which in turn affects the length of simulation) given the desired confidence interval and level, we must first simulate the model for some arbitrary length to estimate the standard deviation of the random variable. This standard deviation value is needed in the confidence interval relationship. The proper sample size is then computed using this initial estimate of the standard deviation and the desired width of the confidence interval.

Sample size computations for given desired confidence intervals for mean, proportion, and difference between means are provided in the following sections. The methods used in these sections may yield reasonable estimates of the sample size in many studies. It should be noted, however, that these methods are based on using crude estimates. The assumption of normality used in these procedures may not provide for reliable results in certain circumstances.

7.6.1 Sample Size Based on Population Mean

Given a desired confidence interval and a confidence level for mean of a certain variable in the system being simulated, an analyst might wish to find the appropriate sample size in order to make a simulation experiment that yields the given confidence interval characteristics. If we let d denote one-half the size of the given confidence interval (i.e., one-half of the difference between the upper and lower bounds), then according to the definition of confidence interval we have

$$P\{\overline{x} - d \leq \mu \leq \overline{x} + d\} = 1 - \alpha \tag{7.13}$$

Equation (7.13) assumes that the confidence interval is symmetrical around the mean. Comparing this equation with the confidence interval equation for mean, the following relationship is obtained:

$$d = Z_{\alpha/2}\frac{S}{\sqrt{n}} \tag{7.14}$$

We may now derive the sample size from equation (7.14):

$$n = \frac{(SZ_{\alpha/2})^2}{d^2} \tag{7.15}$$

Notice that the population standard deviation must be known to determine the sample size. In rare instances and for certain variables in the model the information on the population standard deviation may be obtained using historical data. In most instances, however, there is no applicable historical data available. As previously mentioned, the alternative is to run the model for an arbitrarily chosen sample size. The output of this pilot simulation can provide an estimate (perhaps a crude one) of the value of the standard deviation for the variable in question. This estimate may then be used in equation (7.15) for sample size calculation.

It should be noted that the sample size may be estimated independently of the population standard deviation if the size of the desired confidence interval is expressed in terms of the number of the population standard deviation of the random variable. For example, if d is desired to be two-tenths of the population standard deviation, then substituting the value of d in the sample size equation yields the following:

$$n = \frac{Z_{\alpha/2}^2 S^2}{(2S/10)^2} = 25Z_{\alpha/2}^2$$

Note that the above sample size calculation requires only the value of the standard normal variable for the given confidence level. However, since d is expressed in terms of an unknown standard deviation, the actual size of the confidence interval is not known in this case.

Example 7.8. Suppose that in Example 7.4 we wish to estimate the mean interarrival time of tankers to the gas station such that the probability is 0.95 that our estimate is within 0.1 hour of the population mean. To obtain an estimate for the population standard deviation used in the sample size formula, we take the result of the run made for Example 7.4 as a pilot run with a sample size of 100, as specified in that example. According to the EZSIM output, the standard deviation of system time for the run is 2.39 hours. Substituting this value in the related equation, we obtain the following sample size:

$$n = \frac{[(2.39)(1.96)]^2}{(0.1)^2} = 2194$$

7.6.2 Sample Size Based on Proportion

Given a desired confidence interval and a confidence level for the proportion of a certain occurrence in the system being simulated, an analyst may wish to find the appropriate sample size in order to make a simulation experiment that yields these confidence interval characteristics. Assuming that the confidence interval is symmetric around the population proportion, and denoting one-half the size of the confidence interval by d, then according to the definition of the confidence interval we have

$$P\{\overline{p} - d \leq p \leq \overline{p} + d\} = 1 - \alpha \tag{7.16}$$

$$d = z_{\alpha/2}\sqrt{\overline{p}(1 - \overline{p})/n} \tag{7.17}$$

Solving the above equation for n yields the following sample size formula:

$$n = \frac{Z^2_{\alpha/2}\overline{p}(1 - \overline{p})}{d^2} \tag{7.18}$$

Notice in the above equation that $\overline{p}$ must be known to compute the sample size. A pilot simulation run of an arbitrary length could again provide an estimate for this parameter. A more reliable approach, however, is to take advantage of the fact that the maximum possible value of $\overline{p}(1 - \overline{p})$ is 0.25, when $\overline{p} = 0.5$. Substituting the maximum value of this term in the sample size equation provides the following equation, which usually yields a sample size that is larger than necessary:

$$n = \frac{Z^2_{\alpha/2}}{4d^2} \tag{7.19}$$

The major advantage of equation (7.19) is that it is independent of the population parameters; hence it relieves the user of performing the pilot simulation run.

Example 7.9. In Example 7.6, assume that we wish to determine the proportion of products that are produced in 15 days or less such that the probability is 0.95 that our estimate is within 0.03 of the actual proportion of such product units.

Using equation (7.19), we obtain the following value:

$$n = \frac{(1.96)^2}{4(0.03)^2} = 1067$$

This computation indicates that at least 1067 entities must be specified at the TERMINATE node named COUNT in the corresponding EZSIM model to meet the required specifications for the confidence interval on the proportion of the above product units.

7.6.3 Sample Size Based on Difference between Means

Given a desired confidence interval and a confidence level for the difference between two means of a certain variable, where each mean is associated with a different model condition, we may wish to find the appropriate sample size in order to make a simulation experiment that yields the desired confidence interval characteristics. Assuming that the confidence interval is symmetric around the difference in the population means, and denoting one-half of the confidence interval by d, we get

$$P\{(\overline{x} - \overline{y}) - d \leq \mu_x - \mu_y \leq (\overline{x} - \overline{y}) + d\} = 1 - \alpha \tag{7.20}$$

Comparing the above relationship with that of the confidence interval for the difference between means, we obtain the following equation:

$$d = Z_{\alpha/2}\sqrt{\frac{S_x^2}{n_x} + \frac{S_y^2}{n_y}} \tag{7.21}$$

Using the same size for both samples, we obtain the following equation for simulation sample size:

$$n = \frac{Z_{\alpha/2}^2(S_x^2 + S_y^2)}{d^2} \tag{7.22}$$

Therefore, the simulation run for each of the two modeling scenarios should generate estimates of means based on the above sample size that result in the desired confidence interval characteristics.

As discussed in the case of the confidence interval for difference between means, use of the same random number streams and the same sample sizes for pilot simulation runs of the two scenarios (intended to provide estimates for standard deviations) is preferred in the above procedure.

Example 7.10. Suppose we are interested in determining a sample size for the two simulation scenarios presented in Example 7.7. Let us further assume that the sample size should be large enough to allow for a 95 percent confidence interval with a width of 0.5 year ($d = 0.25$) for the difference between the two mean device lifetimes. Given the standard deviation values obtained from the pilot run, which was based on 100 observations for each of the two scenarios, we may derive the sample size using the corresponding formula:

$$n = \frac{(1.96)^2[(3.74)^2 + (4.18)^2]}{(0.25)^2} = 503$$

This means that each scenario must be simulated for at least 503 replications to provide the desired width for the confidence interval on the difference between the mean lifetimes.

7.7 TEST OF HYPOTHESIS

Often in simulation we may be interested to know if the mean of a certain variable is actually equal to a certain value. We may also be interested in knowing if there is a significant difference between two means obtained from different simulation experiments. Earlier in this chapter we discussed the methods for establishing confidence intervals for mean and difference between means. The *test of hypothesis* is based on similar foundations, assuming normality of sample means.

A hypothesis states a certain relationship which may or may not be true. This statement is called the *null hypothesis*. If we reject a hypothesis that is actually true, we say that a Type-I error has been made. The maximum acceptable risk of making a Type-I error in a test, α, is called the *level of significance* of the test. $1 - \alpha$ is called the *confidence level*.

The test of hypothesis is performed on the information drawn from a statistical sample. If the mean of the random variable is $\overline{x}$ based on a sample of size n, we may accept the null hypothesis which states that $\mu = a$ (recall that μ is the population mean) under a confidence level of $1 - \alpha$ if the following relationship holds:

$$-Z_{\alpha/2} \leq \frac{\overline{x} - a}{S/\sqrt{n}} \leq Z_{\alpha/2} \tag{7.23}$$

If this relationship does not hold, we reject the hypothesis.

We may fail to reject the null hypothesis which states that $\mu > a$ under a confidence level of $1 - \alpha$ if the following relationship holds:

$$\frac{\overline{x} - a}{S/\sqrt{n}} < Z_{\alpha} \tag{7.24}$$

Finally, we may fail to reject the null hypothesis which states that $\mu < a$ under a confidence level of $1 - \alpha$, if we have

$$\frac{\overline{x} - a}{S/\sqrt{n}} > -Z_\alpha \tag{7.25}$$

The Z values in the above relationships may be found using a standard normal distribution table.

The above relationships may be used to test hypotheses on the difference between means. For example, if we state the hypothesis that the mean of a variable is equal to the mean of another variable, then we are actually stating that $\mu_1 = \mu_2$, or $\mu_1 - \mu_2 = 0$. Having taken samples of each type of variable, the corresponding means and standard deviations will be available. The relationship used for testing the single variable mean may now be applied to the difference between means; that is, we accept the above hypothesis if the following holds:

$$-Z_{\alpha/2} \leq \frac{(\overline{x}_1 - \overline{x}_x) - 0}{\sqrt{S_1^2/n_1 + S_2^2/n_2}} \leq Z_{\alpha/2} \tag{7.26}$$

As the following example demonstrates, this method may be used to compare the results of two simulation scenarios.

Example 7.11. Suppose that in Example 7.7 we are interested in knowing if there is actually a significant difference between the mean device lifetimes under each design configuration. Let us assume that we would like to have a confidence level of 95 percent.

The null hypothesis should state that there is no significant difference between the two means. The test score based on the given data may be computed as follows:

$$Z = \frac{4.9 - 4.04}{\sqrt{\dfrac{(3.74)^2 + (4.18)^2}{100}}} = 1.53$$

Since this value is within the range $[-1.96, 1.96]$, which is found from the normal distribution table for the given confidence level, we fail to reject the null hypothesis and therefore conclude that there is no significant difference between the expected lives of the device under the two simulated design scenarios.

Note that in certain situations several alternatives may have to be compared to select the most desirable one. In such situations the above pairwise comparison approach may be used to eliminate one of each pair of candidates at each comparison stage. A more efficient approach uses multivariate analysis, which compares all alternatives in a concurrent manner (see Kirk, 1968; Law and Kelton, 1991; Banks and Carson, 1984).

7.8 THE INDEPENDENT REPLICATIONS METHOD

The materials covered in the preceding sections have addressed some related statistical sampling and analysis issues that apply to the situations in which individual samples are independent of one another. These methods apply to simulation as well as many other types of statistical studies.

As discussed in Section 7.3, there are many situations in which there is an autocorrelation among individual samples (because of queues, conditional branches, etc.). This correlation limits the direct use of the foregoing statistical techniques, since those techniques are based on the assumption that individual samples are independent and are identically distributed.

In simulation models with autocorrelated samples, independent consecutive observations are not collectable within one run; therefore, the independent replications method performs several shorter runs (replications), instead of one long run. In this method, each run starts from the beginning (e.g., from time zero), and thus a replication is not a continuation of its preceding replication. To be able to obtain new information in each replication, different random number seeds are used for each run. Alternatively, in certain situations when the system starts with a nonempty state, a different initial condition (e.g., number in queue) may be used for each replication. The merit of the independent replications method lies in the fact that the results of each run are expected to be independent of other runs.

As shown in Fig. 7.19, each run in the independent replications method contains both transient and steady state conditions. Thus a major advantage of this method over other methods is that it makes it possible to study the transient as well as steady state behavior of the system.

The mean of the data collected in each run serves as an independent observation. The traditional methods of statistical analysis may then apply to these data points. More specifically, sample means of each run that are statistically independent of one another and are identically distributed are found according to the following formula, which assumes N replications:

$$\overline{x}_i = \frac{1}{n_i} \sum_{j=1}^{n_i} x_{ij} \qquad i = 1, 2, \ldots, N \tag{7.27}$$

In equation (7.27) n_i is the number of samples in the ith replication and x_{ij} is the jth observation in the ith replication. Note that all simulation software tools are capable of automatically generating the

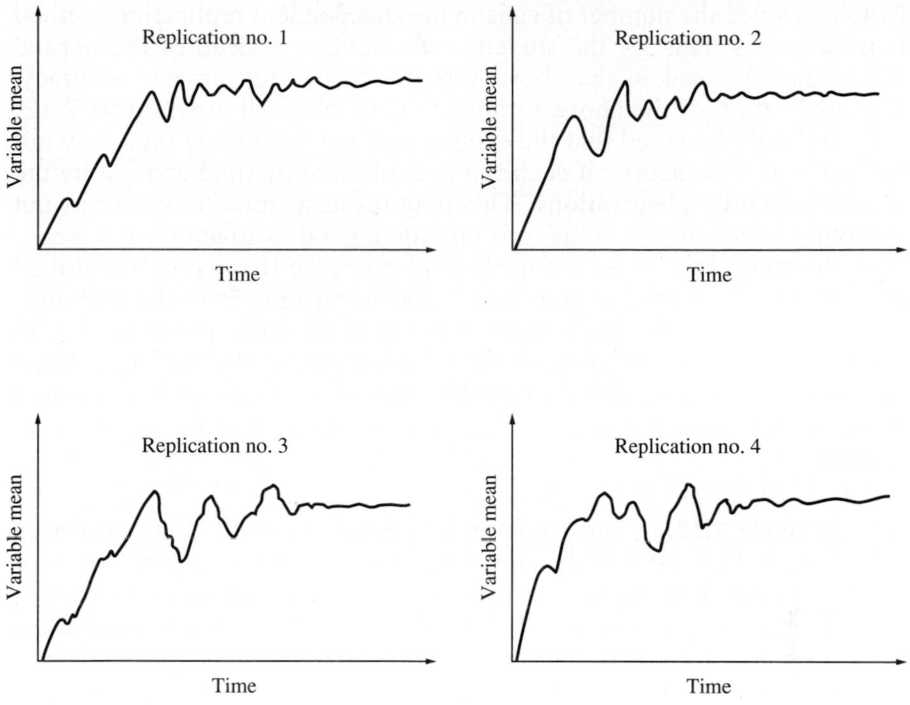

FIGURE 7.19
Illustration of the independent replications method.

corresponding mean $\overline{X}_i$ for each run. The grand mean of the above means is then found using the following equation:

$$\overline{\overline{x}} = \frac{1}{N}\sum_{i=1}^{N} \overline{x}_i \tag{7.28}$$

The following relationship yields the variance of the means:

$$S^2 = \frac{1}{N-1}\sum_{i=1}^{N}(\overline{x}_i - \overline{\overline{x}})^2 \tag{7.29}$$

Using the following relationship, a confidence interval may be built for the population mean:

$$P\left\{\overline{\overline{x}} - \frac{t_{N-1,\alpha/2}S}{\sqrt{N}} \le \mu \le \overline{\overline{x}} + \frac{t_{N-1,\alpha/2}S}{\sqrt{N}}\right\} = 1 - \alpha \tag{7.30}$$

Note that since the number of runs in the independent replication method is usually less than 30, the student t distribution instead of the normal distribution is used in the above formula to provide greater accuracy. The application of the above formula is demonstrated in Example 7.12.

It should be noted that the sample size per individual runs may not be known if termination of each run is controlled by time and not by the number of entity observations. This may result in some replications not receiving large enough samples to provide a good estimate for the mean. The potential error in the estimates will decrease if runs are controlled by the number of observations (e.g., 200 departures from the system).

The independent replications method is the only practical way to study the system behavior during its transient period. In this case, statistics for each run are collected from the start of simulation to any desired point in time (which may be before or after the start of the steady state period).

Example 7.12. A simulation model has generated the data provided in Table 7.3 on the average number of patients in the emergency room of a hospital. Each replication is made with a different stream of random numbers for 100 patients. We are interested in a 95 percent confidence interval on the average number of patients waiting in the room to decide on the number of physicians to be assigned to the emergency room.

Based on the given data, the grand mean and the standard deviation of sample means are 52.7 and 7.07 cars, respectively. From the student t table given in Appendix D, Table D.4, for $10 - 1 = 9$ degrees of freedom and $(\alpha/2) = 0.025$, we read the corresponding t statistic value of 2.26. Using the equation for the confidence interval, we obtain the following:

$$52.7 - \frac{(2.26)(7.07)}{\sqrt{10}} \leq \mu \leq 52.7 + \frac{(2.26)(7.07)}{\sqrt{10}}$$

$$47.65 \leq \mu \leq 57.75$$

TABLE 7.3
Simulation results under various random number streams

Run number	Average no. of patients	Run number	Average no. of patients
1	56	6	43
2	47	7	59
3	52	8	64
4	61	9	45
5	49	10	51

7.9 THE BATCH MEANS METHOD

To study the steady state period using the independent replications method, data collection in each run should start from the end of the transient period. This is viewed as a disadvantage of the method of independent replications, since it requires the detection of the steady state condition and the discarding of the data collected during the transient period in every run. In the batch means method, however, the end of a transient state must be detected only once.

The method of batch means applies only to the study of systems during their steady state. This method is suggested to alleviate the problem of excessive data generation and analysis associated with the transitional periods in each run of the independent replications method.

As shown in Fig. 7.20, the batch means method consists of a single run divided into multiple intervals. The data collected in each interval is considered as one batch. For each batch, the mean of the desired variable is computed. The batch mean values then serve as individual observations for which the grand mean and variance are computed using the same formula used in the independent replications method. The dependency among batch means is expected to be significantly weaker than the dependency among individual observations.

One of the problems in the batch means method is the identification of the length of the interval representing the batch. Shorter intervals result in stronger dependencies between the batch means, because the chance of a given system state during one interval continuing into the next interval is increased. For example, if there are five entities in

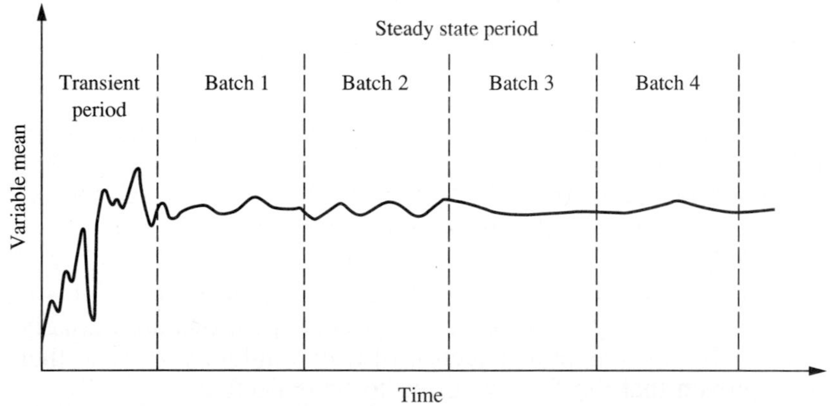

FIGURE 7.20
Illustration of the batch means method.

the system during the ith interval, the same five may remain in the system during the $(i + 1)$th interval. This results in a strong dependency among the data (e.g., number in queue, waiting time) collected for the two batches. Generally, larger batch sizes and fewer batches (5 to 30) are recommended.

When a relatively high precision in estimations is required, or when a specific system is frequently simulated (e.g., daily simulation of shop floor schedules), it may be worthwhile to use the statistical *runs up and down* test, which detects the possible dependencies among the means of batches of various sizes. Several simulation executions may be performed with various batch sizes and tested in this way to find the batch size that results in the minimum amount of data dependency.

7.9.1 Runs Up and Down Test

The runs test is used to detect possible dependencies among statistical data points. These data points are represented as numbers in a sequence. A run is defined as a succession of similar patterns (increasing or decreasing) in the sequence that is followed by a different pattern. The length of the run is the number of data points in the run before the data pattern changes. Consider, for example, the following sequence of single digit numbers: 4, 5, 0, 3, 9, 8, 7, 1, 2, 6. Let us indicate the transition from a lower to a higher number by a + sign placed after the smaller number. Likewise, let us show the transition from a larger to a smaller number by a − sign placed after the larger number. The following may then be written for the given sequence:

$$4+, 5-, 0+, 3+, 9-, 8-, 7-, 1+, 2+, 6$$

The test of runs is unconcerned about the actual value of each data point; it examines only the patterns of changes in the sequence of data. Therefore, we can summarize the above finding as

$$+ - + + - - - + +$$

Note that there are five runs in the above sequence. The lengths of these runs are 1, 1, 2, 3, and 2, respectively. Also note that if the above numbers were arranged as 0, 1, 2, 3, 4, 5, 6, 7, 8, 9 there would have been only one run, which is the minimum possible number of runs for a sequence. On the other extreme the sequence 0, 5, 1, 6, 2, 7, 3, 8, 4, 9 has nine runs (i.e., $N - 1$, where N is the total number of data points in the sequence). Obviously, both of these extreme situations are unlikely. If R is the number of runs in a sequence of truly random variables, then it has been proven that the following are to be expected:

$$\mu_R = \frac{2N - 1}{3} \tag{7.31}$$

and

$$\sigma_R^2 = \frac{16N - 29}{90} \tag{7.32}$$

For $N > 20$ the distribution of R is approximately normal. To test the hypothesis of data independency in a sequence of numbers with R runs, the following test statistic may be computed using the values for mean and standard deviation of the number of runs found from equations (7.31) and (7.32):

$$Z = \frac{R - \mu_R}{\sigma_R} \tag{7.33}$$

Based on the discussion of tests of hypothesis given earlier, we may fail to reject the hypothesis that R is equal to what it is expected to be (the mean number of runs for truly random variables) if the following relationship holds:

$$-Z_{\alpha/2} \leq Z \leq Z_{\alpha/2} \tag{7.34}$$

where α is the level of significance. In other words, if the above relationship holds, we fail to reject the hypothesis which states that the given samples are independent; otherwise, we would reject this hypothesis.

Example 7.13. Suppose 22 batches of a single simulation run have generated the following mean values for the number of cars in the queue of a freeway toll station:

15.7	22.8	17.4	14.9	20.1	24.5	24.7	19.5	16.3	18.9	21.3
23.6	23.9	18.2	18.8	21.3	20.6	17.7	19.4	19.9	16.8	21.2

We would like to use the test of runs to determine under a 95 percent confidence level if the above batch means are independent of each other. The sequence of runs for the above data is as follows:

$$+ - - + + + - - + + + + - + + - - + + - +$$

There are 11 runs in the above sequence. Using equations (7.31) through (7.33) we have

$$\mu_a = \frac{2(22) - 1}{3} = 14.33$$

$$\sigma_a^2 = \frac{16(22) - 29}{90} = 3.59$$

$$Z = \frac{11 - 14.33}{3.59} = -0.93$$

From the standard normal distribution we read $Z_{0.975} = 1.96$ and $Z_{0.025} = -1.96$. Since $Z = -0.93$ falls between these two values, on the basis of this test the hypothesis which states that the above data points are independent cannot be rejected. Further statistical analyses may now be performed on the above data in a manner similar to the independent replications method. Note that since the batch sample size in this case is smaller than 30, we must use the student t table, not the normal table, for confidence interval computation and other related analyses.

To demonstrate the procedure for using the given batch data, let us suppose that we would like to estimate the average number of cars at the toll station such that the probability is 0.95 that our estimate is within ± 2 of the actual mean number of cars waiting in the queue. Computation of the appropriate sample size may now be based on the above data, which has been generated through a pilot simulation run. The sample standard deviation in this case is 2.85 cars. Using equation (7.15) and replacing $Z_{\alpha/2}$ with $t_{N-1,\alpha/2}$, which for $22 - 1 = 21$ degrees of freedom and $\alpha/2 = 0.025$ is 2.08, we obtain the sample size of nine. This indicates that the original 22 batches have been sufficient for obtaining the desired confidence interval, and further simulation for obtaining data from additional batches is not necessary.

7.10 VARIANCE REDUCTION TECHNIQUES

Generally, as the size of a statistical sample increases, the sample variance decreases. As shown in the confidence interval formulas, the estimation precision depends on the sample variance (the square of the standard deviation). Large sample sizes in the simulation of some large-scale systems, however, could be undesirable, since it may take many hours of execution to study various model scenarios. Variance reduction techniques are recommended to reduce the required sample size and still attain a relatively small variance. These techniques were especially popular when computers were expensive and slow.

For a given sample size, the use of variance reduction techniques results in attaining better estimation precision. Stated differently, given a required precision in estimation, analysts need a smaller sample size if they use a variance reduction technique in their simulation studies. Methods such as stratified sampling, correlated sampling, antithetic variates, and Russian roulette have been devised for variance reduction.

It should be mentioned that variance reduction techniques, when used improperly, can result in adverse effects (i.e, they could increase the variance). They do not have much utility for small and simple models for which large sample sizes usually do not take much time to run, and for large-scale models it is difficult to use them properly. Because of these drawbacks, and because of the increased speed and availability of computers, variance reduction techniques are not widely used today. (For further readings on the topic of variance reduction in simulation see Law and Kelton, 1991; Moy, 1971; Pritsker, 1986; Wilson, 1984.)

7.11 OTHER STAGES IN THE SIMULATION PROCESS

The simulation process was briefly discussed in Chapter 3 and several stages (model construction, data acquisition, and output analysis) of this process were described in more detail in Chapters 4 through 7. This section discusses the issues related to the other stages of the simulation process. The discussion of these stages has been deliberately deferred because most of these stages require prior knowledge of the analysis of simulation results.

7.11.1 Verification

Verification is the process of establishing that the computer implementation of the model is error-free and is a correct representation of the logical behavior of the conceptual model built by the analyst. Note that verification is not concerned with the process of establishing whether the conceptual model is a reasonable representation of the system. The latter process is validation, the description of which is in the next section.

In general, a computer program may fail to perform as intended either because of coding errors or because of logical errors. Errors of the first type are usually easier to detect since they often negate the program execution process and are identified by the compiler diagnostic error message system. Mistyping variable and function names is the major cause of coding error. This is especially common in the case of those simulation software tools that employ a conventional programming style in which model logic is written in the form of sequential program statements. Many modern simulator software systems (i.e., special-purpose simulation tools) employ an icon- and menu-based modeling approach that significantly reduces the chance of typing errors. Most current general-purpose simulation tools, however, employ the conventional programming style.

Logical errors are considerably more troublesome than coding errors. These errors occur when the program executes but fails to generate the correct results, or when the program execution encounters unrecoverable numerical errors such as division by zero or pointers pointing at unintended memory locations. The worst scenario is when the programmer unknowingly gives a value to a pointer which corresponds to a memory location in which the program for the computer operating system resides. This error may result in a total system halt, which necessitates rebooting of the computer to rerun the operating system program.

7.11.1.1 POTENTIAL SOURCES OF ERROR IN SIMULATION PROGRAMS. Following are some of the potential sources of error in simulation programs:

Numerical data errors. Often a large amount of data is used in a simulation model. Distribution parameters, probability values at probabilistic branches, number of parallel servers, initial values of variables and resource levels, and so on are some examples. Even when model logic is correctly incorporated in the program, a mistake (such as a misplaced decimal point) in data entry can create significant errors in the results.

Unexpected random variate values. Random process generators, especially those for open-ended probability distributions, may generate extreme values which are unpredicted by the analyst. For example, if normal distribution is used to represent the duration of an activity, it is always possible (regardless of the mean value of the random variate) for the generator to produce a negative number. A negative value for time corrupts the event calendar and creates major execution errors whose source, depending on the choice of programming tool, can be very hard to detect. Certain distribution parameter values make this form of error relatively unlikely to occur. For example, if for a normal distribution the relationship $\mu - 3\sigma > 0$ holds, creation of negative random variates from the distribution is highly unlikely. When this relationship does not hold, a truncated form of the distribution which excludes negative values should be used.

Inconsistency in units of measurement. Another source of potential error is the inconsistency in the assumed units of measurement. This error often takes place in relation to units of time. An example is the case of specifying interarrival times in minutes but service times in hours. As long as the analyst is consistent, any convenient unit of measurement may be used. For example, both interarrival and service times must be specified either in minutes or in hours. The chance of making errors in relation to the units of measurement may increase when multidimensional measures such as speed (distance per time unit) are involved.

Overwriting variables. Variables, attributes, and resource levels are often unintentionally overwritten and their intended values are lost. A common scenario is when a user variable (a global variable) is incorrectly utilized where an entity attribute should have been used instead. For example, if the distribution of service durations differs for various entities, the corresponding durations must be stored in an attribute of the entities and not in a user variable; otherwise, before an entity starts its service, a following entity may overwrite its intended service duration by altering the value of the user variable. Overwriting system variables, such as the master simulation clock, can also create serious execution or output errors.

Problems caused by concurrent events. Concurrent events (i.e., events that take place at the same epoch of time) may result in improper handling of the model logic during program execution. For example, if an entity opens a gate while another entity simultaneously closes the gate, then depending on the sequence of execution of the related logic, entities waiting behind the gate may or may not leave the gate as a result of the gate opening. This type of situation is usually difficult to detect by observing the simulation output alone. Event animation and trace of variable values are helpful in detecting the errors caused by concurrent events.

Entity flow problems. Often due to incorrect incorporation of the model logic or mistakes in specifying conditions for branching entities; entities take routes that are not intended, or do not take some intended routes. Entity tracing and animation are effective methods for detecting the sources of such logical errors.

Entity deadlocks. When entities do not receive the intended resources or service at facilities, excessive growth of queue sizes may be encountered. For long runs or large models this may result in computer memory overflow. Deadlocks can also take place at conditional branches where none of the branch conditions are met. If the branching point does not immediately follow a queue where the blocked entities may be accumulated, an entity loss or execution error takes place. Event animation is especially useful for finding the deadlock points.

Error in statistics specification. Simulation analysts sometimes create simulation programs that do not encounter any of the above errors and run successfully, but the statistics that they provide are incorrect. This error usually takes place when a statistic that is time-based by nature is specified as an observation-based statistic, or vice versa. For example, if observation-based statistics are collected on a variable that represents the number of entities in a certain segment of a model, then the results will be unreliable, especially when the event at which the observation on the variable is made takes place infrequently. Another example is the case of specifying a variable that has a time dimension (such as the time during which a certain resource is unavailable) as a time-based statistic. This type of error is very serious since it is commonly generated by seemingly valid programs that have won the analyst's confidence. It should be kept in mind that if the output of a simulation program is incorrect, then as far as the end user is concerned the entire study is worthless.

7.11.1.2 ERROR PREVENTION AND RECOVERY TIPS. Although the degree of success in the creation of valid simulation programs strongly depends on the skill and experience of individual analysts, the following general tips should help every analyst in the validation process.

Modularize the validation process. Validating a large program is often more cumbersome than creating it. It is always a good practice to generate submodel programs and test each submodel by subjecting it to a rough estimate for the input patterns. The next step is validating the integrated model, which—assuming the submodels perform as intended—becomes a much easier task.

Create readable programs or models. Using meaningful process and variable names, and writing reminder comments either in the body of the program or in a supplementary document, simplifies the task of reviewing the computer representation of the model. If a program has been altered at various points in time there is a good chance that it will be difficult to decipher at a later stage. It is a good practice to re-create these types of programs anew. The new version is very likely to be cleaner and easier to understand.

Use various modeling approaches. There are usually numerous ways of modeling the same system. Examples include use of resources instead of facilities, use of several conditional branching points instead of specifying complicated expressions on one branch, and use of resources to represent entities when applicable. If the outputs of alternative models are similar, then there is a good chance that the model logic has been correctly understood and incorporated. Note, however, that various modeling approaches may use different allocations of random numbers to processes, thereby resulting in some variations in the outputs of the alternative models. In such cases, tests of hypotheses may be conducted to determine whether the variations are due to inherent randomness or to errors.

Use outside analysts. Model builders may become attached to their creations and see what they want to see; this loss of objectivity can cause them to unconsciously avoid certain tests that are likely to detect an error. It is often advisable to get the opinion of an impartial analyst who is familiar with the technique and can take a fresh and critical look at the computer representation of the model.

Perform numerical evaluations. A simulation program may be numerically validated by such approaches as (a) running the program for a short run length and checking the results by hand calculation, (b) eliminating random effects and treating the model as deterministic, and (c) replacing complex probability distributions with simple ones and comparing the simulation results with the results obtained from an equivalent analytical model (such as an exponential queuing model).

Enforce the occurrence of infrequent events. A model may sometimes run successfully and generate correct results for a given set of input conditions or run length but fail under a different condition. For example, if, based on satisfying a certain condition, a small fraction of entities is to be routed to a certain segment of a model, that segment may escape the analyst's attention and never be tested if the condition is not met during the validation run. Infrequent events may be enforced for validation purposes by mechanisms such as increasing arrival rates, increasing probability values on certain branches, relaxing certain branching conditions, increasing simulation run length, and so on.

Use event animation. Event animation is a valuable tool for validity checking, provided that it does not take too much effort to develop. It can be speeded up during uncritical time periods and slowed down during eventful and critical periods to scan the model and detect potential sources of errors.

In closing this section, it should be pointed out that a unique capability of EZSIM is its provision for prevention of coding as well as logical errors in simulation modeling. For example, all names of nodes, variables, attributes, and resources are entered only once when building an EZSIM model. All subsequent references to the names are made through selection windows. By eliminating the need for retyping the assigned names, EZSIM eliminates the possibility of mistyping these names. By incorporating the generic logical rules governing discrete systems, EZSIM also offers some provision for preventing logical error. For example, connections between certain nodes are not allowed, system variables cannot be overwritten, the nature of a considerable number of statistics types are automatically distinguished, the user is aided in building conditional and other forms of expressions with minimal chance for errors, and an event animation module for effective checking of entity flow and possible blockages is provided.

7.11.2 Validation

Validation is the process of establishing that the simulation model correctly represents the important aspects of the system being simulated. Recall that the purpose of the verification process is to establish that the simulation program performs in accordance with the analyst's intent (which is abstracted in the form of a model). The validation process, on the other hand, aims at establishing that the model and the related activities (i.e., the entire simulation study) are in accordance with the end user's problem and intent. Therefore, whereas verification concerns the relationship between the computer program and the model and serves to address the analyst's performance criteria, the validation process concerns the relationship between the model and the system under study and serves to address the end user's performance criteria. An error-free computer program that successfully passes all verification tests does not necessarily pass the tests for validity, since the model (which is the product of the analyst's frame of mind, not the end user's) may not correctly represent the system.

Obviously, a model can be valid only in the context of the purpose of the study for which it is developed. For example, if a simulation model is designed to find the suitable buffer capacities on a factory floor, this model should not be expected to be valid for the purpose of providing the manpower schedules for factory operations.

If the simulation study concerns an existing system, the test of validity is relatively straightforward. In this case, the model parameters and configurations may be tuned to those of the existing system. The performance results generated by the model may then be compared with those of the real system. In the literature of model validity this approach is usually referred to as the *empiricist's approach,* which is characterized by its disregard for the internal makeup of the model; it observes only the results. As an example of this approach consider the simulation of the traffic intersection described in Chapter 3. To validate the corresponding simulation model, the traffic light timing in the model may be set at the current timing of the actual traffic light. The relevant statistics generated by the model (e.g., average number of cars waiting at each direction) may then be compared with the statistics collected from the actual system. If the two statistics are reasonably close, it may be concluded that the corresponding simulation model is valid and further model experimentation may then be performed. It should be noted that even when a model is perfectly valid, the statistics generated by simulation and the system performance statistics are unlikely to be exactly the same. Hypothesis tests, such as the ones described earlier in this chapter, may be used to determine if the difference between the two performance results is statistically significant.

When simulation studies are applied to nonexisting systems for which historical data is unattainable, the problem of validity checking becomes a perplexing one. In such situations the approach known as the *rationalist's approach* is recommended. This approach is based on closely examining the system, the model, and the assumptions that surround the model; if the model can be characterized as an "unquestionable truth," it is considered valid.

Naturally, an analyst who develops a model has the tendency to be easily convinced about the correctness of his or her model, and, as was the case for verification, he or she might unconsciously avoid deliberations that could void the model validity. Consulting with a field expert in the examination process can greatly enhance the reliability of the validation process. Unlike the case for verification, however, the field expert is not likely to know the simulation methodology and technical details of the simulation model. Effective communication between the analyst and the field expert is crucial to the creation of relevant validity tests. Scene animation can be instrumental in facilitating understanding of the model behavior for the field expert. This type of validity checking is called *face validity.*

Following are some objective approaches that analysts can take to increase the degree of the validity of their simulation models before consulting with field experts:

Testing the parameters and relationships. This test involves close examination of the modeling assumptions regarding parameter values and various relationships specified in the model. Statistical tests, such as goodness-of-fit tests, should be performed on all stochastic inputs and stochastic internal processes.

Structural and boundary testing. Through this test it should be established that a reasonable correspondence exists between the real system and the model structure. The model must include all of the components that are within the identified system boundary and should not include components that are nonexistent in the system.

Sensitivity analysis. Slight changes in the model parameters should not result in significantly different results and thereby affect, in major ways, the corresponding conclusions that may be drawn from the results. When models are very sensitive to slight parameter changes, the model validity is usually questionable.

Continuity testing. This test is intended to find out if slight changes in the model parameters do in fact result in slight changes in the model output. For example, if service duration at a facility is slightly increased, a slight increase in the server utilization statistics must be expected. Satisfying this test, however, is not a necessary condition for model validity. For example, if the level of a resource which is already underutilized is slightly increased, no statistic other than those related to the resource itself may be affected.

Test of degeneracy. If a portion of the model is removed, the results should reflect the removal, or the removed portion may not represent any segment of the real system. In the device reliability model discussed earlier in this chapter, if a component is removed from the model, the device reliability should be either increased or decreased (depending on whether the component is in serial or parallel configuration with other components).

Test of absurd conditions. Imposing absurd conditions on the simulation model may reveal some modeling flaws. For example, if the arrival rate to a queuing system is made infinitely large, the departure rate from the system should not demonstrate a similar behavior, because the service rate imposes practical limitations on the entity departure rate.

Elaborating on the model validity process always increases the end user's confidence in the overall study. A simulation project that has not undergone a rigorous validation process may prove to be worthless or detrimental to the end user.

7.11.3 Experimentation

Experimentation is the process of devising relevant and efficient experimental conditions under which the model behavior is examined. This stage may require parametric changes as well as structural changes to the model; that is, model reconstruction may be performed at this stage. Assuming that statistical analysis of the results of each experiment is correctly conducted and the results are correctly compared, the experimentation process can lead to an understanding of the behavior of existing systems and how to improve them, or to the design of future systems with desirable performances.

Identifying the system type with respect to output analysis (terminating, nonterminating, etc.), deciding on the length of each run, systematically changing various model parameters and configuration for each experiment, and pairwise and multivariate comparisons are some of the major tasks that are involved in the experimentation process.

In the experimentation process, a large number of values for some model parameters must often be tried to find the best system performance. For instance, in the freight company example given in Chapter 3, depending on the cost values for leasing the facility and plane downtime, a large number of scenarios—each using a different number of parallel facilities—may have to be examined to find the least-cost alternative (which may correspond to leasing as many as 20 facilities). Heuristic search procedures may be effectively utilized to greatly reduce the number of possible experiments in such cases. The experimental approach to finding the maximum of the demand function described in Chapter 2 is actually a heuristic search process. The efficiency of heuristic search techniques can be increased if instead of a blind search a systematic technique (e.g., linear search, binary search, hashing method) is used for varying the model parameter values in various simulation experiments (Kuester and Mize, 1973 and Bazaraa and Shetty, 1979 are good references for search techniques).

The utility of heuristic search techniques increases as the number of decision parameters increases. For example, to find the best traffic light timing for which the average waiting time of cars arriving from all four directions to an intersection is minimized, various practical values for the durations of red and green lights have to be tried. Simulating the model for all practical combinations of pairs of duration values can be prohibitive in this case. A systematic search technique can dramatically reduce the number of trial values that lead to the best solution. Figure 7.21 demonstrates a typical response surface for the average waiting time of all cars and some representative trial points that in a few iterations converge to provide the minimum point of the response surface. Note that the model performances for the current and previous trial points may be compared using a pairwise test method (such as the test of hypothesis presented in this chapter) to find the proper search direction.

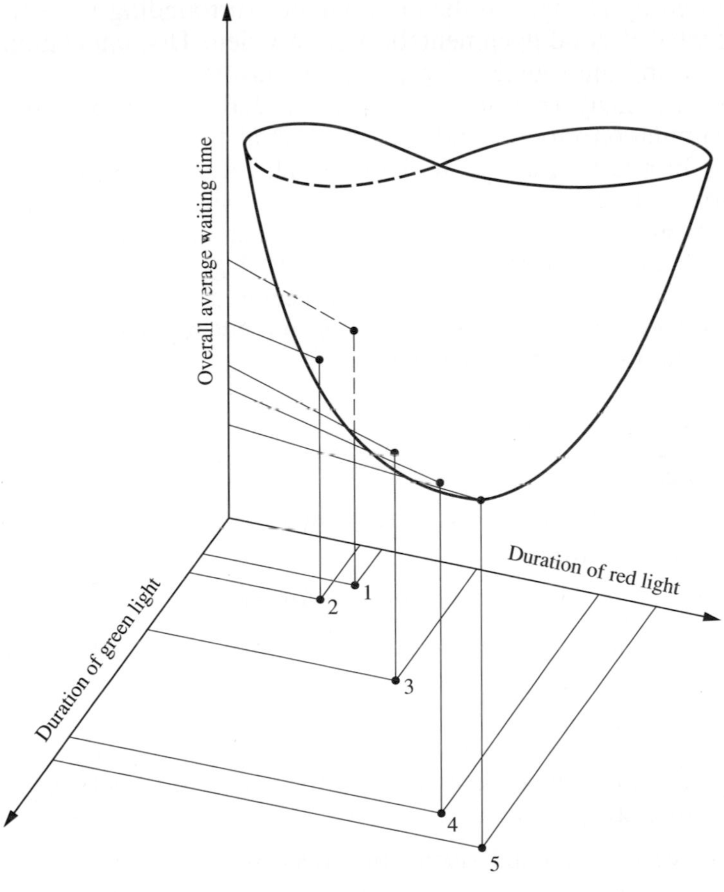

FIGURE 7.21
Illustration of a heuristic search for the traffic light simulation example.

7.11.4 Documentation

Documentation is an important stage in the simulation process. Documentation should include a clear and systematic presentation of the problem, the objectives of the simulation study, all of the activities involved in the simulation process, and the analyst's final recommendation. The results of the study must be summarized in an organized form to relieve the end user from the task of digging the facts out of countless pages of computer printouts. Documentation should be free of technical simulation jargon.

Good documentation can greatly enhance the chance of project acceptance and implementation of the analyst's recommendations. It can also serve as an indispensable source for future reference, especially in the absence of the original analyst. Continued implementation of the simulation results over time, and possible modifications to the original model (as required by changes in the environment surrounding the system), are facilitated if good documentation is provided. Documentation can also serve to find the sources of previous mistakes.

Unfortunately, many analysts spend a great deal of time on other stages of the simulation process and then put together a hasty and incomplete report in a last-minute rush. This can hurt a project's chances of being accepted. Remember that the greatest reward for an analyst should be the eventual implementation of his or her objective recommendations; clear and complete documentation will greatly increase a project's chances of being accepted and implemented.

As a recommended format for simulation project reports, the following list of contents may be considered:

- Problem description
  - Purpose of study
  - System description (preferably supported by diagrams that illustrate the system configuration)
  - Assumptions and the underlying rationale for each
- Data acquisition (source of data, method for collection, analysis, etc.)
- Model description
  - Modeling approach
  - Model overview (accompanied by related flowcharts, flow networks, block diagrams, etc.)
  - Approach to verification and verification results
- Approach to validation and validation results

- Experimentation and analysis of results (supported by tabulated data, bar graphs, plots, histograms, pie charts, etc.)
- Recommendations
- Appendices (selected computer outputs, related letters, technical articles, etc.)

7.12 SUMMARY

This chapter has discussed estimation of performance measures of simulated systems using confidence intervals and hypothesis testing. The method for determining simulation run length based on statistical approaches has been presented. Various types of simulation studies with respect to output analysis have been considered, and certain statistical issues affecting the nature of sampling methods have been discussed in this chapter. Systematic analysis of simulation output is an essential ingredient of any simulation study and should not be overlooked.

All of the stages of the simulation process serve as means to the ultimate objective of obtaining useful results and implementing them for attainment of desirable system performances. Without proper verification, validation, experimentation, analysis of results, and documentation, a simulation study is not likely to benefit the end user.

7.13 EXERCISES

7.1. Generate tables similar to Tables 7.1 and 7.2 using an exponential interarrival time distribution with a mean of 20 time units. Service times are uniformly distributed between 14 and 21. Why are the variations in the results significantly different from those of Tables 7.1 and 7.2?

7.2. In the thief of Baghdad example modify the model to enable you to find a 95 percent confidence interval on the proportion of times that the thief takes the long channel more than three times before he is freed. Choose a sample size of 1000.

7.3. Consider the project network shown in Fig. 7.22, in which the numbers on the branches indicate the upper and lower ranges of uniformly distributed activity durations:

(*a*) Find an appropriate simulation sample size on the basis of a pilot run using a sample size of 50. The sample size should be large enough to allow for the estimation of the project duration within five days under a 95 percent confidence level.

(*b*) Simulate the project with your calculated sample size. Based on the results of your simulation, find the probability of finishing the project within 165 to 170 days after inception.

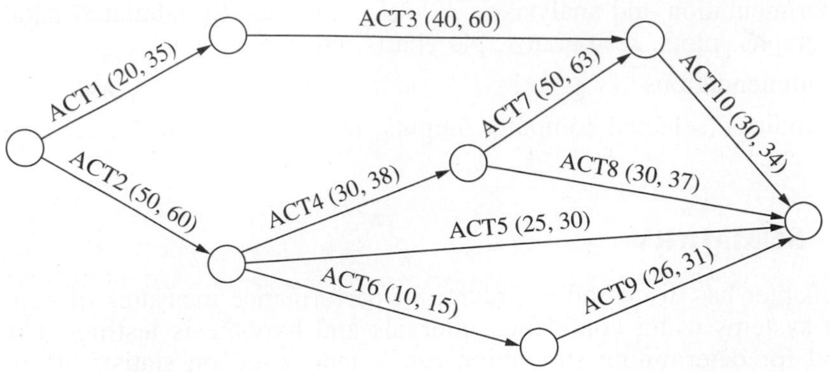

FIGURE 7.22
Project network with uniformly distributed activity durations.

(*c*) Suppose that the net profit of performing the project will be $45,000 if the project is completed in 120 days. For each day thereafter the contractor must pay a $5000 penalty. Determine the expected profit or loss of the contractor.

7.4. Figure 7.23 shows the arrangement of components in a device. The time before failure of each component has an exponential distribution with a mean of 20 years. Make a pilot run of the simulation model of the device using a sample size of 30. Find the sample size required to estimate the mean time before device failure such that there is a 0.95 probability that your estimate is within 0.5 year of the actual mean time before failure.

7.5. Simulate a queuing system with one server in which the interarrival times and service times are distributed exponentially with means of 10 and 8 minutes, respectively.

(*a*) Run your model for 5000 time units.

(*b*) Run your model five times with each run having a length of 1000 time units and a different random number seed for the interarrival times.

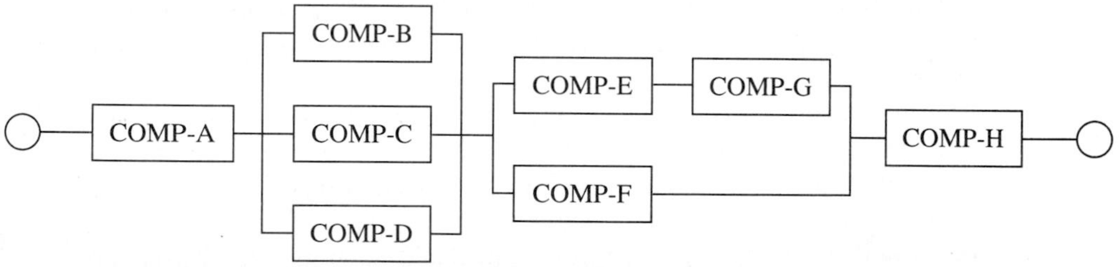

FIGURE 7.23
Arrangement of components in a device.

(*c*) Using the formula for the independent replications method, find the estimate of the mean and standard deviation of the length of queue using the results of part (*b*). Compare these results with those of part (*a*).

(*d*) Compare the above mean values with the value derived from the analytical formula given in Exercise 6.13.

7.6. Using the independent replications method, find a 95 percent confidence interval on the mean waiting time of planes in Example 5.11 on the basis of the simulation of the system for 50 arriving planes. Run your simulation model five times, each time for 100 arriving planes, and build a new confidence interval for the mean waiting time. Compare your results.

7.7. In the system described in Example 5.11, it is proposed to use larger trucks, which can each carry 12 unit loads of cargo. Find a 95 percent confidence interval for the difference between the mean waiting time of planes under the current and proposed scenarios. Using the independent replications method, simulate each scenario five times for 100 arriving planes.

7.8. Suppose that we are interested in finding an estimate of the average waiting time of the planes in Example 5.11 such that our estimate is within 2 minutes of the actual mean waiting time with a 0.95 probability. What is the proper number of replications?

7.9. In Example 5.11, assume that there are initially 1000 cargo units at the terminal. Plot the average time between plane takeoffs after each 50 minutes of simulation to detect the end of the transient period. Rerun the model, but this time identify the proper length of the transient period. Compare the results.

Hint: To plot this variable, select the Interarrival Time at a Node option from the Statistics menu, and specify FLY as the choice of the node. Select the Graph option, select the Plot option, start the plot at time zero, and end it at your desired time.

7.10. In a manufacturing operation 15 percent of machined parts are returned to the machining station for rework. Part interarrival times are distributed exponentially with a mean of 5 minutes. Machining time is uniformly distributed between 2 and 4 minutes. Management is concerned about the effect of reworks on the production throughput (number of parts produced per time unit). It has been proposed by a consultant that the rate of rework can be reduced to 7 percent if, on the average, an additional 1 minute is spent on inspection during the machining process (parameters of machining time will be 3 and 5 minutes). Using the independent replications method, run each scenario five times for 1000 time units. Use the test of hypothesis to find if the mean time between production of completed parts (the inverse of production throughput) is significantly shorter under either the current or the proposed scenario. Use a 95 percent confidence level.

7.11. Suppose that for the system described in Example 5.6 we are interested in finding a 90 percent confidence interval on the proportion of parts that are sent for subcontracting. Using the method of independent replications, find this interval on the basis of five runs, each with a length of one day.

7.12. Assume that the system described in Example 5.4 is a nonterminating system.

(*a*) Run the simulation model of the system for 1000 minutes and plot the average number of trucks in the queue. The plot should start at time zero with an interval size of 10 time units. Based on your observation of the plot, do you think that the system steady state is reached during simulation?

(*b*) Is there a steady state for the operation of the system?

(*c*) Detect the beginning of the steady state period and run the model again, but this time identify the proper length of the transient period. Compare the results of the two runs. Is there a significant difference between the two sets of data? Use a 95 percent confidence level.

Hint: To plot the average queue length, select Expression from the Statistics menu, select System Variable, select Number in Queue, select LINE, and specify the Time Based statistics choice. EZSIM automatically plots the "cumulative average" of the queue length.

7.13. Consider the single-server queuing program given in Appendix A. Assume that the interarrival and service times are exponentially distributed with means of 10 and 8 minutes, respectively.

(*a*) Incorporate the necessary program statements for calculating the standard deviation of waiting time in queue. Run the program to find the average and standard deviation of waiting time in queue on the basis of 200 served entities.

(*b*) Modify the program to print the average waiting time in queue for each batch of 20 entities. Run the program for 20 batches (i.e., for a total of 200 served entities).

(*c*) Perform a test of runs on the average waiting times in queue obtained in part (*b*) to determine if the batch means are independent. Use a 95 percent confidence level. If the test shows that the batches are not independent, try a different starting point for collecting statistics (e.g., after the 100th entity), and/or change the batch size until you obtain 20 independent batch means.

(*d*) Find the average waiting time in queue by using an independent replication method with 10 replications, each having a sample size of 20 served entities.

(*e*) Build three 95 percent confidence intervals on the average waiting time in queue on the basis of the results obtained from parts (*a*), (*b*), and (*d*). Compare these confidence intervals with each other and with the value of the mean waiting time as found using the analytical formula given in Exercise 6.13. Explain your assessment.

7.14. There is a traffic light at the intersection of 2 one-way streets. The interarrival times of cars from the north-south street at the intersection are exponentially distributed with a mean of 2 minutes. The interarrival times of cars to the intersection from the east-west street are exponentially distributed with a mean of 1 minute. It takes 0.15 minute for cars waiting at a red light to accelerate and clear the intersection once the light has turned green. We are interested in finding the traffic light timing for which the average waiting time of all cars at the intersection is minimized. Perform a search procedure to find the best traffic light timing. Assume that observations are independent and use a run length of 1000 time units for each trial.

Hint: Use a variable to represent the total waiting time of all cars. Each time a car that waits at a red light leaves the intersection, add its waiting time to this variable (the waiting time of these cars should include their acceleration time). Dividing this variable by the total number of cars that have passed through the intersection by the end of a run provides the average waiting time of all cars. Note that the total number of cars should also include those cars that do not wait at the intersection. See Chapter 8 for examples of traffic systems.

7.14 REFERENCES AND FURTHER READING

Banks, J., and J. Carson: *Discrete-Event System Simulation,* Prentice-Hall, 1984.

Bazaraa, M., and C. Shetty: *Nonlinear Programming—Theory and Algorithms,* Wiley, 1979.

Hogg, R., and A. Craig: *Introduction to Mathematical Statistics,* 4th ed., Macmillan, 1978.

Kirk, R.: *Experimental Design Procedures for the Behavioral Sciences,* Wadswort, 1968.

Kleijnen, J.: *Statistical Techniques in Simulation—Part I,* Marcel Dekker, 1974.

Kleijnen, J.: *Statistical Techniques in Simulation—Part II,* Marcel Dekker, 1975.

Kuester, J., and J. Mize: *Optimization Techniques with Fortran,* McGraw-Hill, 1973.

Law, A., and W. Kelton: *Simulation Modeling and Analysis,* 2nd ed., McGraw-Hill, 1991.

Mize, M., and G. Cox: *Essentials of Simulation,* Prentice-Hall, 1968.

Moy, W.: "Practical Variance Reduction Procedures for Monte Carlo Computer Simulations," in *Computer Simulation Experiments with Models of Economic Systems,* T. Naylor, Editor, Wiley, 1971.

Naylor, T., and J. Finger: "Verification of Computer Simulation Models," *Management Science,* October 1967, pp. 92–101.

Pegden, D., Shannon, R., and R. Sadowski: *Introduction to Simulation Using SIMAN,* McGraw-Hill, 1990.
Pritsker, A.: *Introduction to Simulation and SLAM II,* Wiley, 1986.
Shannon, R.: *Systems Simulation – The Art and Science,* Prentice-Hall, 1975.
Schmeiser, B.: "Simulation Experiments" in *Handbooks in Operations Research and Management Science, Volume 2: Stochastic Models,* D. P. Heyman and M. J. Sobel, Editors, North-Holland, Amsterdam, 1990, pp. 295–330.
Thesen, A.: *Simulation for Decision Making,* West, 1992.
Watson, H., and J. Blackstone: *Computer Simulation,* 2nd ed., Wiley, 1989.
Wilson, J.: "Variance Reduction Techniques for Digital Simulation," *American Journal of Mathematical and Management Sciences,* Vol. 4, 1984, pp. 227–312.

CHAPTER 8

APPLICATIONS OF SIMULATION

8.1 INTRODUCTION

Simulation can be applied to numerous problem areas in the service and manufacturing industries. Tables 8.1 and 8.2 present a classification of various representative application areas of simulation. The nature of industries, the resources involved, and typical activities are listed for each category. Note, however, that these tables are by no means complete; there are many other application areas in which simulation can be successfully utilized.

The examples presented in the earlier chapters introduced some potential application scenarios. The intention of those chapters was to introduce various features of EZSIM or highlight some issues related to simulation output analysis. The objective of this chapter is to introduce a number of diversified application scenarios to familiarize the reader with realistic problems and typical design or analysis questions that may be answered by simulation. The presentation starts with application examples in the service industries. Several examples in the manufacturing industries are then provided.

Each example includes the problem statement and the specific questions to be answered. The solution section then describes the EZSIM modeling approach used to solve the problem. Specific elements of the network are defined, and the general approach is described. The EZSIM network diagram and a corresponding simulation output are presented for the specified run length. Note that since the main intention of this

TABLE 8.1
Services and construction application areas

Industry	Category	Resources	Activities
Communications	Telephone, radio, computer, circuit switching, packet switching, wide area, local area	Terminals, stations, links, switching devices, satellites, operators	Connection establishment, data transfer, message assembly, compression, decompression, routing update, failure recovery, management of control
Construction	Building, bridge, dam, tunnel, highway, railroad, housing, airport, waterway	Bulldozers, earth movers, trucks, rollers, shovels, loaders, cranes/lifts, concrete, stone, steel, brick, block, lumber, personnel forms, tools, vehicles, pumps, scaffolding, storage area	Surveying, digging, hauling, preparing, cleaning, forming, stripping, paving, pouring concrete, laying brick/pipe, framing, setting steel, planting, positioning, erecting, loading, unloading
Education	Public schools, college/university, day care, special schools	Classrooms, courses, staff, athletic facilities, books, supplies, teachers, cafeteria, students	Registration, attendance, absence, teaching, SSR, graduation
Entertainment	Movie, theater, amusement park	Seats, waiting areas, parking, concessions, personnel	Ticket purchase, length of stay, concession, sales, shows, movies, rides, exit
Financial	Banking, securities, insurance	Cash, reserves, staff, supplies, parking, waiting area	Service, waiting, deposits, payments, withdrawals, purchases, sales, ATM usage, loans
Food service	Full-serve, restaurant, caterer, cafeteria, fast food, carryout	Tables, seats, waiters, waiting area, food, service items, chefs, bussers	Food preparation, customer service, delivery, occupancy
Health care	Hospital, emergency care, doctor, dentist, outpatient	Beds, rooms, medicine, staff, equipment, elevators, parking, waiting areas, restaurants, appointment times	Diagnosis, laboratory testing, serving, cleaning, medicine inventory control, staff and doctors scheduling
Hotel/ hospitality	Resort, hotel, motel, tours, travel agency	Beds, rooms, staff, water, linens, elevators, parking, telephone, TV	Lengths of stay, arrivals, walk-ins, cancellations, changes, no-shows, room occupancy, registration, checkout, food service, cleaning
Retailing	Department store, grocery, drug store	Parking spots, shopping carts, dressing rooms, staff, inventory, shelves, checkout counters	Service, shopping, waiting, checkout, fitting
Transportation	Taxi, shuttle, limo, intercity bus, intracity bus, airline, charter airline, car rental, ships, tankers	Vehicles, seats, cargo space, drivers, parking space, mechanics, staff, spare parts	Travel, idle, maintenance, passenger pickups, dropoffs, fueling, routing, staff and crew scheduling, ticket sales, reservation

TABLE 8.2
Production and manufacturing application areas

Industry	Category	Resources	Activities
Power generation	Steam, fossil fuel, hydro, nuclear, solar, wind	Turbines, pumps, cooling towers, generators, conveyors, scrubbers, capacitors, transmission lines, coal, gas, uranium, windmills, panels, meters, instruments	Firing, cooling, generating, conducting, storing, cleaning, tending, loading, unloading
Natural resource extraction/ harvesting	Strip mining, deep mining, timber harvesting, well drilling, fishing	Explosives, drill bits, cutters, shovels, drag lines, elevators, derricks, trucks, crushers, conveyors, bulldozers, nets, saws, storage yards, tanks, pumps, scales, ships, personnel, tools, winches, lights	Blasting, drilling, digging, scraping, hauling, cutting, boring, crushing, sorting, clearing, draining
Animal farming	Dairy, cattle, fish, poultry, sheep	Tractors, loaders, pens, trailers, barns, tanks, troughs, pumps, shelters, buildings, scales, feed, milkers, chemicals, personnel, breeding stock, cans, tools, buckets, grazing land	Milking, feeding, caching, spraying, cleaning, transporting, watering, weighing, breeding, neutering, branding
Plant farming	Grain, fruit, vegetables, cotton, flowers	Plows, planters, sprayers, disks, pickers, trimmers, ladders, tractors, trucks, tankers, trailers, combines, pumps, pipes, bins, bailers, gins, washers, elevators, scales, barns, greenhouses, fertilizers, pesticides, personnel, seed, boxes	Planting, pickling, baling, fertilizing, crating/packing, caching, spraying, cleaning, thrashing, transporting, irrigating, weighing
Chemical processing	Oil refining, water treatment, sewage treatment, paint making	Pumps, tanks, chemicals, mixers, heaters, pipelines, tankers	Cracking, mixing, settling, fermenting, skimming, testing, filling, pumping, draining, heating, cooling
Food processing	Dairy, grain, meat packing, baking, bottling, frozen foods, packaging	Coolers, heaters, racks, tanks, conveyors, chemicals, tools, personnel, processing equipment, packing materials, carts, tables, ovens, mixers, cutters, mills	Separating, mixing, measuring, sorting, packing, freezing, baking, drying, cutting, forming, raising, cooling, inspecting, cleaning, filling
Mineral processing	Iron/steel, copper, aluminum, glass, ceramics	Loaders, trucks, carts, raw material, chemicals, furnaces, conveyors, cranes, castings, ladles, personnel, tools, workspace, fuels	Hauling, fueling, pouring, firing, sampling, monitoring, tending, loading, cleaning, cooling, casting, rolling
Vehicle assembly	Auto, truck, bus, aircraft, ship, spacecraft, military	Conveyors, handtools, robots, welders, cranes, riveters, personnel, floor space, prints, attachment hardware, parts, cleaners, paint, tables, testers, manuals, monitors	Matching, retrieving, installing, moving, attaching, inspecting, testing, monitoring, painting, cleaning, etching, welding, riveting
Other product manufacturing	Furniture, electronics, tools, machinery, appliances, leisure, personal products	Conveyors, handtools, robots, riveters, personnel, floor space, attachment hardware, product parts, cleaners, paint, tables, testers, manuals, monitors, X-ray, packing materials	Matching, retrieving, installing, moving, attaching, inspecting, testing, monitoring, painting, cleaning, etching, soldering, assembling, packing, storing

chapter is the presentation of various application scenarios and modeling approaches, no effort is devoted to proper experimental design and statistical analysis of the results.

As has been shown in the earlier chapters, a given system may be modeled in many different ways (using resources to represent some entities, using facilities to represent some resources, etc.). The modeling approach may vary depending on the individual analyst's preferred modeling style. Note that the solution given for each example in this chapter represents only one modeling approach. Trying alternative modeling approaches enhances model building capability and hence is strongly recommended. The model files for all of the examples in this chapter are included in the accompanying diskette.

8.2 APPLICATION EXAMPLES IN THE SERVICE INDUSTRIES

As Table 8.1 shows, the application areas of simulation in the service industries are very diversified. A number of interesting capabilities have been incorporated in some of the new simulation tools to meet the modeling requirements for simulation applications in the service industries. Also, several special-purpose simulators have been developed for various forms of communications systems, chain restaurant design, traffic systems, parking lot design, and so on. As recent technical presentations in various simulation conferences indicate, simulation practitioners are discovering some new and unusual applications for simulation in the service sector.

Some representative examples are presented in this section to familiarize the reader with typical simulation problems and the corresponding solution methods in relation to service industry applications.

Example 8.1 Cafeteria. A cafeteria chain is designing a new food service concept for business lunch customers. In addition to the regular cafeteria line, a parallel self-service line with limited food items will be added as an option. Data was collected at a prototype operation. During the peak 45-minute period, customer interarrival times were found to be normally distributed with an average of 15 seconds and a standard deviation of 3 seconds. Half of the customers used the self-service line. The regular line consists of the salad, entrée, dessert, and beverage stations. Each station's service time is exponentially distributed with averages of 15, 30, 10, and 20 seconds, respectively. The entrée station has two servers; all others have one. The self-service line can accommodate up to six customers at a time. Its service time was normally distributed with an average of 45 seconds and a standard deviation of 5 seconds. After obtaining their food, customers pay the cashier. There are two cashiers at the cafeteria. The cashier service time is exponentially distributed with an average of 15 seconds. Assuming that customers stop at each station of the regular line, we would like to simulate the system for 1000 customers served to determine the following:

1. The average waiting time and average queue length at the cashier
2. The station in the regular line that has the largest average queue length

Solution. This system has two parallel service lines made up of queues and facilities fed by a single source node. No variables or statistics are required. Queue and facility data are sufficient to answer the questions (see Fig. 8.1*a*). The source node generates customer entities that follow one of two probabilistic branches to a queue representing the two cafeteria lines. The queue representing the regular cafeteria line begins a series of four sequential queue-facility pairs. Each facility has the number of parallel servers specified in the problem. Entities exit the beverage facility and converge with the self-service entities at a queue representing the cashier line. After passing through the cashier line, the entities are terminated. The self-service queue is followed by a single facility representing the self-service time. Six parallel servers are specified at this facility. Each server represents one of the six locations at which a customer may stand to select food items. Entities exit this facility and proceed to the cashier queue. Figure 8.1*b* shows the simulation output for this model. The reader is advised to change the number of parallel servers in the cashier facility to simulate additional cashiers.

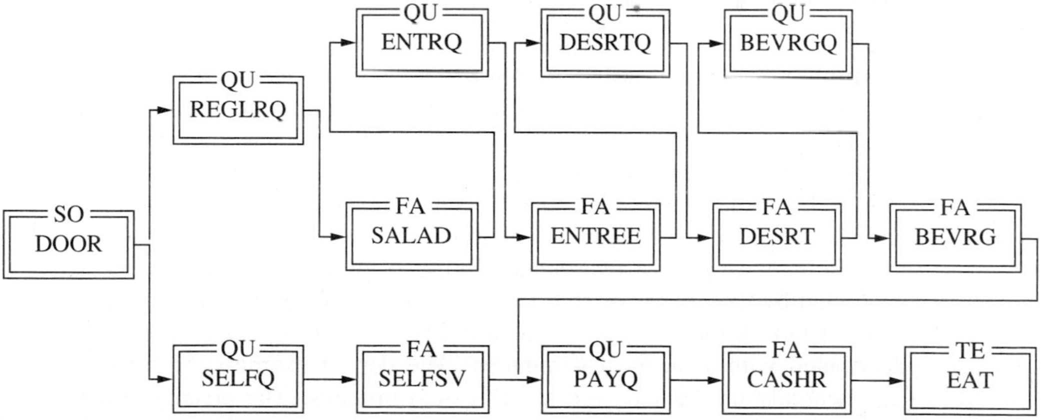

FIGURE 8.1*a*
Network model of the cafeteria example.

```
*** E Z S I M STATISTICAL REPORT ***

Simulation Project: PROBLEMS
Analyst: HILSCHER
Date: 03/15/92
Disk file name: CAFETERI.OUT

Current Time: 15164.38    Transient Period: 0.00

Q U E U E S:
NAME     MIN/MAX/LAST         MEAN     STD      MEAN     STD
         LENGTH               LENGTH   LENGTH   DELAY    DELAY
-----------------------------------------------------------------
REGLRQ    0/ 6/ 0             0.29     0.72     8.35     17.24
SELFQ     0/ 1/ 0             0.00     0.00     0.00     0.00
ENTRQ     0/ 8/ 0             0.34     1.10     9.85     26.98
DESRTQ    0/ 6/ 0             0.18     0.59     5.06     11.69
BEVRGQ    0/14/ 1             1.23     2.15     35.47    49.39
PAYQ      0/ 4/ 0             0.17     0.53     2.63     6.74

F A C I L I T I E S:
NAME     NBR    MIN/MAX/LAST   MEAN     STD      MEAN     MEAN
         SRVRS  UTILIZATION    UTLZ     UTLZ     IDLE     BUSY
-----------------------------------------------------------------
SALAD    1       0/ 1/ 0       0.51     0.50     14.10    14.68
ENTREE   2       0/ 2/ 1       0.99     0.80     29.02    28.53
DESRT    1       0/ 1/ 0       0.33     0.47     19.29    9.54
BEVRG    1       0/ 1/ 1       0.69     0.46     9.01     19.88
CASHR    2       0/ 2/ 0       0.93     0.80     16.21    14.12
SELFSV   6       0/ 5/ 2       1.42     0.96     145.36   44.99
```

FIGURE 8.1*b*
Simulation output of the cafeteria model.

Example 8.2 Dental office. A dentist is planning a layout for her new practice. She wishes to determine how many dental examination chairs and X-ray machines are needed to adequately serve the patients. The typical dental patient is first seated in an examining chair until the X-ray room is available. The patient then has X-rays taken by a dental technician and is returned to the examining chair. Twenty percent of X-rays must be retaken and have priority over others. In the examining area, the patient is examined by the dentist (with a technician's assistance) and then has his or her teeth cleaned by the technician. The X-ray time is exponentially distributed with a mean of 4 minutes. The dentist examination time is exponentially distributed with a mean of 6 minutes. The cleaning time is normally distributed with a mean of 8 and a standard deviation of 2 minutes. Patient appointments are scheduled every 10 minutes for four hours. There are three technicians, two X-ray machines, and four chairs in the dental office. We are interested in finding the statistics on the utilization of the dentist's resources and the amount of time that patients spend for their visit based on a simulation of 1000 served patients.

Solution. This problem uses a series of resource queues and delay nodes to determine total system time and to track resource utilization (see Fig. 8.2*a*). Patient entities are generated and passed to a series of resource queue nodes to acquire dental chair, X-ray machine, and dental technician resources. Note that at node XRAY the X-ray machine and one technician are requested. A delay node is then used to account for X-ray time, after which the X-ray machine and the technician are released. This node is followed by probabilistic branches to the exam resource queue or to the X-ray retake node. At the retake node, another X-ray and technician resource is acquired using priority without preemption. After resource acquisition, the X-ray delay node is again passed as before. After X-ray, entities enter the exam resource queue and acquire technician and dentist resources. A delay node then follows, after which the dentist is released. A second delay node accounts for cleaning time. The technician and chair resources are released after cleaning. The traversal time of patient entities between the source node and the terminate node specified in the Statistics menu provide the desired system time statistics. Figure 8.2*b* shows the simulation output for this model.

Example 8.3 Security screening. When passing through an airport security checkpoint, people first leave handbags or briefcases at the X-ray machine and then walk through the metal detector. The interarrival time of people is exponentially distributed with a mean of 6 seconds. Eighty percent of these people carry handbags. Handbags are placed on the X-ray conveyor, which has room for two bags before entering the machine. When the conveyer feed is full, people with bags wait in line. The X-ray processing time is uniformly distributed between 4 and 7 seconds. People line up to walk through the metal detector, whose processing time

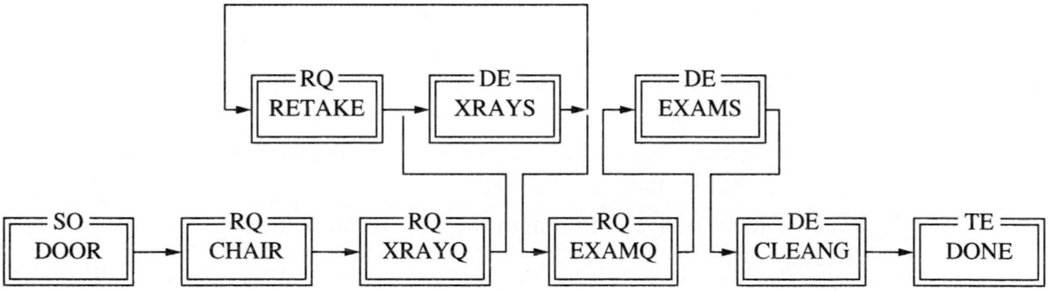

FIGURE 8.2*a*
Network model of the dental office example.

```
*** E Z S I M  STATISTICAL  REPORT ***

Simulation Project: PROBLEMS
Analyst: HILSCHER
Date: 03/15/92
Disk file name: DENTIST.OUT

Current Time: 10011.61    Transient Period: 0.00

Q U E U E S:
NAME      MIN/MAX/LAST      MEAN      STD       MEAN     STD
          LENGTH            LENGTH    LENGTH    DELAY    DELAY
-------------------------------------------------------------------
XRAYQ      0/ 2/ 0          0.00      0.05      0.02     0.27
EXAMQ      0/ 3/ 0          0.34      0.68      3.44     6.84
RETAKE     0/ 1/ 0          0.00      0.00      0.00     0.00
CHAIR      0/ 4/ 0          0.10      0.41      1.04     3.89

R E S O U R C E S:
NAME      INIT   MIN/MAX/LAST     MEAN     STD      MEAN     STD
          LEVEL  USAGE            USAGE    USAGE    LEVEL    LEVEL
-------------------------------------------------------------------
TECHN     3        0/ 3/ 1        0.50     0.59     2.50     0.59
XRAY      2        0/ 2/ 1        0.50     0.58     1.50     0.58
DENTIST   1        0/ 1/ 1        0.60     0.49     0.40     0.49
CHAIR     4        0/ 4/ 3        2.55     0.90     1.45     0.90
```

FIGURE 8.2*b*
Simulation output of the dental office model.

is normally distributed with a mean of 6 and a standard deviation of 1 second. Twenty percent of the people who walk through the metal detector must pass additional screening, which averages 9 seconds and has an exponential distribution. After personal screening, it takes 3 seconds to pick up the handbag and depart. Assume that any times not given are negligible. We would like to simulate the system for 1000 passengers to answer the following questions:

1. What is the average time it takes to pass through the security checkpoint?
2. What capacity is required in the buffer where bags collect after X-ray screening, before they are claimed?
3. What is the average line length waiting to walk through the metal detector?

Solution. This problem utilizes the MATCH-Q node to generate corresponding parallel processing lines for people and their baggage.

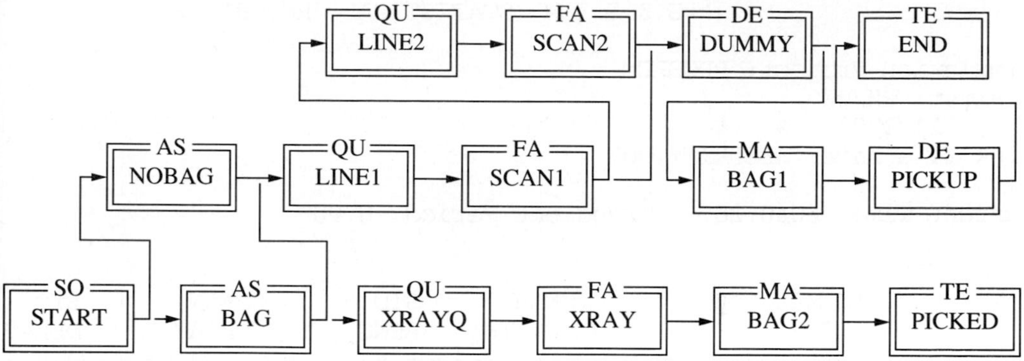

FIGURE 8.3*a*
Network model of the security screening example.

Probability-based branching is also used (see Fig. 8.3*a*). Passenger entities are generated and passed through one of two probabilistic branches to an assignment node where an attribute (BAG) is assigned the values of YES or NO, indicating whether or not the person has a handbag. Entities with a bag are then duplicated using an Always branching option at node BAG. The duplicated entity is the handbag belonging to the person. Passenger entities with and without bags then pass through the queue and facility nodes representing the walk-through metal detector. Entities are then sorted using probabilistic branching to either pick up their bag at the match node or to pass through additional screening. Those requiring additional screening are routed to the SCAN2 queue and facility nodes. Entities exiting these nodes are sorted using conditional branching based on the entity attribute (BAG) to either be terminated or to pick up their bag at the MATCH-Q node. Entities are matched with their bag at the MATCH-Q node and then pass through an associated delay node. Entities are then terminated. Entity count at the terminate node is 1000. Figure 8.3*b* shows the simulation output for this model.

Example 8.4 Crosswalk. Pedestrians use a crosswalk to cross a two-way street. Interarrival times are uniformly distributed between 7 and 12 seconds from side A to side B, and between 10 and 14 seconds from the other direction. The time it takes a pedestrian to cross a single traffic lane (one-half the street width) is uniformly distributed between 2 and 4 seconds. Pedestrians have the right-of-way over vehicle traffic, and the capacity of the crosswalk is 10 pedestrians per lane. Vehicle interarrival times from both directions are exponentially distributed with means of 8 seconds from direction A and 6 seconds from direction B. Vehicle times to drive over the crosswalk are exponentially distributed with a mean of 3 seconds. Assume that vehicles may proceed if their lane of traffic is

*** E Z S I M STATISTICAL REPORT ***

Simulation Project: PROBLEM
Analyst: RH/BK
Date: 11/1/92
Disk file name: SECURITY.OUT

Current Time: 6040.28   Transient Period: 0.00

Q U E U E S:

NAME	MIN/MAX/LAST LENGTH	MEAN LENGTH	STD LENGTH	MEAN DELAY	STD DELAY
XRAYQ	0/ 5/ 0	0.07	0.34	0.51	1.43
BAG2	0/45/42	16.55	7.80	113.87	54.63
LINE1	0/55/53	20.44	9.55	111.18	54.94
LINE2	0/ 4/ 0	0.07	0.32	2.25	6.22
BAG1	0/ 1/ 0	0.00	0.01	0.00	0.02

F A C I L I T I E S:

NAME	NBR SRVRS	MIN/MAX/LAST UTILIZATION	MEAN UTLZ	STD UTLZ	MEAN IDLE	MEAN BUSY
XRAY	2	0/ 2/ 1	0.77	0.76	8.88	5.53
SCAN1	1	0/ 1/ 1	1.00	0.06	0.02	6.01
SCAN2	1	0/ 1/ 0	0.28	0.45	22.26	8.72

V A R I A B L E S:

NAME	MEAN	STD	MIN	MAX	No.OBSRVD
SYSTIME	1.27E+02	5.01E+01	6.61E+00	2.50E+02	1000

FIGURE 8.3*b*
Simulation output of the security screening model.

clear of pedestrians. All times are given in seconds. We would like to simulate the system for 10,000 seconds of operation to determine the average waiting time for vehicles in each direction.

Solution. This problem demonstrates use of resources to control access to the crosswalk. The maximum number of resources for each lane of the crosswalk is 10 (capacity of the crosswalk). Each pedestrian in the lane uses a single resource. These resources are named XWALKA and XWALKB in this model. Each car uses all 10 resources to cross its lane of the crosswalk (see Fig. 8.4*a*). Four separate networks are used in this problem to simulate the four traffic movements—two pedestrians and two vehicles. Pedestrian and vehicle networks are identical except for node names. Pedestrians are created as entities and proceed to a RESOURCE-Q

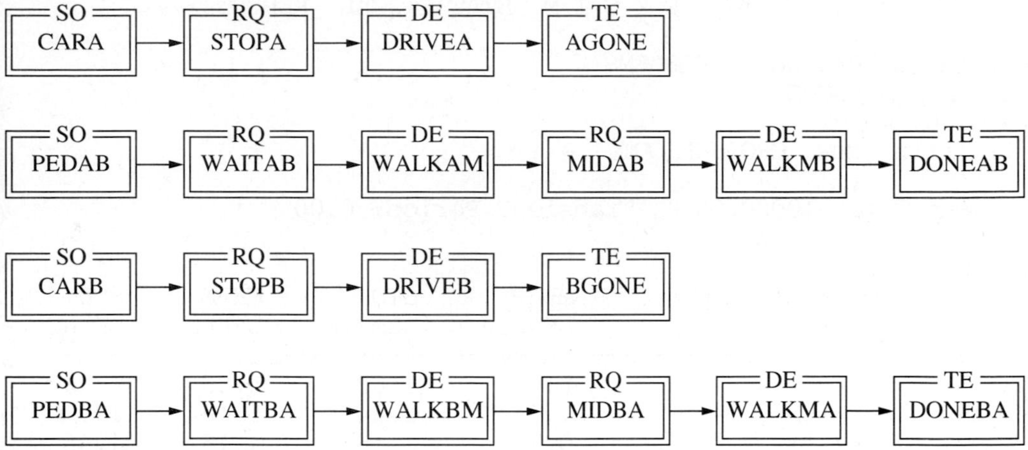

FIGURE 8.4*a*
Network model of the crosswalk example.

node to acquire one of the crosswalk resources in the proper lane. After acquiring the resource, the entity moves through a DELAY node which represents crossing the first traffic lane. The lane resource is released after the delay node. The entity then acquires a resource for the next lane in another RESOURCE-Q node and passes through a corresponding delay node as before. The resource is again released after the delay node, and the entity is terminated. The vehicle networks are similar to the pedestrian networks but have only one set of RESOURCE-Q and DELAY nodes. These represent crossing the crosswalk in their respective lane. To do so, the car must acquire all resources, simulating an empty lane. Pedestrians are given resource priority (right-of-way) without preemption over vehicles. Fig. 8.4*b* shows the simulation output for this model.

Example 8.5 Traffic lanes. Two parallel traffic lanes (A and B) traveling in the same direction widen to three lanes at the approach to an intersection. The third lane (C) is a left turn lane. Interarrival times of cars approaching the intersection are distributed exponentially with an average of 3 seconds. Forty percent of these cars take lane A, and the rest take lane B. Forty percent of the cars in lane B enter lane C to turn left. The traffic signal at the intersection is on a fixed 2-minute cycle. Fifteen seconds of this time is for lane C traffic to turn left. This is followed by thirty seconds of time for lanes A and B to proceed simultaneously. We would like to simulate the system for 500 cars passing through lane C to find the capacity necessary in lane C to eliminate blocking of lane B by cars waiting to make a left turn.

*** E Z S I M STATISTICAL REPORT ***

Simulation Project: PROBLEMS
Analyst: HILSCHER
Date: 03/16/92
Disk file name: CROSSWK.OUT

Current Time: 10000.00 Transient Period: 0.00

Q U E U E S:

NAME	MIN/MAX/LAST LENGTH	MEAN LENGTH	STD LENGTH	MEAN DELAY	STD DELAY
WAITAB	0/ 7/ 4	0.60	1.02	5.66	9.24
MIDAB	0/14/ 0	1.37	2.24	13.04	19.90
STOPB	0/ 9/ 0	0.71	1.15	4.16	4.66
MIDBA	0/12/ 2	0.84	1.77	10.11	19.30
WAITBA	0/16/ 3	2.06	2.63	24.57	31.23
STOPA	0/ 5/ 0	0.32	0.66	2.53	3.18

R E S O U R C E S:

NAME	INIT LEVEL	MIN/MAX/LAST USAGE	MEAN USAGE	STD USAGE	MEAN LEVEL	STD LEVEL
XWALKA	10	0/10/ 0	4.10	4.48	5.90	4.48
XWALKB	10	0/10/10	5.64	4.58	4.36	4.58

FIGURE 8.4*b*
Simulation output of the crosswalk model.

Solution. This problem requires two networks: one for vehicle traffic and one for signal control. Gates and switches are the primary nodes used. Special attention must be paid to queue blocking to properly set up the relationships in lanes B and C (see Fig. 8.5*a*). In the first network, car entities are generated and probabilistically branched to any of the three lanes. Entities passing through the lane gates are counted, and then they proceed to termination. When the two gates are open, entities are counted, and then they proceed to termination nodes. All gates should be set to a closed initial condition. The second network generates a single key entity which first passes through a switch to open the lane C gate. The key is then delayed 15 seconds before passing through switches that close the lane C gate and open the lanes A and B gates. The key entity is then delayed for 30 seconds before passing through switches that close the lane A and B gates. A final 75-second delay occurs, after which the key entity is routed back to the first switch to open gate C. Note that lanes A and B are opened by one switch node named OPENAB. These gates are also closed with one switch node named CLOSAB. This shows the capability of the switch node, which can open and close several

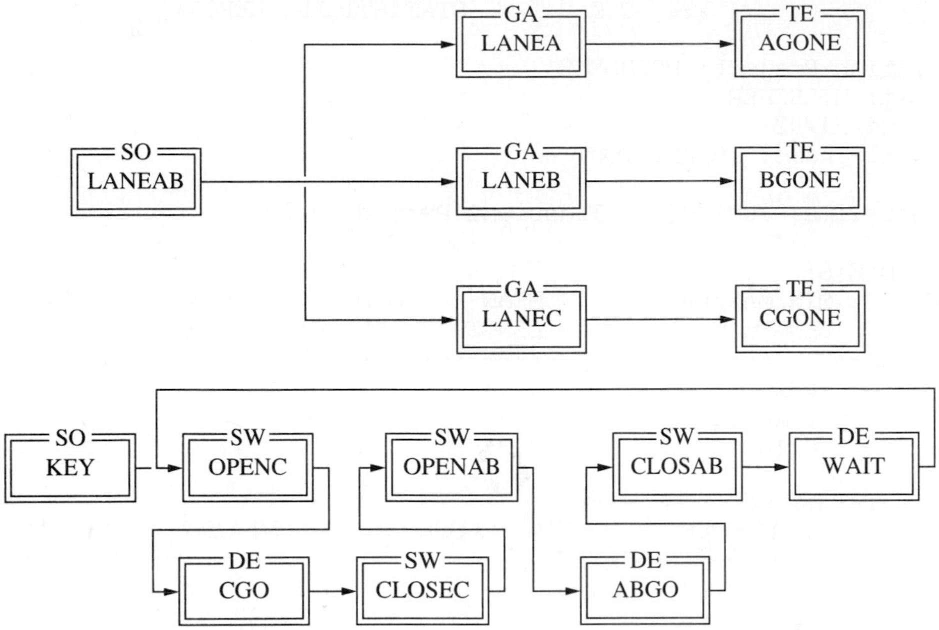

FIGURE 8.5*a*
Network model of the traffic lanes example.

nodes simultaneously. (Note that to specify more than one gate the S key should be pressed in the corresponding window of the SWITCH node.) The simulation output for this model is shown in Fig. 8.5*b*.

Example 8.6 Roller coaster design. The Willey-Nilley Amusement Company is planning a new roller coaster at one of its amusement parks and needs to determine the maximum average speed for a three-train operation. During peak attendance days at the park, people arrive at popular attractions at an exponential rate averaging 4 seconds. Each train has a capacity of 20 people. Each train waits until it is full to capacity. A loaded train cannot leave the station until the preceding trains pass the one-third and two-third points on the track, respectively. Likewise, any train will stop at a one-third point if the preceding train has not crossed the next one-third point. The planned length of the ride circuit is 1500 meters. Assume an equal average speed on all sections of the ride. A simulation model of this system is to be built to answer the following questions:

1. What is the maximum average speed of the ride which will not cause stoppage at any point on the track except the station?
2. What is the average number of passengers per hour that can be accommodated with this ride?

```
                *** E Z S I M  STATISTICAL  REPORT  ***

Simulation Project: PROBLEMS
Analyst: HILSCHER
Date: 04/21/92
Disk file name: TRAFFIC.OUT

Current Time: 7090.00     Transient Period: 0.00

Q U E U E S:
NAME       MIN/MAX/LAST          MEAN     STD       MEAN      STD
           LENGTH                LENGTH   LENGTH    DELAY     DELAY
-----------------------------------------------------------------------
LANEA        0/20/12             4.61     4.54      34.97     30.26
LANEB        0/20/12             4.44     4.55      32.22     29.38
LANEC        0/16/ 0             3.37     3.05      47.47     32.72

GATE  STATISTICS:
NAME         TIME OPEN        TIME CLOSED   PERCENT OPEN   PERCENT CLOSE
-----------------------  ----------------------------------------------
LANEA          1.80E+03       5.30E+03          25.32             74.68
LANEB          1.80E+03       5.30E+03          25.32             74.68
LANEC          8.95E+02       6.20E+03          12.62             87.38

V A R I A B L E S:
NAME       MEAN       STD          MIN          MAX          No.OBSRVD
-----------------------------------------------------------------------
CYCLES     6.00E+01   0.00E+00     6.00E+01     6.00E+01     60
LANEA      9.13E+02   0.00E+00     9.13E+02     9.13E+02     913
LANEB      9.52E+02   0.00E+00     9.52E+02     9.52E+02     952
```

FIGURE 8.5*b*
Simulation output of the traffic lanes model.

3. What is the maximum number in the loading queue during the three hours of operation?
4. Does having more trains reduce the waiting time of customers significantly? Note that no more than three trains can be on the track. Having additional trains can help in loading customers while the three trains travel on the track.

Solution. This problem focuses on carriers, gates, switches, and delay nodes. Some trial and error is required to balance the system speed. A variable can be used to represent the speed value for a single entry during the trial and error period (see Fig. 8.6*a*). Three train entities are generated at the start of the simulation. They then pass to the group node to pick up riders. Rider entities are generated and go directly to the group node,

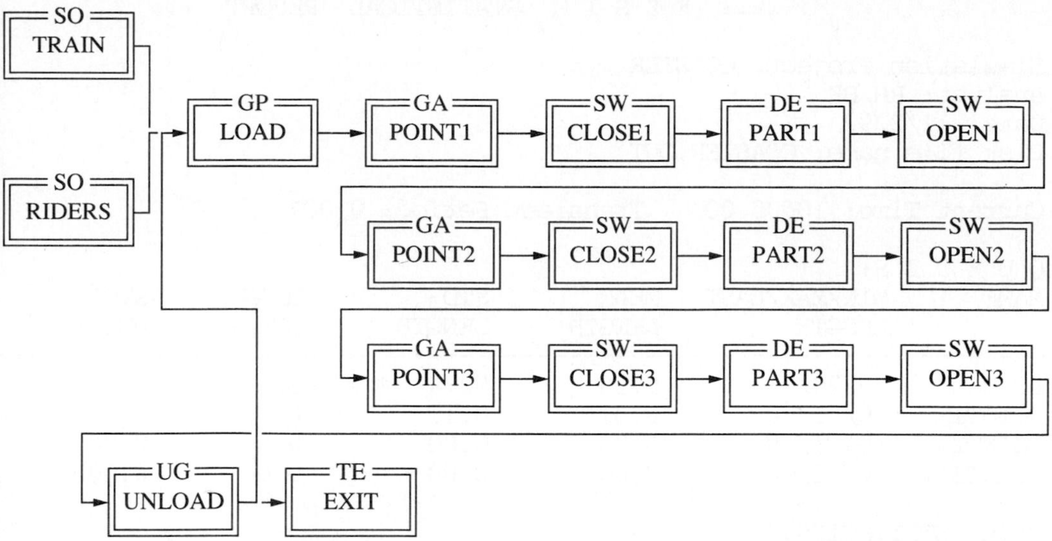

FIGURE 8.6*a*
Network model of the roller coaster example.

where a train waits for a time given by the load time distribution. It then takes all available riders up to its capacity and moves on to the gate queue. The gate queue is set open to start and allows the train to pass on to a switch node that closes the gate behind it. The entity then goes to the delay node representing the first third of the ride time. After the delay node, the entity passes through a gate node and on to a switch that closes the one-third gate behind it and opens the station gate. The entity continues on through similar series of nodes for the middle and end of the ride. When the train returns to the station, riders are unloaded in the ungroup node. The train then returns to the group node for another load of passengers. In the meantime the two other trains have departed the station on the same circuit. The output shown is for a simulation length of 10,000 seconds. Delay times may be changed to study the effects of various train speeds. Fig. 8.6*b* shows the simulation output for this model.

Example 8.7 Ski rental. The management of a major ski resort wants to simulate the rental program during the high season. There are two types of customers: people who pay cash and people who use a credit card. The interarrival times of these customers are exponentially distributed with means of 130 and 60 seconds, respectively. After arrival, customers must fill out a waiver form. The time taken to fill the waiver form is normally distributed with a mean of 60 and a standard deviation of 20 seconds. After filling out the form, customers get in line to pay. Time to pay

```
                    ***  E Z S I M  STATISTICAL  REPORT  ***

Simulation Project: COASTER
Analyst: RH/BK
Date: 11/2/92
Disk file name: COASTER.OUT

Current Time: 10000.00    Transient Period: 0.00

Q U E U E S:
NAME        MIN/MAX/LAST   MEAN        STD         MEAN        STD
            LENGTH         LENGTH      LENGTH      DELAY       DELAY
---------------------------------------------------------------------------
LOAD          0/20/ 1      9.56        5.74        35.80       26.76
POINT1        0/ 1/ 0      0.01        0.11        0.92        2.50
POINT2        0/ 1/ 0      0.00        0.00        0.00        0.00
POINT3        0/ 1/ 0      0.00        0.00        0.00        0.00

GATE   STATISTICS:
NAME          TIME OPEN    TIME CLOSED         PERCENT OPEN    PERCENT CLOSE
---------------------------------------------------------------------------
POINT1          2.44E+03    7.56E+03                24.38             75.62
POINT2          2.46E+03    7.54E+03                24.65             75.35
POINT3          2.52E+03    7.48E+03                25.25             74.75
```

FIGURE 8.6*b*
Simulation output of the roller coaster model.

depends on the method of payment. The time required to pay in cash is uniformly distributed between 30 and 45 seconds; the time required to pay by credit card is uniformly distributed between 90 and 120 seconds. There are three types of rental equipment: boots, skis, and poles, rented in that order. Not everyone rents all three types because some people already have some equipment. After paying, 80 percent of the people rent boots, 10 percent go directly to skis, and the rest just rent poles. Once at the boot counter, it takes a normally distributed time with a mean of 120 and a standard deviation of 5 seconds to obtain the boots. The time to try the boots on is uniformly distributed between 60 and 240 seconds. About 20 percent of the people need a different size boot; the rest go to get their skis. Once in the ski rental area, everyone waits until a resort employee is free to obtain the right size ski, which takes a uniformly distributed time between 45 and 75 seconds. Twenty percent of these people need their bindings adjusted. The binding adjustment process is exponentially distributed with a mean of 180 seconds. Seventy percent of the people go on to get their poles; the rest of the people leave. Ninety percent of the people who get their bindings adjusted go on to get poles, and the rest leave. At the station where poles are rented, service time is normally distributed with a mean of 60 and a standard deviation of 10 seconds. There are currently two employees working in the rental shop. Management wants to use simulation to find answers to the following:

1. How should the employees be allocated so that the queues do not become excessively long?
2. Should additional workers be hired to keep the average queue lengths below 20 for every queue?
3. How many seats are necessary for the boot-fitting area?

Solution. Fig. 8.7*a* shows the network model for this problem. The model starts with two SOURCE nodes: one for customers who pay in cash and the other for customers who pay by credit card. At the register facility, "entity name dependent service time" is specified, and the corresponding service time distributions are assigned to the customer entities named CASH and CREDIT. Probabilistic branching is chosen, and the proper percentages are defined for the three branches that lead the customer entities to various equipment-rental sections. The boot-fitting area is modeled as three nodes: a QUEUE node, a FACILITY node for

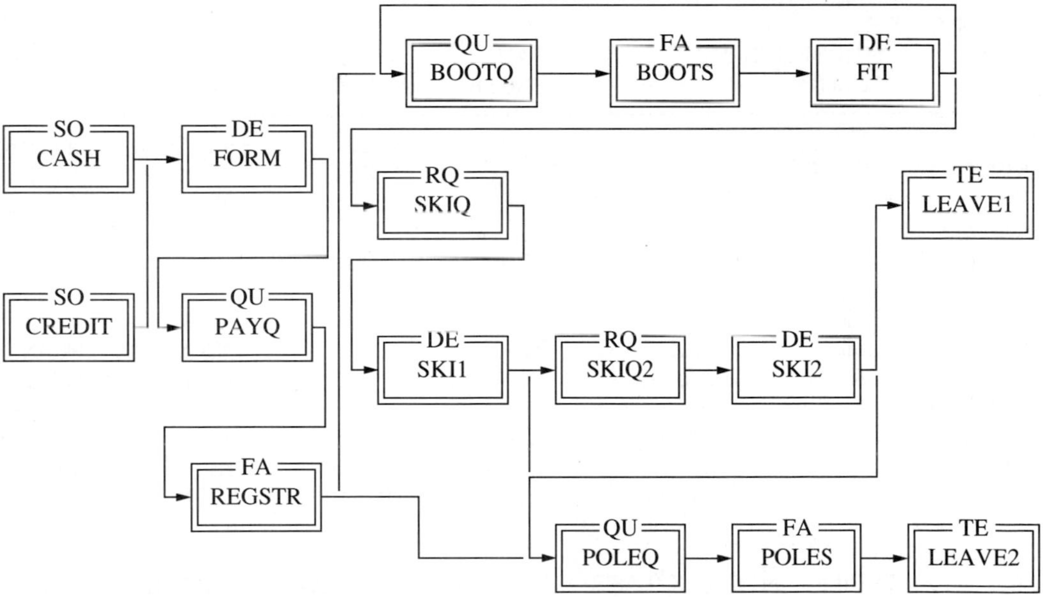

FIGURE 8.7*a*
Network model of the ski rental example.

finding the boots, and a DELAY node to fit the boots. After the DELAY node, probabilistic branching is chosen again, where 20 percent of the customers go back to the boot QUEUE. The ski-rental area is modeled as a RESOURCE-Q node followed by a DELAY node for representation of the process of obtaining skis. Another RESOURCE-Q/DELAY node combination is used to represent the binding adjustment process. Note that resources are used here as opposed to facilities, because the employee must stay with the customer if the bindings need adjustment. After both DELAY nodes, the proper probabilistic branching is defined. The last rental area is for the poles. This is simply modeled by a QUEUE/FACILITY node combination. The output shown in Fig. 8.7*b* is for a run length of 10,000 seconds.

*** E Z S I M STATISTICAL REPORT ***

Simulation Project: SKIRENTAL
Analyst: DERUITER & PRIDE
Date: 4/4/93
Disk file name: SKI.OUT

Current Time: 10000.00 Transient Period: 0.00

Q U E U E S:

NAME	MIN/MAX/LAST LENGTH	MEAN LENGTH	STD LENGTH	MEAN DELAY	STD DELAY
PAYQ	0/13/ 9	3.53	3.20	144.50	127.05
BOOTQ	0/ 9/ 1	1.44	1.86	63.44	74.02
SKIQ	0/11/ 8	3.98	3.30	195.74	173.41
POLEQ	0/21/16	9.71	7.13	513.25	384.46
SKIQ2	0/ 1/ 0	0.00	0.00	0.00	0.00

F A C I L I T I E S:

NAME	NBR SRVRS	MIN/MAX/LAST UTILIZATION	MEAN UTLZ	STD UTLZ	MEAN IDLE	MEAN BUSY
REGSTR	2	0/ 2/ 2	1.86	0.45	6.26	83.42
BOOTS	3	0/ 3/ 3	2.70	0.70	13.43	119.31
POLES	1	0/ 1/ 1	0.91	0.29	5.78	58.73

R E S O U R C E S:

NAME	INIT LEVEL	MIN/MAX/LAST USAGE	MEAN USAGE	STD USAGE	MEAN LEVEL	STD LEVEL
SERVCMN	2	0/ 2/ 2	1.73	0.61	0.27	0.61

FIGURE 8.7*b*
Simulation output of the ski rental model.

Example 8.8 Capital investment. An investment firm is considering an investment opportunity in a four-year project. It is estimated that the revenue for each of the project years is distributed uniformly between \$1000 and \$1500. Assuming a 10 percent rate of return, the firm is interested in finding the expected value (mean) and variance of the present value of the project for risk analysis purposes based on a simulation of the project.

Solution. The modeling approach for this problem is straightforward. Fig. 8.8*a* shows the network model in which ASSIGN nodes YEAR1 through YEAR4 assign uniformly distributed values between 1000 and 1500 to user variables X1 through X4, respectively. The ASSIGN node named PW calculates the total present worth of annual revenues. The relationship used is

$$X = (0.91)X1 + (0.82)X2 + (0.75)X3 + (0.68)X4$$

The above relationship is based on $P = F(1 + i)^{-n}$, where P is present worth, F is future value, i is the interest rate, and n is the number of periods.

The source node generates entities that trigger the assignment processes. Each entity sent by the SOURCE node results in one replication (one observed value for X). By specifying the maximum number of creations in the SOURCE node, the sample size can be specified (100 is used for the simulation output shown in Fig. 8.8*b*). Note that the time between creation of entities may be arbitrarily chosen because it does not have any meaning in the context of this model (i.e., this is a timeless model). However, choosing a value of 1 for the intercreation time makes the master simulation clock correspond to the number of replications. This may be desirable in an animation of this model. To compute the mean and standard deviation of the observations of the present value, X, observation-based statistics are collected on this user variable. The statistics are observed each time an entity hits the TERMINATE node.

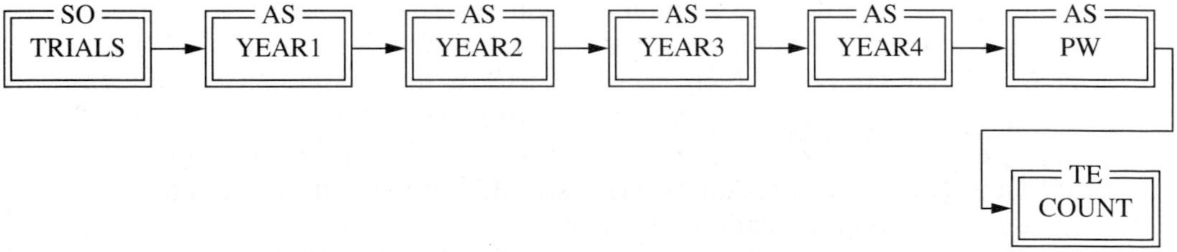

FIGURE 8.8*a*
Network model of the capital investment example.

```
                    *** E Z S I M  STATISTICAL  REPORT ***

Simulation Project: CAPITAL BUDGETING
Analyst: BK
Date: 6/23/93
Disk file name: CAPITAL.OUT

Current Time: 99.00     Transient Period: 0.00


V A R I A B L E S:
NAME       MEAN        STD         MIN          MAX        No.OBSRVD
--------------------------------------------------------------------
PVALUE    3.97E+03   2.20E+02    1.00E+03    4.45E+03    100
```

FIGURE 8.8*b*
Simulation output of the capital investment model.

8.2.1 Communication Systems

As mentioned in Chapter 1, simulation is extensively used in the analysis and design of communication networks, thousands of which are deployed worldwide. These networks connect telephones, computers, and other devices by means of various media such as copper wires, optical fibers, coaxial cables, and radio waves. Communication systems can have different sizes. They can interconnect units within a building or within a university campus, or they can span the entire globe.

The increasing use of digital communication to transmit computer data, digitized voice communication, and digitized images (e.g., fax machines) necessitates the systematic design and utilization of communication networks which provide for maximum communication channel availability and communication speed. Recent methodological developments such as the packet-switching concept, which is based on nonsequential and distributed transmission of packetized chunks of data, have led to effective but complicated communication control systems that require sophisticated analysis and design tools. Consequently, analytical approaches such as queuing theory as well as simulation are being used extensively in the field of communication systems. Complicated control strategies, which are frequently implemented in wide area networks and local area networks (LANs), limit the use of the analytical techniques in communication systems design and analysis. Hence, simulation is an especially attractive tool in this application area. Several commercial simulators (i.e., special-purpose simulation software tools) are now available for communication systems.

Generally, in communication networks the users (terminals, computers, etc.) may access the network for transmission or reception of data by one of two distinct methods. In the first method a dedicated access line is available for each user. In this method there is no network access contention. Switching stations (multiplexers) are used to connect the user lines to the main network lines (trunks).

In the second access method, a common channel or medium that is shared by all users is utilized. Examples of this popular method are local area networks using coaxial cables, interactive CATV (cable television), packet radio (which is applied to cellular telephone networks), and satellite radio schemes. There is contention for the common medium in all of these applications. Provisions are therefore necessary for providing fair access to the common channel to all users. There are two basic strategies for achieving this. In *polling*, users are scanned sequentially either by a centralized approach or by a decentralized method (e.g., the token-passing method used in token-ring LANs). In the *random access* method, users transmit at will. Since, in the latter method, one or more users may decide to use the common channel at the same time, contention resolution methods are needed. A special case of the random access strategy is applied in the popular Ethernet protocol used in some LANs.

In this section, two representative examples are provided which serve to demonstrate the application of simulation in the analysis of polling and random access strategies in communication networks.

Example 8.9 Network random access. Suppose three computers are connected to a network. The times between data transmission requests are exponentially distributed with means of 3, 4.5, and 6 seconds for computers 1, 2, and 3, respectively. When a computer intends to send a message, it routinely checks the line availability for data transfer. (This method is called the pure Aloha method; it was developed at the University of Hawaii in the early 1970s.) The time between line checkings is 0.1 second (in the Aloha method, this time may have a random length). When the line is available, data is transmitted. Data transfer time is uniformly distributed between 0.5 and 1.5 seconds. We are interested in finding the average waiting time of messages before transmission in each computer. Various lengths of time between line checkings are to be tried to find the timing that results in the smallest overall average message waiting time.

Solution. The network model for this example is shown in Fig. 8.9*a*. The three SOURCE nodes represent the computers which send message entities to the corresponding GATE-Q nodes, where messages await line availability. The message entity that leaves one of these GATE-Q nodes immediately closes the gate behind it so that no more than one message entity at a time is received by the RESOURCE-Q node. At this node, the resource called LINE is seized and kept for the duration of the message

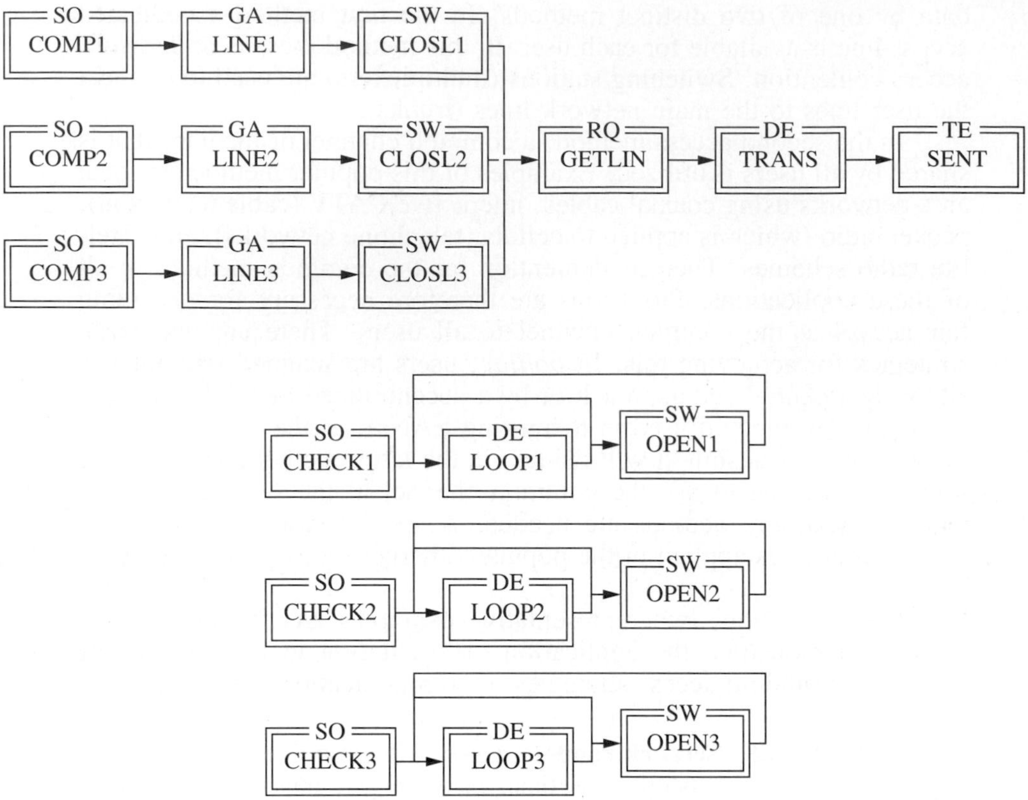

FIGURE 8.9*a*
Network model of the network random access example.

transfer process, which takes place in the DELAY node named TRANS. The LINE resource is released after this DELAY node.

Each of the lower disjoint networks has a circulating entity which checks the availability of the line and a message in the corresponding computer. For example, in the DELAY node LOOP1, conditional branching to the SWITCH node OPEN1 is specified. The entity proceeds to OPEN1 if the size of the queue in the LINE1 gate is larger than zero (i.e., there is a message at computer 1) and the LINE resource is available. Otherwise, the checking entity loops back to the DELAY node, which has a duration of 0.1 second. OPEN1 opens gate LINE1. This procedure ensures that no message is sent at any time other than the line-checking time and that no more than one message occupies the LINE resource at any time. The output shown in Fig. 8.9*b* is for simulation of 1000 messages sent. Note in this output that the duration of open gate status for all GATE-Q nodes is zero. This is true because gates open and close in one

```
*** E Z S I M  STATISTICAL  REPORT ***

Simulation Project: COMMUNICATION NTWRK
Analyst: BK
Date: 8/3/1993
Disk file name: ALOHA.OUT

Current Time: 1324.78    Transient Period: 0.00

Q U E U E S:
NAME        MIN/MAX/LAST              MEAN      STD        MEAN       STD
            LENGTH                    LENGTH    LENGTH     DELAY      DELAY
-------------------------------------------------------------------------------
LINE1        0/ 4/ 1                  0.32      0.67       0.90       1.01
LINE2        0/ 4/ 2                  0.37      0.75       1.66       2.36
LINE3        0/ 8/ 1                  0.58      1.15       3.27       4.81
GETLIN       0/ 1/ 0                  0.00      0.00       0.00       0.00

GATE  STATISTICS:
NAME         TIME OPEN          TIME CLOSED    PERCENT OPEN    PERCENT CLOSE
-------------------------------------------------------------------------------
LINE1         0.00E+00           1.32E+03          0.00              100.00
LINE2         0.00E+00           1.32E+03          0.00              100.00
LINE3         0.00E+00           1.32E+03          0.00              100.00

R E S O U R C E S:
NAME       INIT    MIN/MAX/LAST     MEAN      STD       MEAN      STD
           LEVEL   USAGE            USAGE     USAGE     LEVEL     LEVEL
-------------------------------------------------------------------------------
LINE       1       0/ 1/ 1          0.76      0.43      0.24      0.43
```

FIGURE 8.9*b*
Simulation output of the network random access model.

epoch of time, each time releasing one entity, which immediately closes the gate after it exits.

Example 8.10 Network polling. Three computers are connected to a local area network. The times between data transmission requests are exponentially distributed with means of 5, 6, and 7 seconds for computers 1, 2, and 3, respectively. Using a polling strategy, each computer is checked in a cyclic order for its possible transmission request. The status of a computer is checked only when there is no message transmission in progress. If at the time of checking a computer it is found to have a message to send, the message is transmitted. Status checking is not performed during message transmission. Checking starts as soon as a message transfer is completed and the common channel (line) is made available. The time between two consecutive checkings of the status of computers (referred to as the *walk time*) is 0.1

second. Message transmission times are uniformly distributed between 0.5 and 3 seconds for computer 1, between 0.5 and 2 seconds for computer 2, and between 1 and 4 seconds for computer 3. We are interested in finding the average waiting time of messages in each computer before transmission.

Solution. Fig. 8.10*a* shows the simulation network model of this communication network. In the upper disjoint network one entity circulates in a closed loop. As the entity reaches the ASSIGN node called POLL1, it resets the level of two resource types called LINE and LINE1 at 1 and sets the value of the user variable X to zero. X is the duration of transfer, which is set in the lower disjoint network. Resource LINE is the common channel. LINE1 will be explained later. As the polling entity leaves the ASSIGN node, it is delayed by the duration of walk time followed by the duration of transmission (which may be zero, if no message entity is sent by the lower network). The process continues for ASSIGN nodes POLL2 and POLL3 in which X is set at zero, LINE is reset to 1, and LINE2 (in POLL2) and LINE3 (in POLL3) are reset to 1.

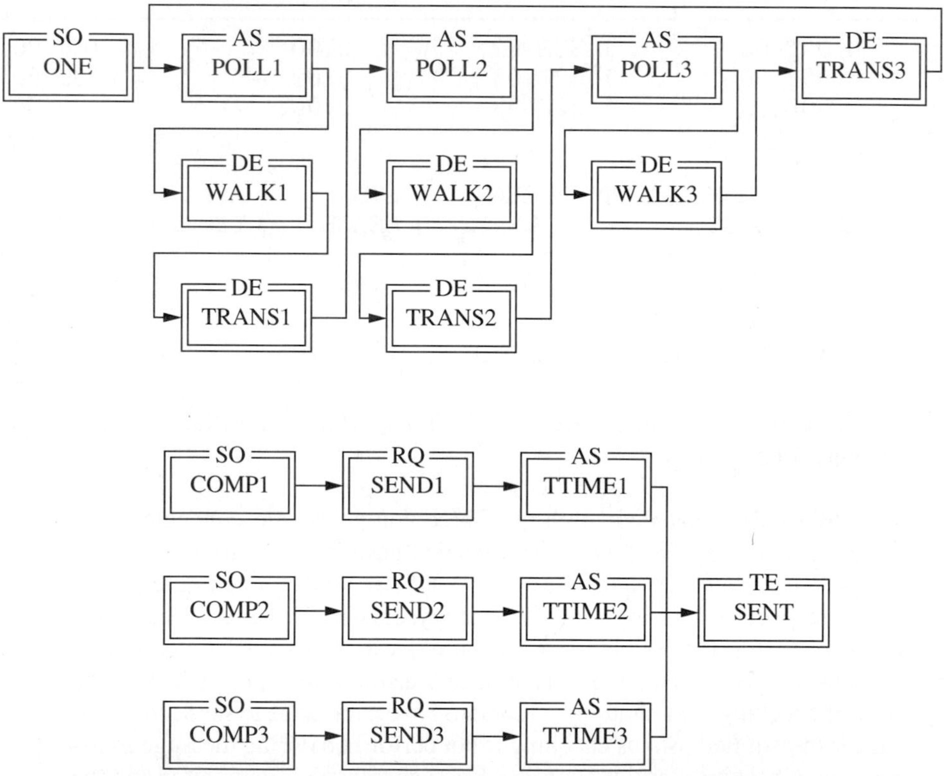

FIGURE 8.10*a*
Network model of the network polling example.

In the lower disjoint network, each SOURCE node generates its respective messages according to the given interarrival time distributions. The message entities are sent to the RESOURCE-Q nodes, in which one unit of each of the two types of resources is requested. For example, in SEND1 one unit of resource LINE and one unit of resource LINE1 are requested. Likewise, in SEND2 one unit of LINE and one unit of LINE2 are requested. Note that resource LINE is used to ensure that no more than one message is transmitted at a time, and LINE1 is used to make sure that computer 1 sends a message only when it is polled and not when another computer ends its transmission and releases the common channel (LINE). LINE2 and LINE3 are used for similar controls of computers 2 and 3, respectively. The ASSIGN nodes that follow the RESOURCE-Q nodes assign the respective transmission durations to variable X, which is used in the upper disjoint network. Note that in this model the seized resources are never released. The resources are reset to their original level of 1 by means of the ASSIGN nodes. This prevents the unauthorized seizure of the common channel at times other than polling times. The output shown in Fig. 8.10*b* is for simulation of 1000 transmitted messages.

```
                 ***  E Z S I M  STATISTICAL  REPORT  ***

Simulation Project: POLLING NETWORK
Analyst: BK
Date: 8/7/1993
Disk file name: POLLING.OUT

Current Time: 1957.52    Transient Period: 0.00

Q U E U E S:
NAME       MIN/MAX/LAST          MEAN      STD       MEAN     STD
           LENGTH                LENGTH    LENGTH    DELAY    DELAY
-------------------------------------------------------------------
SEND1        0/13/ 4             1.93      2.91      9.41     11.96
SEND2        0/36/18             5.60      7.75      30.01    39.85
SEND3        0/18/ 4             5.19      4.99      34.46    37.37

R E S O U R C E S:
NAME     INIT    MIN/MAX/LAST      MEAN      STD       MEAN     STD
         LEVEL   USAGE             USAGE     USAGE     LEVEL    LEVEL
-------------------------------------------------------------------
LINE1    0       0/ 1/ 1           0.68      0.47      0.32     0.47
LINE     0       0/ 1/ 0           0.89      0.31      0.11     0.31
LINE2    0       0/ 1/ 0           0.45      0.50      0.55     0.50
LINE3    0       0/ 1/ 0           0.59      0.49      0.41     0.49
```

FIGURE 8.10*b*
Simulation output of the network polling model.

8.3 APPLICATION EXAMPLES IN MANUFACTURING

Manufacturing systems involve complex and dynamic interrelationships between their many components. Materials, tools, fixtures, machines, parts, work-in-process, finished parts inventories, material-handling equipment, storage systems, operators, and various types of information are some of the components of manufacturing systems. The systems analysis and design tasks in manufacturing can be performed using one or more of the following:

- Intuitive management decisions, which rely on human experience and judgment. This method can be highly subjective and inconsistent. Intuitive decisions may work in one setting but fail in others.
- Mathematical decision rules that usually apply only to extremely simplified and unrealistic scenarios.
- Utilization of expert systems that usually apply to a limited range of problems. Knowledge acquisition and truth maintenance in these systems are generally difficult.
- Computer simulation models of manufacturing systems, which provide for the most realistic problem representation, even for very complicated scenarios.

Simulation is an indispensable tool for successful planning and design of new plants and processes as well as for studying the behavior of existing systems and creating reliable decision policies at many levels (e.g., top management strategy policies, middle management planning and control policies, and operating management resource allocation and scheduling policies). Because of the widespread application of simulation in manufacturing, several special-purpose tools (simulators) for simulation of manufacturing systems have been developed and commercialized in recent years.

Because of the vast scope and diverse nature of manufacturing systems, it is not practical to present every possible manufacturing application of simulation in this book. The following examples present only a small set of possible application examples; they also illustrate the useful features of EZSIM as applied to manufacturing systems simulation.

Example 8.11 Eyeglass manufacturing. An eyeglass manufacturer receives orders daily via fax. Order interarrival times are uniformly distributed between 5 and 18 minutes. Orders are filled in a manufacturing process which includes the following steps and associated processing times and rework rates:

Process	Duration (minutes)	% rework
Plain grinding	Exp(14)	20
Bifocal grinding	Exp(18)	20
Tinting	Exp(5)	15
Frame installation	Uniform(3,5)	0

Bifocal lenses are ordered on 38 percent of all glasses, and tinting is ordered 50 percent of the time. Assume that each operation in the process includes both lenses in a pair of glasses, and use only one grinding operation for each order, either plain or bifocal. A pair of lenses can be reworked only once before being scrapped. There are two grinding and two tinting machines available. We would like to simulate the system for 80 hours of operation to answer the following questions:

1. What is the average time to process a single order?
2. How many orders are scrapped during an eight-hour period?
3. What is the percent utilization of the grinding and tinting operations?

Solution. This problem suggests a straightforward serial manufacturing process with cumulative processing times and branching based on probabilities and conditions (see Fig. 8.11*a*). Orders are generated as specified and then routed using probabilistic branching through assigned nodes to determine the time required to grind the type of lenses ordered. This time is assigned as an attribute (GRIND). Orders then proceed through the grinding queue and facility nodes using the attribute to determine the processing time. Probabilistic branching is used, leaving the grind facility to determine the reworks. Reworks are counted in an assign node using an attribute (REWORK). On the second rework, the entity exits to an assignment node in which the attribute is reset to zero, and a variable (SCRAP) is incremented to track the statistics on scrapped lenses. Conditional branching based on the number of reworks controls branching to the SCRAP node. The entity leaving SCRAP node must not be terminated or the corresponding order would in effect be cancelled; therefore, the entity leaving SCRAP and returning to GRINDQ represents the original unsatisfied order transaction and not the scrapped lens. Those orders not requiring rework are routed to the tinting queue and facility. Probabilistic branching follows the tinting facility to determine reworks. Reworks are routed back to the queue and reprocessed. Counting is not required because there is no limit on the number of tinting reworks.

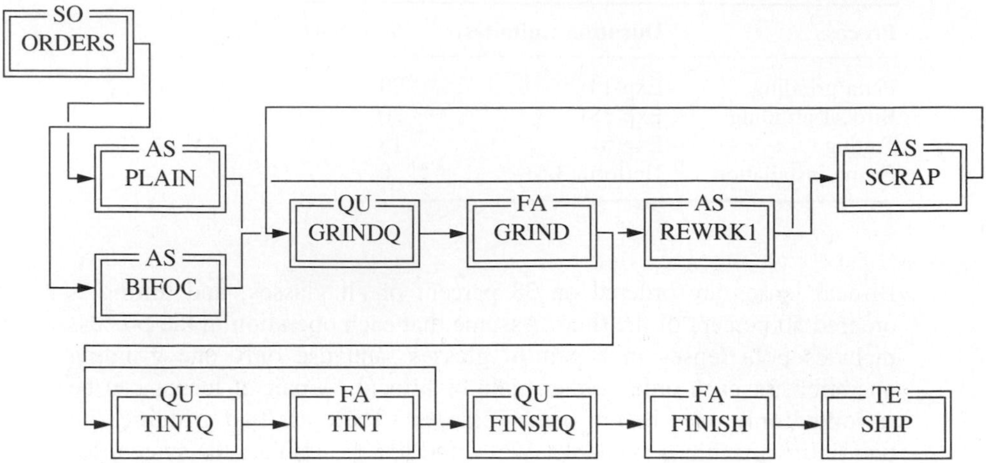

FIGURE 8.11*a*
Network model of the eyeglass manufacturing example.

Following tinting, the order is sent to the finishing queue and facility. After finishing, the entity is counted and terminated. Fig. 8.11*b* shows the simulation output of this model for 4800 minutes (80 hours) of operation. Dividing the number of scraps given in the output by 10 provides an estimate for the number of scraps in an eight-hour operation period.

Example 8.12 Plywood manufacturing. A lumber mill produces one-eighth inch veneer sheeting at the rate of one sheet per second to make plywood. The 4×8 sheets are graded into classes A (10%), B (40%), and C (50%). Class A sheets are used as is for external layers on AC plywood. Class B sheets are plugged (wood holes are mended) and used for external layers on BC plywood. The plugging operation for class B sheets is exponentially distributed with an average of 20 seconds. An average of 20 percent of the class B sheets are damaged during plugging and must then be downgraded to class C. Class C sheets are used as is for internal layers only. Two bonding machines are used to make $\frac{5}{8}''$ plywood from the veneer sheets. Each machine makes only one kind of plywood. The bonding process time has a triangular distribution (40, 48, and 60 seconds for AC; 30, 36, and 42 seconds for BC). Assume that all production operations run concurrently. We are interested in simulating the system for 1000 time units of operation to answer the following questions:

1. What portion of the production process is AC plywood?
2. How many sheets of $\frac{5}{8}''$ plywood will be produced in two hours?

*** E Z S I M STATISTICAL REPORT ***

Simulation Project: PROBLEMS
Analyst: HILSCHER
Date: 04/18/92
Disk file name: GLASSES.OUT

Current Time: 4800.00 Transient Period: 0.00

Q U E U E S:

NAME	MIN/MAX/LAST LENGTH	MEAN LENGTH	STD LENGTH	MEAN DELAY	STD DELAY
GRINDQ	0/15/13	2.86	3.97	25.89	36.18
TINTQ	0/12/ 3	0.95	1.77	10.09	15.90
FINSHQ	0/ 3/ 0	0.08	0.31	0.95	1.82

F A C I L I T I E S:

NAME	NBR SRVRS	MIN/MAX/LAST UTILIZATION	MEAN UTLZ	STD UTLZ	MEAN IDLE	MEAN BUSY
GRIND	2	0/ 2/ 2	1.62	0.63	3.68	15.91
TINT	2	0/ 2/ 2	1.31	0.77	7.37	14.01
FINISH	1	0/ 1/ 0	0.32	0.47	8.28	3.97

V A R I A B L E S:

NAME	MEAN	STD	MIN	MAX	No.OBSRVD
COMPLETE	3.92E+02	0.00E+00	3.92E+02	3.92E+02	392
SCRAP	1.20E+01	0.00E+00	1.20E+01	1.20E+01	12

FIGURE 8.11*b*
Simulation output of the eyeglass manufacturing model.

Solution. This problem involves two disjoint networks, one for resource stocking and one for plywood production (see Fig. 8.12*a*). Equal product demand for the two grades of plywood is controlled through the use of a double Always branching option from the SOURCE node in the second network. Parallel resource queues control allocation of sheets for the two products. In the first network, entities representing veneer sheets are generated and then routed to one of three ASSIGN nodes, which represent the three grades of plywood. Here, each respective resource stock is increased as the entity passes. Entities entering the class B branch first enter a DELAY node representing the time delay for plugging. Probabilistic branching is used after this node to sort out the downgraded veneer sheets, which are routed to the class C ASSIGN node. Other entities pass through the class B node to increase the resource. All entities are

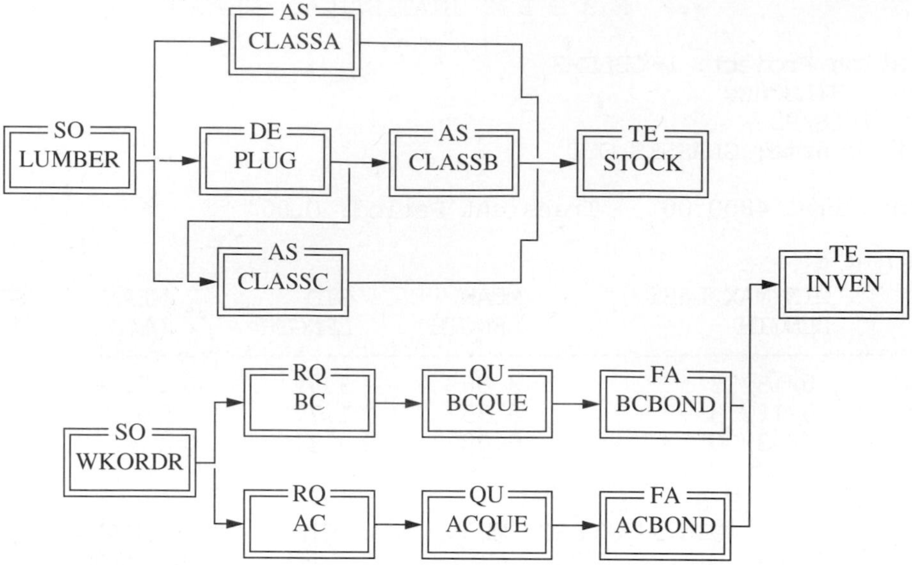

FIGURE 8.12*a*
Network model of the plywood manufacturing example.

terminated after the ASSIGN nodes. In the second network, work-order entities are generated and branched to both resource queues for material acquisition. Veneer resources generated in the first network are acquired in the numbers necessary to build a sheet of the respective plywood type. The entities then enter QUEUE and FACILITY nodes representing the manufacturing process. The resources are released and not returned to stock after exiting the FACILITY node. Entities of each type of plywood are then counted and terminated. The initial levels of all resources in this model are set to zero. Figure 8.12*b* shows the simulation output for this model.

Example 8.13 Focal plane assembly. A high-tech experimental production operation makes remote sensors for geodetic survey sensors. Focal plane sensing units are produced from substrate chips made in a chemical deposition process. Substrate chips are tested for uniformity before being cut into focal planes for sensor cards. This test results in a 20 percent rejection rate. Each acceptable substrate film chip is cut into 14 focal planes. The focal plane cutting process has a 50 percent scrap rate. Useable focal planes are then installed on the sensing card. Each card requires 6 focal planes. The finished card is then tested for pixel integrity. Ten percent of the finished cards are rejected when tested. We would like

```
*** E Z S I M  STATISTICAL  REPORT ***

Simulation Project: PROBLEMS
Analyst: HILSCHER
Date: 04/20/92
Disk file name: PLYWOOD.OUT

Current Time: 5000.00   Transient Period: 0.00

Q U E U E S:
NAME      MIN/MAX/LAST      MEAN     STD      MEAN     STD
          LENGTH            LENGTH   LENGTH   DELAY    DELAY
---------------------------------------------------------------
AC         0/71/71          24.56    17.65    288.84   329.50
BC         0/ 1/ 0          0.01     0.09     0.22     1.92
BCQUE      0/29/28          14.67    8.03     368.77   274.89
ACQUE      0/18/ 0          9.03     5.88     470.07   275.74

F A C I L I T I E S:
NAME      NBR    MIN/MAX/LAST  MEAN    STD     MEAN    MEAN
          SRVRS  UTILIZATION   UTLZ    UTLZ    IDLE    BUSY
---------------------------------------------------------------
BCBOND    1        0/ 1/ 1     1.00    0.07    0.17    35.81
ACBOND    1        0/ 1/ 0     0.95    0.21    2.51    49.58

V A R I A B L E S:
NAME      MEAN       STD        MIN        MAX        No.OBSRVD
---------------------------------------------------------------
PLYSHTS   2.34E+02   0.00E+00   2.34E+02   2.34E+02   234
ACPLY     9.60E+01   0.00E+00   9.60E+01   9.60E+01   96
BCPLY     1.38E+02   0.00E+00   1.38E+02   1.38E+02   138

R E S O U R C E S:
NAME     INIT    MIN/MAX/LAST   MEAN     STD      MEAN     STD
         LEVEL   USAGE          USAGE    USAGE    LEVEL    LEVEL
---------------------------------------------------------------
SHEETA   0       0/192/192      117.99   63.47    0.65     0.48
SHEETC   0       0/792/792      431.44   238.69   291.74   147.78
SHEETB   0       0/334/334      167.66   96.25    207.98   110.94
```

FIGURE 8.12*b*
Simulation output of the plywood manufacturing model.

to determine the proportion of the acceptable sensor cards based on the following processing times:

Process	Time
Chemical deposition	Uniform(25,34) min/substrate
Uniformity test	Exp(8) min/substrate
Cutting	Normal(12,4) min/focal plane
Assembly	Exp(4) min/focal plane
Pixel testing	Exp(10) min/card

Solution. This is a relatively simple sequence of process networks having substantial rejection rates and employing entity multiplication (see Fig. 8.13*a*). In the first network, substrate chip entities are generated and passed through a QUEUE and a FACILITY node representing the testing information. Probabilistic branching sorts out rejected parts on the downstream side for termination. Accepted entities pass on to the QUEUE and FACILITY nodes representing the cutting station. After cutting, the entity enters an ASSIGN node where the focal plane resource is increased by 7 units (50 percent of 14). The entity is then terminated. In the second network, sensor card entities are generated and routed to a RESOURCE-Q node for acquisition of focal plane resources. Six planes are requested by each entity at this node. Next, the entity passes through the assembly and test delay nodes. (It is assumed that sufficient resources are available

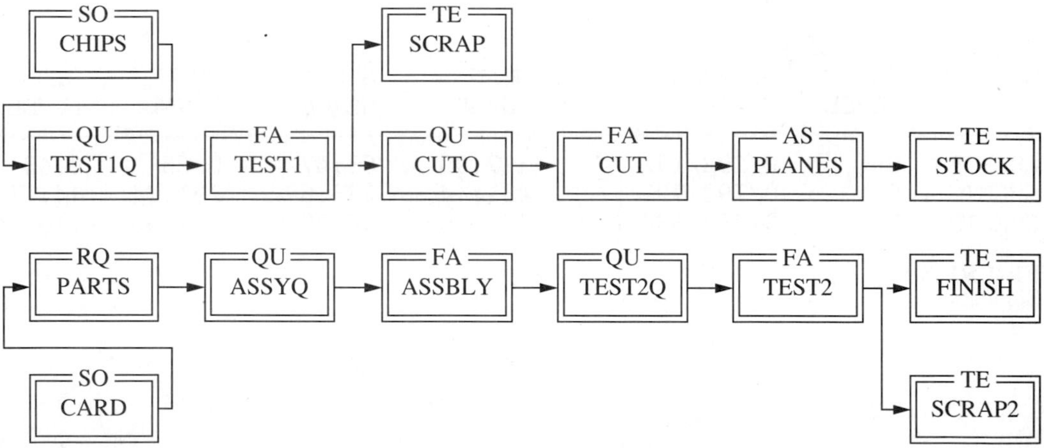

FIGURE 8.13*a*
Network model of the focal plane assembly example.

for assembly and testing; therefore, a simple delay, instead of a facility, is used.) Probabilistic branching sorts out the defective card assemblies, which are counted and terminated. Acceptable cards are also counted and terminated. Fig. 8.13*b* shows the output for this model based on the simulated production of 1000 focal planes.

```
                  ***  E Z S I M  STATISTICAL  REPORT  ***

Simulation Project: PROBLEMS
Analyst: HILSCHER
Date: 04/20/92
Disk file name: FOCAL.OUT

Current Time: 34761.39    Transient Period: 0.00

Q U E U E S:
NAME       MIN/MAX/LAST       MEAN      STD          MEAN        STD
           LENGTH             LENGTH    LENGTH       DELAY       DELAY
------------------------------------------------------------------------
PARTS       0/66/57           30.69     18.54        879.31      580.59
TEST1Q      0/ 1/ 0           0.01      0.08         0.18        1.44
CUTQ        0/ 1/ 0           0.01      0.09         0.30        1.50
ASSYQ       0/ 2/ 0           0.02      0.14         0.63        2.19
TEST2Q      0/ 3/ 0           0.07      0.30         2.32        6.61

F A C I L I T I E S:
NAME       NBR    MIN/MAX/LAST  MEAN       STD        MEAN       MEAN
           SRVRS  UTILIZATION   UTLZ       UTLZ       IDLE       BUSY
------------------------------------------------------------------------
TEST1      1        0/ 1/ 0     0.26       0.44       21.75      7.66
CUT        1        0/ 1/ 1     0.33       0.47       24.69      12.06
ASSBLY     1        0/ 1/ 0     0.13       0.33       27.49      4.06
TEST2      1        0/ 1/ 0     0.32       0.47       21.54      10.00

V A R I A B L E S:
NAME       MEAN        STD         MIN         MAX          No.OBSRVD
------------------------------------------------------------------------
GOOD       1.00E+03    0.00E+00    1.00E+03    1.00E+03     1000
BAD        1.02E+02    0.00E+00    1.02E+02    1.02E+02     102

R E S O U R C E S:
NAME      INIT    MIN/MAX/LAST     MEAN      STD        MEAN      STD
          LEVEL   USAGE            USAGE     USAGE      LEVEL     LEVEL
------------------------------------------------------------------------
PLANE     0       0/6612/6612      3294.76   1903.79    2.53      1.72
```

FIGURE 8.13*b*
Simulation output of the focal plane assembly model.

Example 8.14 Paint factory. A paint factory receives three types of paint orders—small, medium, and large. The interarrival times of these orders are exponentially distributed with means of 2, 3.5, and 5 days, respectively, for each order size. After an order is received, it is queued until a tank is ready for making the paint. The factory has two tanks. Once a tank is available, large orders take 2.5 days to make, medium orders take 2 days to make, and small orders take 1.5 days to make. After processing, the paint is submitted to a color test, in which there is a 10 percent chance of failing the test. If the paint fails to pass the test, the required ingredients are added automatically, and the test is applied again until the color is approved. Each cycle of testing and color adjustment delays the order for one day. After a paint order passes the color test, it goes into another queue to wait until a pump is ready to can it off into 5-gallon containers and send it out for shipment. The can-off time pertains to the size of the order. The small orders take 2 days to can off, the medium orders take 2.5 days to can off, and the large orders require 3 days to can off. There are two pumps in the factory. The factory operation is to be simulated for 500 completed orders of paint. The variable of interest is the waiting time of orders in the queues for tank and pump availability. This information is important because long waiting times result in long production lead times, which in turn make it hard to meet due dates for orders.

Solution. The model network for this example is shown in Fig. 8.14*a*. The three SOURCE nodes create the three different types of order entities with their respective interarrival distributions. These entities then enter a QUEUE where they wait until a tank is available for production operations. The entities then enter a FACILITY node in which the

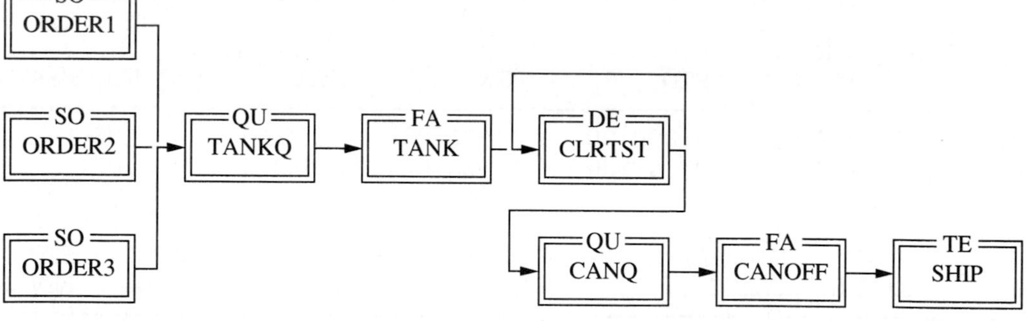

FIGURE 8.14*a*
Network model of the paint factory example.

time that each order takes to be produced is specified. The number of parallel servers at this facility is two, which corresponds to the number of available tanks. After passing through this node, entities enter a DELAY node that represents the color test time. Probabilistic branching out of this node returns 10 percent of the orders to the node. Note that since no information is given on the possible limitations on the availability of test equipment and operators, a DELAY process is used instead of a FACILITY process. After the color test process, the entities enter another QUEUE where they wait until a pump is available to can off the produced paint. They then enter a FACILITY node which represents this canning process, and finally leave for shipment via the TERMINATE node. Note that in the FACILITY nodes for processing and canning, the "entity name dependent service time" option is used to assign the respective delay times to each order size. The simulation output for this model is shown in Fig. 8.14*b*.

Example 8.15 Automatic guided vehicle. An automatic guided vehicle (AGV) is used to pick up finished parts from three machines. The parts

*** E Z S I M STATISTICAL REPORT ***

Simulation Project: PAINT
Analyst: DERUITER AND PRIDE
Date: 2/27/93
Disk file name: PAINT.OUT

Current Time: 593.00 Transient Period: 0.00

Q U E U E S:

NAME	MIN/MAX/LAST LENGTH	MEAN LENGTH	STD LENGTH	MEAN DELAY	STD DELAY
TANKQ	0/22/ 2	7.03	7.35	7.14	6.83
CANQ	0/84/78	28.73	26.17	23.77	24.12

F A C I L I T I E S:

NAME	NBR SRVRS	MIN/MAX/LAST UTILIZATION	MEAN UTLZ	STD UTLZ	MEAN IDLE	MEAN BUSY
TANK	2	0/ 2/ 2	1.81	0.51	0.19	1.84
CANOFF	2	0/ 2/ 2	1.99	0.15	0.01	2.35

FIGURE 8.14*b*
Simulation output of the paint factory model.

from each machine are dropped into standard-sized carrier bins. These bins are automatically transferred to the AGV when it arrives for pickup. Each machine has a limit of five full bins, and will spill parts on the floor if the limit is reached. The AGV travels to each machine in sequence and stops long enough to pick up all full bins. The AGV then travels to a stock room and unloads the bins. The following data describes specific details of the AGV route and machine production (all times are in seconds):

Station time	Time to fill bin	Bin loading time	Travel time to next station
Machine A	60	Exp(20)	45
Machine B	80	Exp(30)	20
Machine C	35	Exp(40)	40
Stockroom	NA	NA	75

The system is to be simulated for 10,000 time units to answer the following questions:

1. For various bin capacities of the AGV, how many parts will be dumped on the floor? What is the required bin capacity of the AGV such that all machined parts are accommodated?
2. What is the total number of transported parts for each machine?

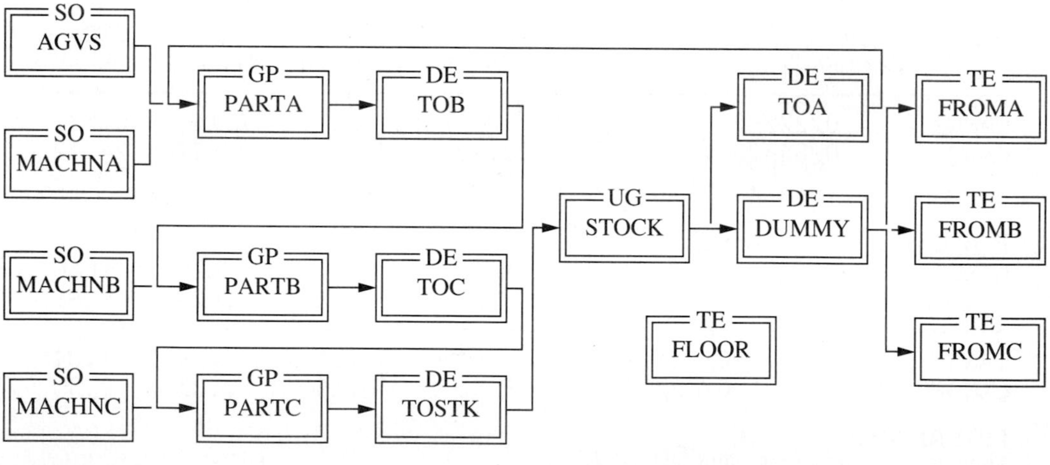

FIGURE 8.15*a*
Network model of the automatic guided vehicle example.

Solution. This problem focuses on GROUP-Q and DELAY nodes with entities being transported by a carrier (see Fig. 8.15*a*). A single carrier entity is generated; it passes through three sets of GROUP-Q and DELAY node pairs that represent the stops at each machine. The AGV then moves to the UNGROUP node representing the stockroom unloading process. At each GROUP-Q node the carrier loads all available bins. The AGV then returns to machine A and starts the circuit again. The AGV capacity is set to a large number (15) in order to determine its maximum required capacity through this simulation. The capacities of the queues at the GROUP-Q nodes are set to five. If the queue at a GROUP-Q node is full, the arriving bin balks to the TERMINATE node called FLOOR. Bin entities are created in three nodes representing the three machines. The bins for each machine are grouped at their respective GROUP-Q nodes and then picked up by the AGV on its route. After ungrouping, bin entities are sent to a dummy DELAY node to be branched on the basis of their name attribute to either FROMA, FROMB, or FROMC nodes. Entity counts are requested for each TERMINATE node in the Statistics menu. Notice that parameters such as the speed of the AGV, its capacity, and the number of AGVs may be varied in this model to find the desirable system configuration. The simulation output of this model is shown in Fig. 8.15*b*.

*** E Z S I M STATISTICAL REPORT ***

Simulation Project: PROBLEMS
Analyst: HILSCHER
Date: 04/14/92
Disk file name: AGV.OUT

Current Time: 10000.00 Transient Period: 0.00

Q U E U E S:

NAME	MIN/MAX/LAST LENGTH	MEAN LENGTH	STD LENGTH	MEAN DELAY	STD DELAY
PARTA	0/ 3/ 2	1.97	0.84	87.98	67.41
PARTB	0/ 3/ 1	1.06	0.76	58.16	58.14
PARTC	0/ 5/ 1	2.62	1.53	76.59	58.68

V A R I A B L E S:

NAME	MEAN	STD	MIN	MAX	No.OBSRVD
FLOOR	8.00E+00	0.00E+00	8.00E+00	8.00E+00	8
MACHA	1.62E+02	0.00E+00	1.62E+02	1.62E+02	162
MACHB	1.23E+02	0.00E+00	1.23E+02	1.23E+02	123
MACHC	2.72E+02	0.00E+00	2.72E+02	2.72E+02	272

FIGURE 8.15*b*
Simulation output of the automatic guided vehicle model.

8.3.1 Production and Inventory Control

Production and inventory control models apply to service as well as manufacturing operations. For example, a grocery store follows an inventory policy for ordering the items on its shelves from suppliers, and an operations manager of a manufacturing department orders the required components from another department within the factory according to a certain inventory policy. Production and inventory control models can become too complicated to study analytically, especially when demand and/or lead time have various stochastic patterns. Simulation has proven to be an effective tool in devising effective production and inventory control policies.

One of the popular inventory control methods is the *periodic review* method. In this method, the inventory level is reviewed at constant time intervals. A measure called *inventory position* (IP) is usually used to determine the order size (if any) after each review. The following relationship determines the inventory position:

$$\text{Inventory position} = \text{Stock on hand} + \text{Stock on order} - \text{Backlog}$$

Stock on hand (STOCK) is the actual physical inventory level. Stock on order is what has been ordered but not yet received because of a nonzero delivery lead time. Backlog represents the number of stock units promised to customers, who wait for their purchased units until they become available. At each review time, the actual STOCK is determined and IP is calculated. IP is then compared with a predetermined value called the *reorder level* (RL). If IP is greater than RL, no action is taken; otherwise, an order is placed to bring IP up to another predetermined level called the *stock control level* (SCL). In other words, the order quantity is $OQ = SCL - IP$, if $IP < SCL$, and it is zero otherwise. When an order is placed, IP immediately reaches the SCL, even though the actual order is received after the delivery lead time. The following example demonstrates the application of simulation to a periodic review policy.

Example 8.16 Periodic review inventory control. Suppose that the sales department of a computer manufacturing company periodically reviews the inventory of a particular microcomputer model and, when needed, places an order to the factory for the production of the required units. The production lead time is six days. The customer demand for the computer has an exponential interoccurrence time with a mean of 0.4 day; customers agree to wait if they face an out-of-stock situation. In this inventory policy, the stock control level is 15 units and the reorder level is 5 units. We are interested in simulating this inventory system for 1000 days of operation to determine the following:

1. The average inventory level
2. The number of backlogs and the statistics on the waiting time of customers before receiving their orders
3. Statistics on the safety stock level (the actual stock level before receiving the shipment from the factory)
4. Statistics on the inventory position

Solution. Fig. 8.16*a* shows the EZSIM model for this system. There are two disjoint networks in this model. The lower network shows the flow of customer demand transactions. Each entity representing demand goes to the assign node POSDWN, at which the user variable IP is reduced by one unit. Entities then proceed to the DEMAND node, at which one unit of the resource called STOCK is requested. If STOCK is zero, the entities wait at the node until the next shipment is received. After receiving their demanded resource, entities leave the system through the TERMINATE node SATSFD. The branch to POSDWN is always taken, regardless of the availability of stock (the maximum number of branches that may be taken is set at two in the branching window of CUSTMR). The branch to BAKLOG is conditional and is taken if the current level of resource STOCK is zero.

The upper network represents the review transactions flow. The SOURCE node creates entities representing review transactions once every 7 days. The branch from REVIEW to ORDER is conditional and is taken if IP is less than 5 units. The branch to NOORDR is of the Last choice type. At ORDER, the production order is placed and the inventory position is updated. The variable PQ (production quantity) is defined by the expression PQ = 15 − IP, and then IP is set to 15. The entity now

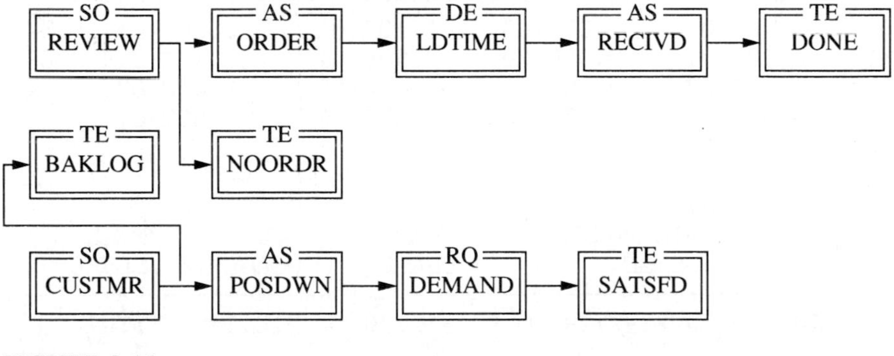

FIGURE 8.16*a*
Network model of the inventory control example.

represents the order transaction and travels through LDTIME representing the lead time delay. At RECIVD the order is received, which increases the resource named STOCK by PQ units. The initial level of stock is set at 10 units at the beginning of simulation. The safety stock statistics STCLVL is requested in the Statistics menu. The statistics type is the number of available resources (a system variable), and it is collected as an observation-based statistic at the RECIVD node. The statistics called INVPOS are time-based statistics collected on the user variable IP. The simulation output of this model is shown in Fig. 8.16*b*. Note in this output that the inventory position (INVPOS) may become negative, as expected.

```
               *** E Z S I M  STATISTICAL  REPORT ***

Simulation Project: INVENTORY
Analyst: BK
Date: 11/6/92
Disk file name: INVENTRY.OUT

Current Time: 1000.00    Transient Period: 0.00

Q U E U E S:
NAME     MIN/MAX/LAST          MEAN      STD        MEAN      STD
         LENGTH                LENGTH    LENGTH     DELAY     DELAY
------------------------------------------------------------------
DEMAND    0/10/ 0              0.46      1.34       0.64      1.64


V A R I A B L E S:
NAME     MEAN        STD         MIN          MAX          No.OBSRVD
------------------------------------------------------------------
STCLVL   1.38E+01    2.05E+00    1.10E+01     2.00E+01     54
NOORDR   8.90E+01    0.00E+00    8.90E+01     8.90E+01     89
LOST     1.47E+02    0.00E+00    1.47E+02     1.47E+02     147
INVPOS   8.88E+00    4.13E+00   -5.00E+00     1.50E+01


R E S O U R C E S:
NAME     INIT    MIN/MAX/LAST    MEAN      STD       MEAN      STD
         LEVEL   USAGE           USAGE     USAGE     LEVEL     LEVEL
------------------------------------------------------------------
STOCK    10      0/729/729       355.30    212.15    5.01      4.03
```

FIGURE 8.16*b*
Simulation output of the inventory control model.

8.3.2 Manufacturing Quality Control

Inspection for quality control purposes is carried out at many stages in manufacturing. Inspection of incoming materials and parts, process inspection at various points in the manufacturing operation, and final inspection of the manufactured product are some of the common inspection stages. Since some inspections involve destructive tests (e.g., burning out some bulbs to find the maximum electric current that the bulb can sustain), and since in many instances 100 percent inspection is costly (e.g., detailed inspection of one million small screws made in a production run), statistical sampling methods are usually used for acceptance or rejection of a batch of manufactured parts. Single sampling and double sampling are two popular sampling methods used for lot inspection purposes. The following examples demonstrate the application of simulation in the analysis and design of these lot inspection plans.

Example 8.17 Lot inspection using single sampling. When the decision to accept or reject the lot is made on the basis of the inspection of one sample of a given size, the inspection plan is called a single-sampling plan. A single-sampling plan requires that three numbers be specified. One is the number of parts in the lot from which the sample is to be drawn. The second is the sample size, N. The third is the acceptance number, A. The acceptance number is the maximum allowable number of defective parts in a sample. If there are more than A defective parts in a sample, the corresponding lot is rejected; otherwise, it is accepted.

Let us consider a specific single-sampling plan in which samples of size 12 are to be taken from each lot. The acceptance number is one (that is, if the number of defective parts in a sample is zero or one, the sample is accepted; otherwise, it is rejected). We would like to simulate the inspection process to find the percentage of lots that are accepted under this sampling plan. Let us assume that 10 percent of the parts in the lots are defective.

Solution. The generalized network model for simulation of single-sampling plans is shown in Fig. 8.17*a*. The SOURCE node generates entities that represent individual parts to be expected. The time between creation of these parts may be arbitrarily chosen, since this is a timeless model. Selection of one time unit as intercreation time is suggested, since the simulation clock in this case would represent the total number of inspected parts. At node XPLUS the user variable X, which represents the total number of inspected parts for the current lot, is incremented by one unit. In this specific example the probabilistic branching out of XPLUS sends 10 percent of the inspected parts to node FAIL, in which the user variable F is incremented by one unit. User variable F represents the

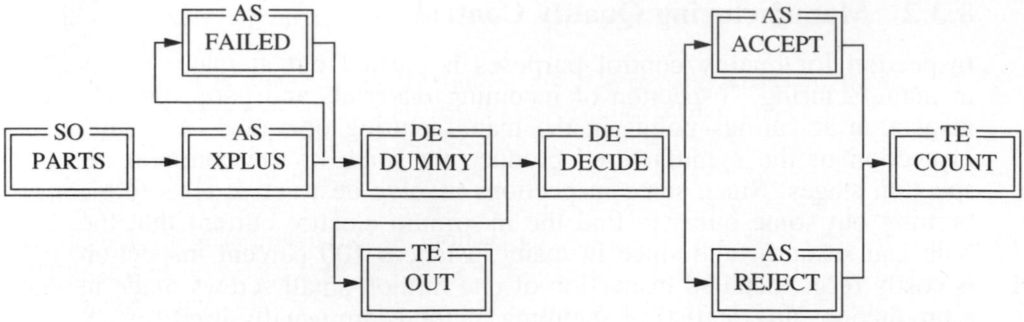

FIGURE 8.17*a*
Network model of the single-sampling inspection plan example.

total number of failed parts in the current sample. Entities leaving nodes XPLUS and FAILED enter the dummy DELAY node. The branching option out of this DELAY node is conditional. The condition for going to node DECIDE is $X = 12$, meaning that when 12 parts are inspected we must decide to either accept or reject the lot on the basis of the total number of failed parts. The branching option from node DECIDE to node ACCEPT is also conditional. This branch is taken if $F \leq 1$. The Last Choice branching option is specified for the branch that connects node DECIDE to node REJECT. At nodes ACCEPT and REJECT, user variables X and F are reset to zero. Entities leaving nodes ACCEPT and REJECT enter the TERMINATE node, in which a termination count of 100 is specified. Thus, simulation stops when 100 lots are inspected. The entity count statistics are specified for nodes ACCEPT and REJECT in the Statistics menu. Fig. 8.17*b* shows the simulation output of this model.

```
                    ***  E Z S I M  STATISTICAL  REPORT  ***

Simulation Project: QUALITY CONTROL-1
Analyst: BK
Date: 9/3/93
Disk file name: QUALITY1.OUT

Current Time: 1199.00    Transient Period: 0.00

V A R I A B L E S:
NAME       MEAN        STD         MIN         MAX          No.OBSRVD
----------------------------------------------------------------------
ACCEPTED   6.90E+01    0.00E+00    6.90E+01    6.90E+01     69
REJECTED   3.10E+01    0.00E+00    3.10E+01    3.10E+01     31
```

FIGURE 8.17*b*
Simulation output of the single-sampling inspection plan model.

Example 8.18 Lot inspection using double sampling. Double sampling includes the possibility of putting off the decision on the lot until a second sample has been taken. A lot may be accepted at once if the first sample is good enough or rejected at once if the first sample is bad enough. If the first sample is neither good enough nor bad enough, then the decision is based on considering the first and the second sample combined. In general, double sampling involves less inspection than single sampling for any given quality protection. In this plan a sample of size $N1$ is first inspected. If the number of defective parts is not greater than $A1$, the lot is accepted. If the number of defective parts is greater than $A2$, the lot is rejected. If the number of defective parts is between $A1$ and $A2$, a second sample of size $N2$ is drawn. If the total number of defective parts in both samples is not greater than $A2$, the lot is accepted; otherwise, it is rejected.

Let us consider a specific double-sampling plan in which $N1 = 12$, $A1 = 1$, $N2 = 15$, $A2 = 4$, and 10 percent of the parts in the lots are defective. Suppose that we would like to simulate the inspection process to find the percentage of lots that are accepted under this sampling plan and the percentage of time that a second sample is needed.

Solution. Fig. 8.18*a* shows a generalized network model for simulation of double-sampling plans. This model is a modified version of the model for the single-sampling case. In the new model, the branching condition from the DELAY node named DUMMY to node DECIDE is $X = N$ where N is the sample size, which may be 12 if the lot is accepted at the first sampling stage or 27 (i.e., 12+15), if a second sample is to be drawn. The condition for the branch connecting node DECIDE to node ACCEPT is $F \leq A$. The condition for the branch connecting node DECIDE to node REJECT is $F > 4$. The Last Choice branching option is assigned to the branch that connects node DECIDE to node SECOND. User variables X

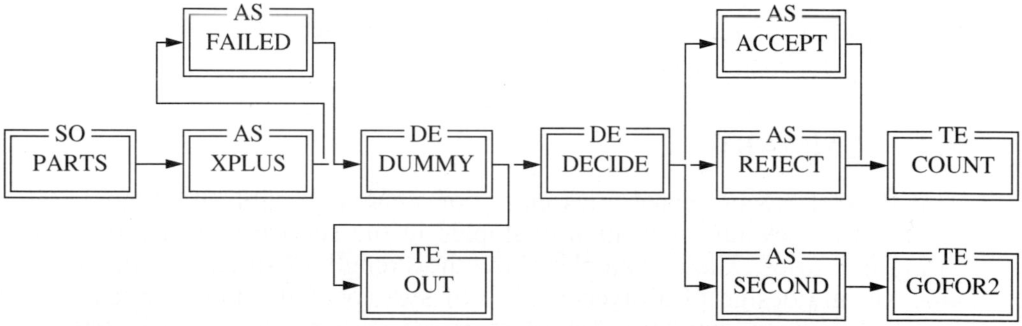

FIGURE 8.18*a*
Network model of the double-sampling inspection plan example.

```
              *** E Z S I M  STATISTICAL  REPORT  ***

Simulation Project: QUALITY CONTROL-2
Analyst: BK
Date: 9/3/93
Disk file name: QUALITY2.OUT

Current Time: 1649.00      Transient Period: 0.00

V A R I A B L E S:
NAME      MEAN        STD         MIN         MAX          No.OBSRVD
------------------------------------------------------------------------
ACCEPTED  8.70E+01    0.00E+00    8.70E+01    8.70E+01     87
REJECTED  1.30E+01    0.00E+00    1.30E+01    1.30E+01     13
SECOND    3.00E+01    0.00E+00    3.00E+01    3.00E+01     30
```

FIGURE 8.18*b*
Simulation output of the double-sampling inspection plan model.

and F are the current number of parts inspected and current number of defective parts in the sample, respectively. In the ASSIGN node named SECOND, user variable A, which represents the acceptance number, is given a value of 4, and user variable N is given a value of 27.

The initial values of the user variables are $X = 0$; $F = 0$; $N = 12$; $A = 1$. The ASSIGN nodes named ACCEPT and REJECT reset these user variables to their original values. Note that in this model if the lot is either accepted or rejected based on the results from the first sample, no more sampling is performed and the variables are reset for a new lot. It is only when a second sample is needed that the sample size is increased to 27 and the acceptance number is changed to 4 in preparation for the second sample inspection. In the Statistics menu, the entity count statistic is also requested for the node named SECOND to provide the number of times that a second sample is drawn. The simulation is stopped after 100 entities reach the TERMINATE node named COUNT (i.e., 100 inspected lots). Fig. 8.18*b* shows the output of this model. Observing the changes in values of user variables as this model is animated effectively clarifies the modeling procedure.

8.4 SUMMARY

This chapter has presented a taxonomy of simulation application areas, followed by several application examples in the service and manufacturing industries. These examples show the strength of simulation in the analysis and design of a diverse range of systems. The study purposes accomplished by simulation for most of the systems described in this chapter are almost impossible to accomplish with other methods.

8.5 EXERCISES

8.1. An automated manufacturing line has two parallel machines, which produce parts A and B, respectively. Two part As and one part B are then used by a third machine to produce part C. Part As are produced at an exponential rate averaging 15 seconds per part. Part Bs are produced at an exponential rate averaging 30 seconds per part. Five percent of part As and 8 percent of part Bs are defective and are therefore scrapped. Part C production time is exponential 25. How long will it take to produce 100 good part Cs?

8.2. A hospital receives three types of patients during the course of a regular day with interarrival times that are distributed exponentially with means of 120, 45, and 25 minutes, respectively. The service times for these patients are uniformly distributed with the parameters $(60, 80)$, $(25, 35)$, and $(10, 20)$ minutes, respectively. Simulate this system to determine the minimum number of doctors needed to ensure that the mean time spent by the patients in the waiting room is not more than 35 minutes.

8.3. Many customers of Local Bank have been complaining that it takes too long to simply deposit money, cash checks, or perform other small transactions. This might be due to the fact that currently all customers wait in the same line until one of three tellers are available. It is known that loan applicants' interarrival times are exponentially distributed with a mean of 65 minutes, and their service time is uniformly distributed between 10 and 80 minutes. All other customers' interarrival times are exponentially distributed with a mean of 5 minutes, and their service time is uniformly distributed between 4 and 9 minutes. The bank wants to minimize the queue waiting time for each type of customer, but they are faced with the constraint of having only three tellers available at the bank. Simulate this system for 100 customers to observe the statistics of the queues.

Now, simulate the following system and see if it improves the queue waiting times. The regular customers can arrive and enter a corresponding line specifically designated to nonloan customers, and loan customers enter their own line. Regular customers may enter the loan queue if no one is being served by the loan teller. Additionally, in the two regular lines, customers may switch between lines if the other line has at least two less people in it. Create different models that meet the constraint and see if queue waiting times are reduced. Make a proposal to the management of the bank on how to modify the existing situation.

8.4. A city is planning to build a road across two parallel train tracks. No more than seven cars can fit between the first and second set of tracks. The planners want to determine the number of parallel lanes that are needed for the segment of the road between the two tracks so that cars will not back up onto the first set of tracks while they wait for a train to pass. Both sets of tracks have trains arriving uniformly between 20 and 30 minutes. Trains take between 5 and 12 minutes, distributed uniformly, to pass and clear the crossing road. Cars arrive exponentially every 1 minute and take 12 seconds to go from the first track to the second. Assume that arriving cars always pick the lane with the least number of cars in it. Note that only one direction of the road needs to be simulated.

8.5. A hotel chain has two buses to pick up and drop off people at a local airport and at two separate hotels. The buses travel from the airport to hotel 1, then to hotel 2, and back to the airport to continue this pattern. It takes a normally distributed time with a mean of 20 and a standard deviation of 2 minutes to travel between each location. Travelers' inter-arrival time from their flights is exponentially distributed with a mean of 2.5 minutes. Fifty percent of the people get off at the first hotel, and the bus picks up people from this hotel who want to go to the airport. Another 50 percent of the people get off at the second hotel, and the bus again picks up people. At the airport, everyone gets off. At both hotels, people arrive at the bus stop to go to the airport with exponentially distributed interarrival times which have a mean of 5 minutes. Simulate the system where the first bus leaves the airport at the beginning of simulation and the second one leaves 30 minutes after the first bus. Determine the amount of seating required on both buses so that everyone waiting can be picked up.

8.6. A bread-baking operation involves the following steps and their associated times for each batch prepared:

Step	Duration	Baker requirement
Mixing	Exp(15)	All mixing time
Raising	Norm(45,5)	None
Baking	Norm(22,3)	5 minutes
Cooling	60	5 minutes
Bagging	Exp(6)	All bagging time

The baker is needed for all of the mixing and bagging time but only for the first 5 minutes of the baking and cooling time (5 minute return time). Only 10 batches of bread are made each day. The number of loaves produced in each batch is uniformly distributed between 15 and 30.

(*a*) What is the percent utilization of a single baker in this operation?

(*b*) How long will it take to finish ten batches with one baker?

(*c*) How will answers (*a*) and (*b*) change if two bakers are working?
(*d*) What is the necessary oven capacity (in batches) for a single and double baker operation?

8.7. A lumber mill and furniture-manufacturing operation is considering introducing a new product line made from maple wood. Maple logs will be delivered and then sawed into boards. The boards will then be dried before they are used for manufacturing a new maple table set. The logs vary in length between 18 and 27 feet (uniformly distributed). They are sawed into 1×6 boards and then into 5-foot lengths before drying. Pieces less than 5 feet long are sorted out for other uses. The number of 1×6 boards produced from each log is normally distributed with an average of 22 and a standard deviation of 6. The manufacturing process requires seven 1×6 boards for each table. Model the sawmill operation to study material requirements and scrap rates.
(*a*) Determine the amount of lumber required to produce 1500 tables.
(*b*) What percentage of the boards cut from the logs will end up in the tables produced?

8.8. A recent graduate from a technical school is going to open a small electronics repair shop. He will be the only person in the operation and will perform both counter transactions and repair work. He plans to work only on TVs and VCRs. TV repair times are normally distributed with an average of 31 minutes and a standard deviation of 5 minutes. VCR repair times are also normally distributed with an average of 45 minutes and a standard deviation of 5 minutes. The owner plans to open for business eight hours each day, but will work an additional two hours on repairs before going home each night. Repairs are evenly divided between VCRs and TVs. Customers arrive uniformly between 20 and 55 minutes to drop off (50 percent) and pick up (50 percent) items for repair. Customer transaction times average eight minutes and are exponentially distributed. Model the repair shop to study the owner's time utilization. Assume there are two TVs and two VCRs already in the shop for repair.
(*a*) Determine how much time the owner can expect to spend performing repairs each day.
(*b*) How many items can he expect to repair each day if he alternates between working on TVs and VCRs?
(*c*) How many can he repair if he hires another person to perform repairs on the same work schedule?

8.9. A movie theater parking lot is being sized by a consulting engineering firm. The complex will have eight theaters. Each theater will seat 250 persons. Movie run times are anticipated to be triangularly distributed (85, 105, 115 minutes). The show start time in each theater is staggered 15 minutes apart (e.g., theater 1 starts at 00:00, theater 2 starts at 00:15, etc.). Twenty percent of all movie patrons are dropped off and picked up after the show. The remaining 80 percent arrive in private autos and require parking space. Observations of these cars showed that each car in the parking lot carries an average of 2.2 people. Assume attendance at any given movie showing is uniformly distributed between 200 and 250. Model the theater to determine parking capacity requirements. Disregard the time spent walking to and from the parking lot.

(*a*) Determine the average number of cars requiring parking.

(*b*) Determine the maximum number of cars requiring parking during a six-month simulation.

8.10. Modify the model for Example 8.9 so that the line availability is checked only when there is a message to be sent by the corresponding computer (the modified model should execute considerably faster).

8.11. An optometrist is planning a new office operation and needs to determine his support staff requirements. He also wants to determine how many patients he can examine per day. The examination process includes four sequential steps: registration, general eye testing, doctor's examination, and eyewear fitting. The table below gives the distribution data for each activity and the required personnel. Only 60 percent of the patients need to see the optician for eyewear fitting.

Activity	Distribution	Personnel
Registration	Exp(2)	Receptionist
Testing	Normal(12,2)	Assistant
Examination	Normal(9,2)	Doctor and asst.
Fitting	Uniform(4,8)	Doctor

Model the optometrist's operation to study staff requirements and service times. Assume that patients arrive every 12 minutes and only 60 percent of them need to see the optician. Only one doctor is available.

(*a*) What is the minimum number of staff members by type to ensure that no patient waits longer than a total of ten minutes during the process?

(*b*) What is the average time a patient spends in the office?

8.12. A grocery store has recently experienced a shortage of shopping carts during peak business hours. The manager wishes to determine how many additional carts are needed to meet the demand. A peak period of two hours was identified, and data on customer shopping activities was collected during this period. Customer arrival times were found to be distributed exponentially with a mean of 30 seconds. Shopping time for customers requiring the use of a shopping cart was normally distributed with a mean of 10 minutes and a standard deviation of 2 minutes. It was also determined that 65 percent of the shopping carts were left in the parking lot and the remainder were immediately returned to service. Those shopping carts left in the parking lot were periodically (every 15 minutes) collected and returned to the store. Model the grocery store system to study shopping cart usage patterns. Determine the maximum and average numbers of carts in use once the store has reached a steady state condition. A cart is considered to be in use until it is returned to the store and is made available for the next customer.

8.13. A major oil company is designing a new layout for their gasoline service stations. The station will offer four types of auto fuel: 87 octane, 89 octane, 95 octane, and diesel. Each pump will dispense the three grades of gasoline. Only one out of every four pumps will also dispense diesel fuel. A recent customer survey indicates the following fuel preferences: 40 percent prefer 87 octane, 30 percent prefer 89 octane, 20 percent prefer 95 octane, and 10 percent prefer diesel. During peak periods customer arrival times are uniformly distributed between 15 and 45 seconds. Fuel is pumped at the rate of five gallons per minute. The time required to start and finish fueling is uniformly distributed between two and seven minutes. Model the service station to study pump usage patterns. Start with eight pumps, two of which offer both diesel and gasoline.

(*a*) What is the maximum number of cars waiting to be serviced?
(*b*) What is the average number of pumps in service?
(*c*) Model the system as a full service station with 3 attendants and 12 pumps available. Assume each attendant can service two cars at once. Compare the results of this operation with the answers to (*a*) and (*b*).

8.14. A hotel chain is studying the elevator system for a new five-story hotel they wish to build. During peak demand, guests on any one floor will arrive at the elevator exponentially with an average time of 15 seconds. Each of the two elevators has a capacity of eight persons. Travel time between floors is 4 seconds, and stops are made only at floors with guests waiting for service. The time required for loading of guests is 5 seconds, plus 2 seconds per person. Empty elevators will be cycled to the top floor after they unload at the ground floor. Assume all guests travel from their floor to the lobby. Model the elevator system to study usage patterns.

(*a*) Determine the longest waiting mean delay time for an elevator.
(*b*) Determine the average number of persons carried during each trip.

8.15. A highway excavation project must move 10,000 cubic yards of material. Ten earth movers with a capacity of six cubic yards each will be used to scrape and haul material. The material will be deposited 1.2 miles from pickup. It requires one bulldozer to push each earth mover in the removal process. The operation time is normally distributed with a mean of 25 seconds and a standard deviation of 4. Three bulldozers will be available at the excavation site. The earth movers travel 30 mph unloaded and 20 mph loaded. Unloading time is exponentially distributed, averaging 20 seconds. The earth-moving operation will be conducted around the clock to minimize completion time. A traffic signal located halfway to the deposit site operates on a two-minute cycle. During ninety seconds of the cycle, earth movers may pass. Empty earth movers returning to the excavation after unloading do not pass the signal.

(*a*) How long will it take to complete the project?

(*b*) What percentage of the time will the bulldozers be idle?

8.16. A continuous-review inventory policy constantly monitors the inventory level (rather than periodically reviewing it). Whenever the inventory level reaches the reorder point, an order of a fixed quantity is placed. Assuming a reorder point of four product units, model the inventory example under a continuous-review policy. We are interested in finding an order quantity such that the average inventory level does not exceed 10 product units and the average waiting time of customers for their backlogged orders does not exceed four days.

8.17. In the AGV example (Example 8.15), assume that there are two AGVs on the same closed-loop track. An AGV may not pass the one that is ahead of it on the track. Both AGVs carry machined parts to the warehouse. All other assumptions of the example hold in this scenario.

(*a*) Model the system to find the average round trip time of each AGV.

(*b*) Assume that after delivering the machined parts to the stockroom, the AGVs carry the initial material for machining from the stockroom back to the shop floor. The number of bins of initial material for each machine is the same as the number of bins of finished parts just unloaded at the stockroom for that machine. Loading and unloading time of bins of initial materials are 20 seconds each. A machine may not process parts if the initial material is not available. Model the new scenario to find the average round trip time for AGVs, the average number of bins of initial materials for each machine, and the machine utilization.

8.18. At an intersection, the traffic light timing for cars coming from the east and west is 15 seconds for the green light and 10 seconds for the red light. The timing for the south and north directions is the opposite. Interarrival times of cars are exponentially distributed with means of 4 and 6 seconds for directions EW/WE and NS/SN, respectively. It takes 2 seconds for each car to clear the intersection.

(*a*) Simulate the system to find the average waiting times of cars on each side of the intersection.

(*b*) Assume that 30 percent of cars arriving from each direction intend to make a left turn. There is no left turn signal in the traffic light. Cars make a left turn if the first approaching car from the opposite direction is at least 7 seconds away from the intersection. Simulate the system to find the average number of cars waiting to make left turns.

(*c*) If you were to suggest a left turn signal for the traffic light at this intersection, what would you recommend for the duration and sequence of red and green lights for each direction and for the left turns? Base your recommendation on a detailed simulation analysis of the system.

(*d*) Suppose that the left turn lanes can each accommodate only 7 cars, and that when they are full, they block all traffic in the corresponding straight direction. Model the system to determine the number of such blockages during a one-hour period.

8.19. Modify the model given for the single-sampling inspection plan for quality control to determine the percentage of defective parts in the accepted lots.

8.20. Perform Exercise 8.19 for the double-sampling inspection plan.

CHAPTER 9

SIMULATION TOOLS AND THE CRITERIA FOR THEIR SELECTION

9.1 INTRODUCTION

The purpose of this chapter is to familiarize the reader with the basic capabilities of some of the representative available simulation tools. General purpose simulation tools are categorized under the event-oriented and the process-oriented categories. A few examples of special purpose tools are also briefly presented. A guideline for selection of a simulation tool is then provided. The presentation does not intend to cover all of the many available simulation tools. The July 1993 issue of *Industrial Engineering* lists 45 simulation software products offered by 20 different vendors, but even this list does not include all of the available simulation tools. The brief discussions of the representative simulation software systems given in this chapter should provide the reader with a comprehensive understanding of the types of simulation tools that are currently available.

9.2 GENERAL PURPOSE EVENT-ORIENTED SIMULATION LANGUAGES

Event-oriented languages are those that provide the basic capabilities typically needed in a simulation program and require the user to construct the logic of events and their effect on the system state. The capabilities provided usually include the modules discussed in Chapter 3 (event calendar, statistics routine, output routine, etc.). These relatively low-level languages have the advantage of being flexible and the disadvantage of being hard to learn and use. Some representative languages are presented in this section.

9.2.1 GASP IV

GASP IV is a software package that provides the user with a collection of subroutines written in FORTRAN. These subroutines perform the basic tasks of event scheduling and triggering, placing and removing entities and their attributes in and from queues, generating random variates from various probability distributions, collecting statistics on observation- and time-based variables, and creating formatted output reports. The user can access these capabilities using single-statement subroutine calls. The major user tasks are in the writing of the individual event subroutines (such as the arrival and departure routines used in the example program) and in linking them with GASP subroutines and compiling the program using a FORTRAN compiler. GASP IV library routines perform the major programming task for the user, the efficient accomplishment of which requires a knowledge of data structures and an advanced programming skill. GASP IV is in the public domain; however, it is no longer supported or maintained.

9.2.2 SIMSCRIPT

One of the oldest and most widely used simulation languages that is still being supported is SIMSCRIPT. It was originally developed at RAND Corporation in the early 1960s. The current version of the software is SIMSCRIPT II.5, which is a full-function computer programming language, marketed by CACI Products Company. The system comes with built-in simulation graphics capabilities. Unlike GASP, which uses FORTRAN, SIMSCRIPT has a special structure and syntax that make the language suitable for simulation applications. A program written in SIMSCRIPT consists of English-like statements in free-format style. This makes the program self-documenting and relatively easy for nonprogrammers to understand. Figure 9.1 shows a code segment for a factory model written in SIMSCRIPT II.5. Since it is possible to define and frequently use several processes in various applications, SIMSCRIPT may also be considered a process-oriented language.

```
PROCESS WORK.ORDER
      LET ARRIVAL.TIME=TIME.V
      UNTIL ROUTING.SET IS EMPTY
      DO
              REMOVE THE FIRST TASK FROM THE ROUTING.SET
              REQUEST 1 UNITS OF PRODUCTION.CENTER(TASK.DOER(TASK))
              WORK.TASK DURATION HOURS
              RELINQUISH 1 UNITS OF PRODUCTION.CENTER(TASK.DOER(TASK))
              DESTROY THE TASK
      LOOP
      LET CYCLE.TIME=TIME.V - ARRIVAL.TIME
  END PROCESS WORK.ORDER
```

FIGURE 9.1
An example of a SIMSCRIPT program.

9.2.3 MODSIM

A modern simulation tool that could be considered as being in the category of event-oriented languages, but with far more advanced features, is MODSIM, which is another CACI product. This language is based on the relatively new object-oriented programming concepts and has its own language syntax and compiler. The modular structure of MODSIM, its object-oriented properties, and the wealth of library objects that it provides allow for the construction of sophisticated simulation models with impressive output report formats and animation features. Some of the library objects in MODSIM perform the standard tasks of a simulation engine as well as other tasks similar to the ones performed by GASP IV subroutines. The expandable object library of MODSIM includes many other objects that may be used in applications ranging from the creation of dedicated windows for user input information specification to the creation of special animation effects. Since some modules may be constructed to represent commonly occurring processes, MODSIM may be considered as a process-oriented language as well. A new simulation software called SIMOBJECT has been developed by CACI. SIMOBJECT is built using MODSIM and has high-level objects that represent some general purpose processes as well as some special purpose ones for application areas such as communication systems. Some discussions regarding object-oriented simulation are presented in Chapter 10.

9.3 GENERAL PURPOSE PROCESS-ORIENTED LANGUAGES

General purpose process-oriented simulation languages incorporate a number of modules that represent common processes, each of which is expected to be present in the systems studied by their users. Although there is a great overlap among the choices of the processes provided by all of these languages, each language differs from the others with respect to the specific features of each process; a few additional processes may be considered by one language and not by the others, and the syntax of program statements may differ. The syntactic differences are usually related to the names of the processes, such as the use of words like generate, create, and source to represent the process that sends entities to the system.

In designing a process-oriented simulation language, the designer should decide on the number of modules and the capabilities of each module, with the intention of (a) being able to represent as many application scenarios as possible, (b) not creating an unreasonably large number of modules, and (c) not creating complicated modules that demand excessive user time and effort to configure these modules to their special applications.

Process-oriented languages generally recommend the use of some form of entity flow diagram prior to writing the actual program statements. The flow diagram in some languages is represented in the form of a network with branches and nodes, and in other languages it may be represented with a vertical block diagram. Some languages have recently introduced accompanying graphics software modules that allow the building of the network or block diagram on the computer screen. The corresponding source program is then generated by these front-end systems. Thus these mechanisms greatly reduce typing errors and relieve the users of memorizing the specific position of each data field for various program statements. Following is a description of some representative process-oriented simulation languages.

9.3.1 SLAM

SLAM (simulation language for alternative modeling) is a simulation language that has evolved from GASP and provides a process-oriented modeling environment represented in a network flow format and an event-oriented environment in which the user may write specialized event subroutines in FORTRAN, which will be linked to the network model. The network model provides a choice of several predesigned processes. The branches in a SLAM network represent the time-consuming activities through which the entities travel (such as a service operation). The nodes represent other processes such as entity source, sink, and queues.

Recent versions of SLAM are supported by animation, presentation graphics, specialized modules for manufacturing applications, and simulation database management software. SLAMSYSTEM is a new mode of the software that provides an interactive graphics environment for network construction and automatic program generation. Pritsker and Associates, Inc. markets and supports SLAM.

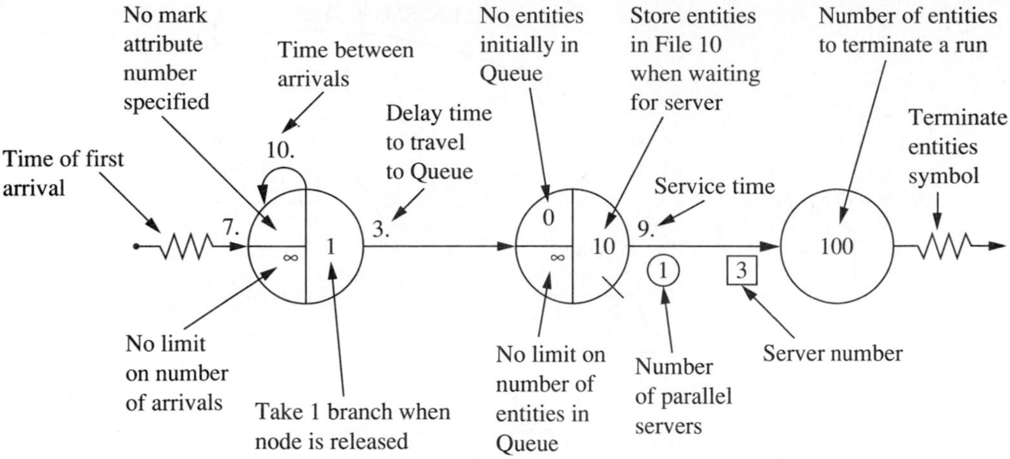


```
 ;EXAMPLE OF A SLAM STATEMENT MODEL
NETWORK;                START OF NETWORK STATEMENTS
      CREATE,10.,7.;    TIME BETWEEN ARRIVALS = 10
      ACTIVITY,3;       TIME TO REACH QUEUE NODE IS 3
      QUEUE(10);        USE FILE 10 FOR QUEUE
      ACTIVITY(1)/3,9;  SERVICE TIME = 9
      TERMINATE,100;    RUN MODE FOR 100 ENTITIES
      ENDNETWORK;       END OF NETWORK STATEMENTS
```

FIGURE 9.2
Structure of SLAM network for a simple queuing system. *Source:* A. Pritsker, *Introduction to Simulation and SLAM II*, Wiley, 1986.

Figure 9.2 shows an example of a simple queuing system, as modeled by a SLAM network. In this example an entity-creation process, a delay process, a queue process, a service process, and a termination process are used. The figure shows the position of each field in the network and in the corresponding program. Entity branching in a SLAM network model is shown in Fig. 9.3. The model simulates a system in which parts arrive at an inspection station with two parallel servers. Eighty-five percent of parts pass inspection and go for packaging. The rest of the inspected parts are found to need adjustment. After the adjustment operation, which is performed by one operator, the parts go back to the queue for inspection. The interarrival time of parts, inspection, and adjustment times are all uniformly distributed with the indicated parameters. The model as shown by the figure is self-explanatory. The node called DPRT in this figure is an auxiliary node that does not represent a realistic physical process. Its function is to collect the statistics on the time spent in the system by each part. Attribute number 1 of the entity is marked with its creation time at the create node, and at node

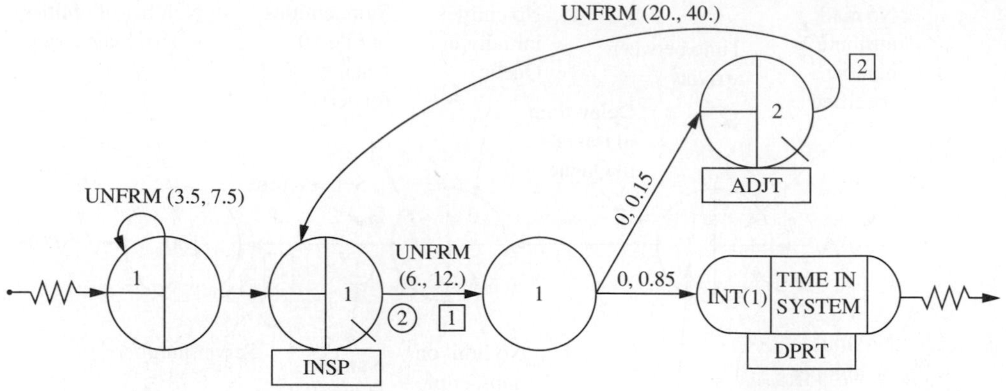

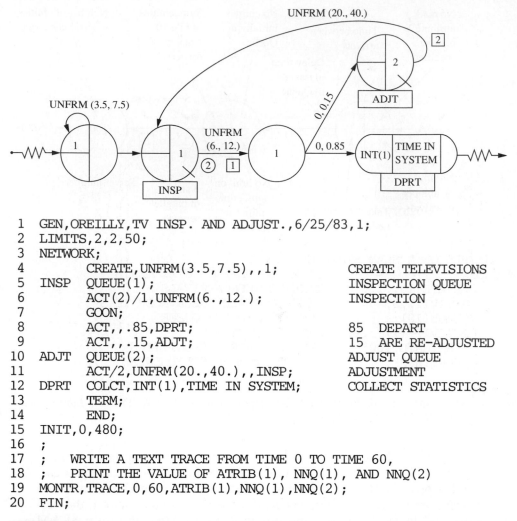

```
GEN,OREILLY,TV INSP. AND ADJUST.,6/25/83,1;
LIMITS,2,2,50;
NETWORK;
        CREATE,UNFRM(3.5,7.5),,1;         CREATE TELEVISIONS
INSP    QUEUE(1);                         INSPECTION QUEUE
        ACT(2)/1,UNFRM(6.,12.);           INSPECTION
        GOON;
        ACT,,.85,DPRT;                    85  DEPART
        ACT,,.15,ADJT;                    15  ARE RE-ADJUSTED
ADJT    QUEUE(2);                         ADJUST QUEUE
        ACT/2,UNFRM(20.,40.),,INSP;       ADJUSTMENT
DPRT    COLCT,INT(1),TIME IN SYSTEM;      COLLECT STATISTICS
        TERM;
        END;
INIT,0,480;
;
;    WRITE A TEXT TRACE FROM TIME 0 TO TIME 60,
;    PRINT THE VALUE OF ATRIB(1), NNQ(1), AND NNQ(2)
MONTR,TRACE,0,60,ATRIB(1),NNQ(1),NNQ(2);
FIN;
```

FIGURE 9.3
An example of a SLAM network model and the related program. *Source:* A. Pritsker, *Introduction to Simulation and SLAM II*, Wiley, 1986.

DPRT this recorded time is subtracted from the current simulation time to provide the entity system time.

9.3.2 GPSS

The General Purpose Simulation System, originally developed at IBM in the early 1960s, is still a popular process-oriented simulation language. GPSS uses a block diagram representation for its program structure. It has more than forty blocks, each representing a common discrete process. A number of blocks are provided for directing nonsequential flow of entities through the block diagram (branching).

Wolverine Software, Inc., one of the distributors of GPSS, is now providing a generic animation software called Proof that can be used with GPSS or any other language that can generate and store a trace file on the magnetic disk. Proof does not run in real-time; that is, for a user-specified period of simulated time, a trace of events during simulation is recorded on a disk and later fed to Proof to show the animation related to the recorded events. As in any other scene animation package, the user can build individual icons and backgrounds using paint software. A sample of a GPSS diagram and the related program are shown in Fig. 9.4. This model represents a single-server queuing system and is self-explanatory as presented.

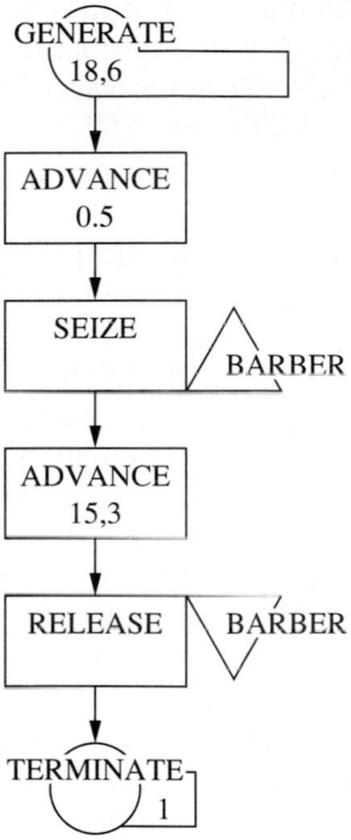


```
ONE-LINE, SINGLE-SERVER QUEUING MODEL

GENERATE          18,6          ARRIVALS EVERY 18 +- 6 MINUTES
ADVANCE           0.5           HANG UP COAT
SEIZE             BARBER        CAPTURE THE BARBER
ADVANCE           15,3          HAIRCUT TAKES 15 +- 3 MINUTES
RELEASE           BARBER        FREE THE BARBER
TERMINATE         1             EXIT THE SHOP
```

FIGURE 9.4
A GPSS block diagram and the related program.

9.3.3 SIMAN

SIMAN is another process-oriented simulation language with a gateway to its base procedure-oriented language, which allows for event-oriented programming. Model representation is in a block diagram format. A graphical editor is available that translates the block diagram input into the equivalent program statements. There are more than forty process blocks in SIMAN. The program structure in SIMAN is similar to that of GPSS. An advantage offered by SIMAN is that it keeps the program logic statements and some data that is typically changed in various simulation experiments (e.g., length of run) in separate files called the *model file* and *experiment file,* respectively. This allows for the changing of some parameters after each run and the running of the original program without recompilation. Also, there are several processes in this language that represent material-handling equipment. This increases the utility of the language in applications such as manufacturing. A scene animation software called Cinema is also available for SIMAN. Systems Modeling, Inc. markets SIMAN and the related software, including ARENA, which is a new simulation environment with some unconventional capabilities.

Figure 9.5 shows a sample SIMAN block diagram and the associated program. The system modeled concerns parts that arrive with an exponentially distributed interarrival time to be processed on a machine with a triangularly distributed processing time. It should be noted that the program has two segments. The first segment, called the *model source file,* directly corresponds to the block diagram and is self-explanatory. The second program segment augments the first segment and is called the *experiment source file*. The DISCRETE statement specifies the maximum allowable number of entities that can exist in the system concurrently. If during the course of simulation this number is exceeded, an error is generated. The QUEUES statement identifies the ranking procedure (e.g., first-in-first-out) and other information related to the queue statements used in the model. In this particular example, since the optional values are not specified, all queue characteristics are set at their default values (e.g., FIFO is chosen as the ranking rule). The RESOURCES statement identifies the resources name, number, and available quantity. Since the default number of resources is one, no specific field values are identified in the example. The COUNTERS statement identifies the name, limit, and other information related to a statistics counter. The REPLICATE statement specifies the number of simulation runs, their start times, the length of each run, as well as initialization-related information. Other fields for the latter two statements are set at their default values and therefore are not shown in the listing.

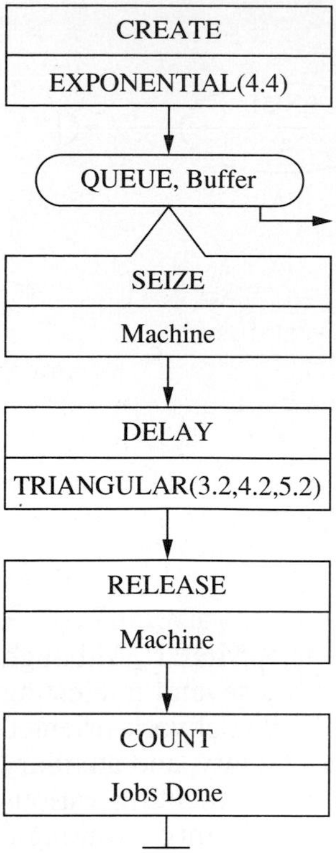


```
BEGIN;

CREATE:   EXPONENTIAL(4.4);               Enter the system
QUEUE,    Buffer;                         Wait for the machine
SEIZE:    Machine;                        Seize the machine
DELAY:    TRIANGULAR(3.2,4.2,5.2);        Delay by the proc.time
RELEASE:  Machine;                        Release the machine
COUNT:    JobsDone:DISPOSE;               Count completed jobs
END;
```

```
BEGIN;

PROJECT,          SampleProblem3.1,SM;
DISCRETE,         100;
QUEUES:           Buffer;
RESOURCES:        Machine;
COUNTERS:         JobsDone;
REPLICATE,        1,0,480;
END;
```

FIGURE 9.5
A SIMAN block diagram and the related program. *Source*: D. Pegden, R. Shannon, and R. Sadowski, *Introduction to Simulation Using SIMAN*, McGraw-Hill, 1990.

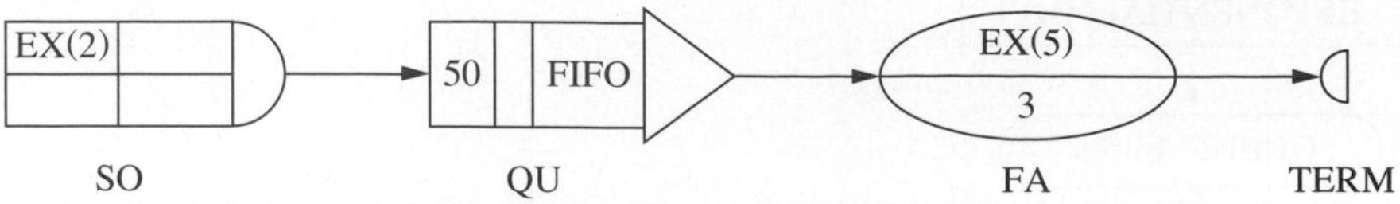


```
$BEGIN:
        SO    *S;EX(2):
        QU    *Q;50;;FIFO:
        FA    *F;;EX(5);3;*TERM:
$END:
```

FIGURE 9.6
A SIMNET network model and the related program. *Source:* H. Taha, *Simulation Modeling and SIMNET*, Prentice-Hall, 1988.

9.3.4 SIMNET

A recently developed process-oriented language is SIMNET. Although this language has not found a wide market base, it has several interesting aspects. SIMNET assumes four basic node types in its network-oriented modeling framework. These are the source, queue, facility, and auxiliary nodes. The queue node may have different modes representing various reasons for entities forming queues (needing to use a facility, awaiting a permission to move on, etc.). SIMNET also provides a relatively free-form mathematical expression possibility within the network modeling framework, and a wide choice of branching possibilities out of each network node. These capabilities give SIMNET enough flexibility to model various scenarios without resorting to a separate event-oriented programming environment. Figure 9.6 shows a simple SIMNET network and program for modeling a queuing system in which entities arrive with an exponentially distributed interarrival time with a mean of 2 minutes. The queue can accommodate a maximum of 50 entities, and its ranking rule is FIFO. Service duration is exponentially distributed with mean of 5 minutes, and there are 3 parallel servers at the facility.

9.4 SPECIAL PURPOSE SIMULATION ENVIRONMENTS

The foregoing presentations concerned general purpose simulation languages that may be applied to a variety of application domains. Special purpose simulation environments are process-oriented simulation tools that are devoted to specific domains of application such as manufacturing systems, communication networks, computer systems, and traffic flows. Frequently reoccurring processes within these specialized domains of application have resulted in the recent creation of these tools. Generally, these tools do not require programming and allow model building using graphics icons and selection menus. This capability is easier to build into the special purpose simulation software, because unlike general purpose languages they are targeted toward specific and narrow possibilities, and need not have unlimited flexibility to deal with unpredictable processes. In the following sections some special purpose simulation software tools will be presented, with an emphasis on manufacturing and communication network domains.

Two major factory simulation software systems are SIMFACTORY of CACI Products Company and WITNESS of AT&T. These special purpose simulation tools are powerful means for analysis and design of manufacturing facilities. They provide a graphics interface for model construction. Several constructs are made available for representing various processes such as machining stations, buffers for work-in-process, various material handling equipment, and so on. The user may choose or construct graphics icons that represent individual stations, parts, background views, and other components of the factory under study. Various domain rules are provided by these environments and can be readily selected by the user.

Animation capabilities that can show parts, workers, and material-handling equipment in motion add to the utility of these tools. System bottlenecks, equipment, and other resource-type utilization are made apparent in animated scenes. Both SIMFACTORY and WITNESS provide for dynamic interactions with the model while it is executing, thus the user is able to interrupt the execution process and change a parameter or reconfigure a portion of the model to see the effect of the modifications. Reports of throughput, part level and delay statistics, cycle time, equipment and personnel utilizations, and system delays are demonstrated in text as well as in graphs by both tools. Another popular factory simulation software is ProModel, which has similar capabilities.

COMNET is a generalized communications network simulator that can be used for voice communication networks, satellite, or radio channels. The network topology and traffic may be represented without programming using a graphics interface. Nodes perform store-and-forward or switching functions and serve as sources and sinks for network traffic. The network work load is defined by end-to-end traffic volumes. Traffic is routed from origin to destination nodes according to user-selected algorithms. Many standard network architectures are considered and their specific properties are defined in the form of ready-to-use modules in COMNET. Simulation shows the network performance using utilization plots and reports that demonstrate response times, blocking probabilities, call queuing and packet delays, and network throughput. COMNET also animates the network during the course of simulation showing routing choices and changing levels of utilization of various parts of the network under study. CACI Products markets COMNET as well as other communication network simulators.

9.5 SELECTION OF A SIMULATION TOOL

Several alternatives for building computer simulation models have been presented in the preceding sections. These are generalized computer programming languages such as BASIC and C, general purpose simulation languages such as MODSIM, GPSS, and SLAM, and special purpose simulation software tools such as SIMFACTORY and COMNET. Any problem that can be simulated by one class of tools may also be simulated by the preceding classes. For example, problems simulated by special purpose simulation tools may also be simulated by general purpose process-oriented, event-oriented, and generalized programming languages. Given all these alternatives, how should a user select a tool? To answer this question, consider first the advantages of using each class of tool.

Following are the advantages of using generalized computer programming languages in creating simulation programs:

1. Maximum modeling flexibility and complete control over the output format may be attained using these languages.
2. A compiler for these languages may be preowned and available, which results in avoiding the additional cost of acquiring a simulation software.
3. No special training for learning a simulation language is needed. A general knowledge about simulation programming logic (such as the one presented by the example) and knowledge of the chosen computer programming language should suffice.
4. Simulation programs written in generalized computer programming languages can be efficient in size and speed because they do not have the overhead (features available but not used in a specific application) of simulation languages.
5. Since compilers of generalized computer programming languages are available on most hardware platforms, simulation programs written in these languages are more portable to other computer types.

The advantages of general purpose simulation languages over generalized computer languages are as follows:

1. Creating simulation programs using the modeling framework and high-level commands provided by simulation languages is significantly easier.
2. Model development, maintenance, and refinement are much faster when simulation languages are used.

3. Simulation programs written using a simulation language are better communicated since the modular structure of these programs is much easier to understand by others.

The advantages of special purpose simulation tools over general purpose languages are as follows:

1. Special purpose simulation tools lack several features of general purpose languages that are not applicable to their particular domain of concern, so they are simpler in structure and are easier to learn.
   Special purpose simulation tools provide powerful constructs that represent complicated processes within their domain of concern.
2. Generally, no programming is done when using a special purpose simulation tool. Graphical icons and selection menus result in minimal model development time.
3. Individual processes and other model components are expressed in terms of the domain names and jargon. Likewise, simulation results are shown in reports that use these expressions. These characteristics make the models and outputs of special purpose simulation tools more understandable to end users.

On the basis of the above considerations it may be concluded that when other alternatives are unavailable, or when a single simulation program is frequently used (such as a daily scheduling model for a particular shop floor configuration) without major structural changes of the program over its usage life, it is better to write the program in a generalized computer programming language that offers higher execution speed and better memory utilization.

On the other hand, if it is necessary to deal with diversified problem scenarios, the use of a general purpose simulation language is recommended. Simulation analysts who are frequently called upon to perform various types of systems analysis and design studies, as well as simulation consultants who are active in diversified domains of application, will find general purpose simulation language more applicable to their needs.

Simulation analysts and consultants working in a certain domain should be familiar with the available special purpose simulation tools for that domain and learn to use their specific tool. In selecting software from the available choices in one of the above classes, several factors should be considered. These include: price, availability of good documentation, availability of educational workshops, nature of maintenance and support agreements, choice of computer hardware, and prior familiarity of the analyst with a tool.

9.6 SUMMARY

A broad overview of some representative simulation tools has been presented in this chapter and a guideline for selection of suitable types of tools for specific situations has been provided. There are numerous simulation tools available, and the number of these tools is expected to increase. Selection of the proper simulation tool will always be a challenging task. Consideration of the logical guidelines provided in this chapter should help in the selection process.

CHAPTER 10

FUTURE DIRECTIONS OF SIMULATION

10.1 INTRODUCTION

Although today's high-level simulation languages offer many advantages as compared with earlier tools, many of them still require the user to possess detailed knowledge about their structure, syntax, commands, and modeling tricks in order to work around uncommon processes. This requires the user to spend numerous hours on becoming proficient in these languages by reading through extensive manuals and writing several trial programs. Often, debugging these programs is a cumbersome process, as syntactic errors are usually the only kind detected by these simulation languages. Logical errors typically result in the system being hung up with an obscure compiler or operating system message.

One of the most crucial catalysts in the successful utilization of simulation in a given application domain is the analyst's knowledge of the details of the domain. Thus, no one is better suited for conducting a simulation study than an expert in the field of the problem area. Therefore, a simulation tool that requires considerable expertise in computer programming and in syntactic details of a language is not likely to be used directly by the application domain experts. If a better choice is not provided to these users, they will have to rely on the services of simulation modelers and programmers even for simple application cases. This reliance naturally results in significant time lags due to the need for essential communication between the domain expert and the simulation analyst at various stages of system modeling, validation, experimentation, analysis of results, and generation of reports that should be understandable by the end users. This practice is not only costly, due to dependency on the expensive services of simulation modelers, but it may also, in certain classes of applications, result in information that is not timely enough to be of any significant use. Modern tools are needed to automate the stages involved in the simulation process. These models will be valuable not only to novice users; they will also be useful to expert simulation analysts, who usually deal with complex application scenarios, in saving modeling and analysis time through the reduction of various forms of errors and provision of easily accessible capabilities.

10.2 INTELLIGENT SIMULATION ENVIRONMENTS

One frequently noted limitation of simulation is the required knowledge and time devotion of the analyst in creating models, specifying alternative parametric and structural instances, designing the sampling experiments, and making sense out of simulation output. Any effort in the direction of automating these tasks will enhance the utility of this powerful tool.

Ideally, all that should be expected of a simulation analyst are knowledge about the system under study (its components, relationships, and boundary), the goals of the study, and the degree of precision required in the results. An intelligent simulation environment should be capable of providing the user with an automatically constructed model of the system, a corresponding computer program based on the logical structure of the model, a fast and error-free execution of the program, an easy to use verification option, an automatically performed analysis of the output, an automatic experiment setting, user-desired output statistics in readable formats, and the capability to rapidly modify the model and generate new results. A truly intelligent simulation tool should minimize the task of the human analyst and should support him or her with capabilities similar to those of a team of expert human simulation specialists.

A truly intelligent simulation system should be able to infer new rules from experience with specific examples. In the process it should analyze cause and effect relationships throughout the simulation, thus discovering hidden loops, undefined variables, or redundancies. In keeping with current progress in the field of expert systems, an intelligent simulation environment should be able to alter its deduction rules from such experiences. Consequently, major research efforts have recently been directed at giving simulation systems the ability to learn. The learning component may be coupled to an explanation facility that describes the flow of activities.

An important current research area in expert systems is concerned with feedback from the system to the user. Diagnostic information should not simply indicate a problem; it should provide an explanation in natural language of the causes and possible remedies. This feature, if incorporated in a simulation system, can provide invaluable assistance to systems analysts who routinely spend significant amounts of time in studying system behaviors for the purpose of improving system performance and productivity.

Intelligent simulation tools are expected to significantly reduce simulation project overheads, which are usually caused by extensive use of human analysts, and will be able to generate more timely information, which is so crucial in today's competitive business environment.

Several attempts have been made in recent years to create intelligent simulation environments. The objective has been to reduce the time spent in the simulation life cycle and to help the simulationist and end users in different stages of the simulation process. Following are some examples of these efforts.

The simulation language EZSIM, designed by the author and presented in this book, is intended to assist the user in building and specifying models by means of menus and graphics. EZSIM creates codes for some available simulation languages or executes them independently. The system also checks consistency while building the model. Handling and presentation of user variables, system variables, attributes, branching options, expression construction, statistics specification, and event animation are some of the unique features of EZSIM.

Another system that attempts to provide assistance with model building and automate the programming task is NATSIM (Khoshnevis and Austin, 1989), a system which creates simulation models of macro level production/distribution systems. NATSIM builds models out of purely natural language (English) descriptions of the systems being studied. It parses each sentence to recognize its critical building blocks and then links the key components in the context of the problem description using a system dynamics worldview (the level/rate relations described in Chapter 2). NATSIM then checks for consistency and completeness and automatically produces executable simulation codes in DYNAMO (a continuous system simulation language).

Cochran and Mackulak (1990) have developed a simulation package called GUIDES. The system contains several frequently used generic models of manufacturing (implemented in SIMAN). Users are assisted in defining their models through a menu-driven interface. If the defined model matches one of the generic models, GUIDES will continue to make that model specific to the user goals. A metamodel is used to predict the performance of simulation models using optimization and statistical analysis.

Mellichamp and Park (1989) have proposed an expert system, SESSA, as a statistical advisor for simulation; it provides assistance with statistical design for parameter estimation and experiment setting. They recognize eighteen statistical issues in simulation methodology, ten of which are incorporated in SESSA.

The *knowledge-based simulation* system, KBS, developed at Carnegie-Mellon University (Fox et al., 1989), analyzes the simulation output to evaluate the approximation of desired goals, and it suggests changes, determines causal relationships, and checks the sensitivity of output to input parameters. KBS uses some artificial intelligence techniques in the simulation process. Frames are used to represent the objects and their relationships, and rules represent the procedural behaviors of the objects. The system combines explanation techniques using rule-based systems with the simulation process.

Mellichamp and Wahab (1987) have developed an expert system for FMS design; their system analyzes simulation output results. The expert system determines whether objectives are met and proposes changes in the model for improvement. Objectives of an FMS design are categorized and ranked as part production goals and financial goals. In the production goals category, it first checks if the desired production output level per unit time is achieved; it then controls the equipment utilization and queue length. The system contains heuristics to look for bottlenecks locally and their causes globally. In the financial goals category, it first checks the total capital investment goal, and then the cost per part is investigated. In order to achieve financial goals, "replacement strategies" are used for equipment. To develop the model, SIMAN is used as the simulation language, and *knowledge engineering environment,* KEE, is used as the expert system shell.

Prakash et al. (1989) have described a goal-directed simulation environment called GDSE which consists of three main components: a model construction module, a simulation language, and an analysis module. Output results of a simulation run are used by an analysis module that contains some knowledge about the system behavior to suggest necessary changes in input parameters. It can be seen from the above review that none of the existing intelligent simulation environments incorporate any learning elements.

10.2.1 Machine Learning and Simulation

One of the most recent advancements in creating intelligent simulation environments concerns the application of machine learning (see Khoshnevis and Parisay, 1993). Machine learning refers to computer programs capable of improving their performance on a specific task over time as they interact with the environment. Basically, learning involves analyzing training examples to extract functions or rules; problem solving involves applying the learned functions or rules to solve new problems.

Machine learning is being used for controlling many physical systems. The learning instances created in these applications are represented by the actual data received through sensing the real system performance, a classic example of this being the pole-balancing problem. In this problem an expert system (a computer program) is initially given a few crude rules for balancing a single pole in a vertical position. The pole base is mounted on a cart driven in either of two directions by a computer-controlled motor. Through a number of trials, the system learns from its failures and successes and creates increasingly robust rules that keep the pole in balance. One may conceive of the same concept applied to scenarios such as controlling a production and inventory system in which the learning instances are created by simulation rather than by actual experience. The resulting rules may then be used in real-world experience for effective operation of the system.

An expert system may be used for simulation model creation, modification, and refinement to achieve a set analysis or design goal. Machine learning may be used to extract heuristic rules to support the knowledge base of such an expert system.

As has been demonstrated by several examples in this book, simulation is a powerful systems analysis and design approach; however, it requires careful setting of experiments by expert analysts to generate reasonable results in a reasonable time. In addition, because of the complexity of the problems studied by simulation, no human expert can truly know the nature of the complicated interrelationships in the model well enough to effectively design the required experiments and guide the simulation process in the direction of generating results that may be used in finding the best system design configurations. Machine learning may be used in conjunction with simulation in order to detect the hidden relationships and extract the unknown rules that govern the behavior of a complicated system.

The prediction rules derived from an integrated environment of simulation and machine learning are expected to be more robust and consistent than those acquired from human experts. Machine learning is expected to introduce a new dimension to the realm of computer simulation. When equipped with a learning module, simulation can generate many instances, learn from these instances, generate decision rules that are superior to human expert rules, and guide itself in the direction of the goals set by the analyst.

10.3 OBJECT-ORIENTED SIMULATION

Another new development in the field of simulation modeling benefits from the concepts developed in the general field of object-oriented software technology. Object-oriented programming (OOP) is currently receiving a considerable amount of attention because of the benefits that it offers. Generalized computer programming languages such as Small Talk and C++ are being widely used in various application domains. The major feature of object-oriented programming is the modularity of program segments. These segments may encapsulate data models that contain the attributes of objects and the object methods (object model), procedures representing functional relations among objects (function model), or representations that maintain the correct sequence of actions that objects create (dynamic model). The major advantages of an object-oriented computer program are the ease of development, documentation, maintenance, modification, and reusability of program segments in other programs.

In the field of simulation, the application of object-oriented programming concepts is resulting in new tools referred to as *object-oriented simulation environments*. These environments can provide a set of preconstructed objects (like the EZSIM nodes) with generic capabilities. The user may modify each of these objects to perform the desired actions. New objects may be built by the user to create an expanded library of application-oriented objects. The objects created by various users may be shared through a public library of the objects. These exciting developments should ease the simulation process and result in more sophisticated applications by a wider range of users. The following is a brief description of some specific elements of object-oriented software technology as applied to simulation:

Objects. Objects are dynamic data types with *fields* and *methods*. Fields define object properties (such as entity attributes) and state (such as the location of the entity in the system or the status of a server), which may be changed only by the object itself, possibly at the request of other objects. A field may point at a dynamic record; for example, a given entity attribute may be a list of properties that could change during the course of simulation. In this case, an object field would point at this dynamic list. It should be noted that objects are not limited to entities. A node in the model network, an animation primitive, a routine that collects statistics, and a routine that prints the simulation output can all be defined and constructed as objects. Several instances of an object (e.g., entity multiplication) may be created easily. Methods describe actions that objects can perform. For example, for an airplane to fuel, it must first find a runway, land, taxi, and fuel. This is the fueling method of the airplane object.

Message passing. Objects can invoke other objects' methods by sending messages for synchronous or asynchronous actions. For example, a flight controller object on the ground may send a message to a flying airplane object requesting it to fuel. The airplane object then executes the fueling method that it encapsulates within. A message may "ask" for an action, in which case the sender object waits until the asked method is executed, or it may "tell" a message, in which case as soon as the message is sent the sender object may go on doing other things. The sent message in the latter case may be executed immediately, or it may be placed in the queue of messages to be executed by the object to which the message is sent.

Inheritance. New object types may be defined from existing types with all properties and methods inherited. Derived objects may override portions of the inherited properties and methods. Most importantly, derived objects may be defined using library objects. Inherited properties may be extended for new objects. Multiple inheritance is possible.

For example, if a vehicle object is defined as one that is powered and can move about carrying things, then new objects called *ship* and *aircraft* may be defined which inherit the properties of the vehicle in addition to their own properties (e.g., ship can move in two dimensions, aircraft can move in three dimensions, etc.). See Fig. 10.1. A multiple inheritance situation in which the helicopter object inherits the properties of the aircraft object as well as the medical emergency object is shown in Fig. 10.2.

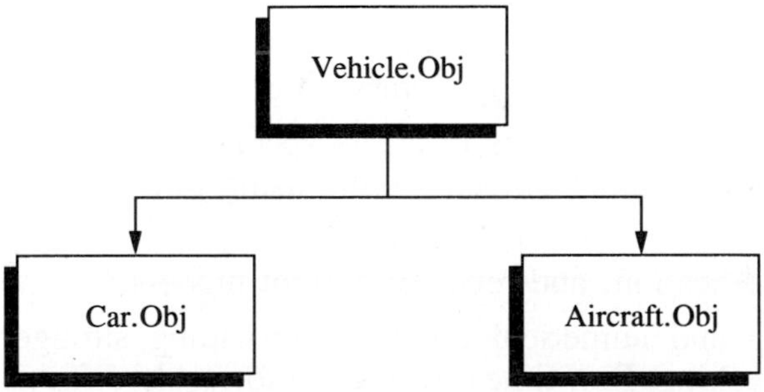

FIGURE 10.1
Inheritance in the object hierarchy.

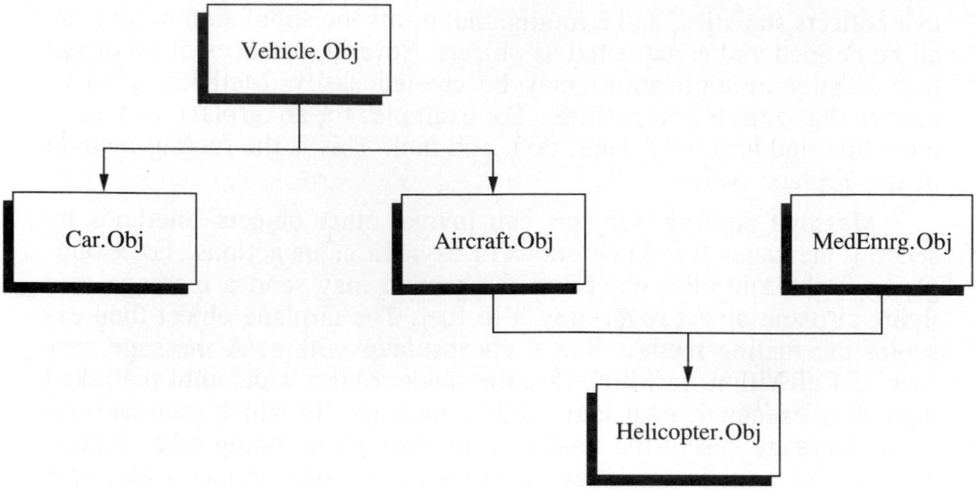

FIGURE 10.2
Demonstration of multiple inheritance.

Polymorphism. Different objects may share the same method name but have a different procedure under that name, making them behave differently in response to the same message. For example, a message to fuel may be sent to the car object, the plane object, and the mule object. All objects receive the same message but respond differently (i.e., the car object moves to a gas station, stops, and fuels; the plane lands, taxis, stops, and fuels; and the mule object goes straight to its hay bucket).

10.3.1 Advantages of Object-Oriented Simulation

Object-oriented simulation shares many advantages of general object-oriented programming. More specifically, the following may be listed as the advantages of this new approach:

- Ease of process-oriented modeling using library objects.
- Possibility of modifying library objects to create specialized objects, and the possibility of contributing objects to and using objects from public-object libraries.
- Ease of extension, modification, and reuse of current models.
- Reduction in modeling and numerical errors by providing stronger syntax and structure checking. Run-time errors are notoriously difficult

to find in simulation. Object-oriented simulations allow for detecting the sources of such errors.

- Ease of creating integrated environments that encompass elements such as modeling activities, execution, output generation and analysis, animation, databases, statistical routines, and expert systems.
- Reusability of previously developed models in constructing new models.

10.4 PROJECTIONS FOR FUTURE ADVANCEMENTS IN SIMULATION

Based on the foregoing discussions and the opinions of various simulation experts, the following developments are expected to take place in the exciting field of computer simulation in the near future:

- Managers will more frequently request simulation to be performed for various design and analysis projects.
- Ease of interface and usage will allow a wider range of users to develop and use simulation models.
- Input data handling, data graphics, and statistical fitting routines to several common models will be automated.
- General statistical support functions offered by such well-known statistical packages as SAS, SPSS, and MINITAB, and database capabilities such as those offered by dBase and Exell, will be offered in the simulation environments.
- Experiment setting, including activities such as determining the number and length of batches for use in the batch-means method and implementation of variance reduction techniques, will be done automatically.
- Object-oriented simulation will be the dominant approach. There will be libraries of standard and specialized objects for various application domains to which simulation users may subscribe. Various end users may contribute their newly developed objects to these libraries.
- Progress in the applications of expert systems and machine learning will eventually result in the automation of output analysis and experimentation stages of the simulation process.
- Progress in the applications of animation and virtual reality will result in simulation systems whose applications will extend beyond the traditional areas. These advancements will result in unusual applications, such as computer-created movies. Traditional application areas

will also benefit from these developments. For example, a factory may be represented by a set of solid models (graphical images of three-dimensional objects) in which a user wearing a hand glove point-and-pick device can rearrange the layout of the factory and create various simulation scenarios.

10.5 SUMMARY

In this chapter the importance of intelligent simulation environments and their potential effects have been described. Some important issues, such as applications of machine learning in simulation and object-oriented simulation methodology, have been discussed in detail, and future projections for advancements in the field of computer simulation have been highlighted.

10.6 REFERENCES AND FURTHER READING

Austin, W. M., and B. Khoshnevis: "Qualitative Modeling with Natural Language: Applications in System Dynamics," in *Qualitative Simulation,* P. Fishwick and P. Luker, Editors, Springer-Verlag, 1991.

Banks, J.: "Future Environments for Simulation Software," *Proceedings of 1st Industrial Engineering Research Conference,* pp. 145–148, May 1992.

Carbonell, J.: "Paradigms for Machine Learning," *Artificial Intelligence,* 40:1–9, 1989.

Chochran, J. and G. Mackulak: "Generic/Specific Modeling: An Improvement to CIM Simulation Techniques," in *Optimization of Manufacturing Systems Design,* D. Shunk, Editor, Elsevier Science Publishers, 237–260, 1990.

Dietterich, T. G.: "Limits of Inductive Learning," *Proceedings of the Sixth International Conference on Machine Learning,* Morgan Kaufmann, 1989.

Dietterich, T. G., and J. W. Shavlik: *Readings in Machine Learning,* Morgan Kaufmann, 1990.

Fayyad, Usama M.: *On the Induction of Decision Trees for Multiple Concept Learning,* Ph.D. Thesis, Stanford University, 1991.

Fox, M. S., N. Husain, M. McRoberts, and Y. V. Reddy: "Knowledge-Based Simulation: An Artificial Intelligence Approach to System Modeling and Automating the Simulation Life Cycle," in *Artificial Intelligence, Simulation and Modeling,* K. A. Loparo, N. R. Nielsen, and L. E. Widman, Editors, John Wiley & Sons, 1989.

Khoshnevis, B., and A. Chen: "An Expert Simulation Model Builder," *Proceedings of Conference on Intelligent Simulation Environments, SCS Simulation Series,* 17:129–133, 1986.

Khoshnevis, B., and W. Austin: "Intelligent Simulation Environment for Production Distribution System Modeling," *Journal of Engineering Costs and Production Economics,* Vol. 17, No. 4, pp. 351–357, 1989.

Khoshnevis, B., and S. Parisay: "Potential of Machine Learning in Simulation," *Proceedings of 1st Industrial Engineering Research Conference,* pp. 149–153, May 1992.

Khoshnevis, B., and S. Parisay: "Machine Learning and Simulation—Application in Queuing Systems," *Simulation,* Vol. 61, No. 5, pp. 294–302, 1993.

Mellichamp, J. M., and Y. H. Park: "A Statistical Expert System for Simulation Analysis," *Simulation,* 52:134–139, 1989.

Mellichamp, J. M., and F. Wahab: "An Expert System for FMS Design," *Simulation,* 48:201–208, 1987.

Mize, J.: "Basic Concepts in Object Oriented Simulation Modeling," *Proceedings of 1st Industrial Engineering Research Conference,* pp. 139–143, May 1992.

Pierreval, Henri, and Henri Ralambondrainy: "A Simulation and Learning Technique for Generating Knowledge about Manufacturing Systems Behavior," *Journal of Operations Research,* 41:461–474, 1990.

Prakash, Subramanian, R. E. Shannon, and Sallie Sheppard: "Goal Directed Simulation Environment: A Prototype," *Proceedings of the Summer Computer Simulation Conference,* pp. 545–549, 1989.

Pritsker, A. B.: "Simulation: The Premier Technique of Industrial Engineering," *Industrial Engineering,* Vol. 24, 7:25–26, 1992.

Quinlan, J. R.: "Induction of Decision Trees," *Machine Learning,* 1:81–106, 1986.

Rendell, L.: "A New Basis for Stat-Space Learning Systems and a Successful Implementation," *Artificial Intelligence,* 1:177–226, 1983.

Rumbaugh, J., M. Blaha, W. Premerlani, F. Eddy, and W. Locrensen: *Object-Oriented Modeling and Design,* Prentice-Hall, 1991.

Schmeiser, B.: "Modern Simulation Environments: Statistical Issues," *Proceedings of 1st Industrial Engineering Research Conference,* pp. 139–143, May 1992.

APPENDIX A

PROGRAM LISTINGS FOR THE SINGLE-SERVER QUEUING SYSTEM

```
'                 SINGLE SERVER QUEUING SIMULATION PROGRAM
'                              (BASIC VERSION)
'                 (Uniform Interarrival and Service Times)
'
'
' SYMBOL DEFINITIONS:
'
' ALQ             Average Length of Queue
' ASULTIL         Average Server Utilization
' ATTRIB(I)       Attribute value (arrival time) of the Ith entity
'                 in the queue
' BUSY            A binary (0,1) variable representing server status
' CLQ             Cumulative Length of Queue
' CSUTIL          Cummulative Server Utilization
' LQ              Current Length of Queue
' LS              Length of Simulation in time units
' MINIAT          Minimum Inter-arrival Time
' MAXIAT          Maximum Inter-arrival Time
' MINST           Minimum Service Time
' MAXST           Maximum Service Time
' NEE             Total Number of Entities Entered the system
' NES             Total Number of Entities Served
```

```
' QLIMIT          Maximum Queue Capacity
' TLE             Time of Last Event
' TNA             Time of Next Arrival
' TND             Time of Next Departure
' TNOW            Current Simulation Time (Master Clock)
' TQT             Total Queue Time
'
'
'
'************************ THE MAIN ROUTINE ***************************
'
gosub INIT
gosub INPT
gosub EVENTS
gosub OUTPT
stop
'
'****************** THE INITIALIZATION ROUTINE ********************
'
INIT:
'
cls
QLIMIT=100
DIM ATTRIB(QLIMIT)
BUSY=0 : LQ=0 : NEE=0 : NES=0 : CLQ=0 : CSUTIL=0 : TQT=0
'
'  Schedule the first arrival at time 0. Set the Time of Next Departure
'  to a large number.
'
TNA=0 : TND=99999
return
'
'********************** THE USER INPUT ROUTINE **********************
'
INPT:
print "INPUT PARAMETERS:": print
input "Length of Simulation in Time Units"; LS
input "Minimum Inter-arrival Time"; MINIAT
input "Maximum Inter-arrival Time"; MAXIAT
input "Minimum Service Time"; MINST
input "Maximum Service Time"; MAXST
return
'
'
'********************** THE EVENT TIMING ROUTINE ********************
'
EVENTS:
'
TOP:
'
'   Set the Time of Last Event equal to TNOW prior to advancing TNOW to
'   Time of Next Event
'
```

```
TLE=TNOW
'
'   Compare the two possible Next Event times and set TNOW equal to the
'   smaller time. Call the associated event subroutine. If TNOW exceeds
'   the simulation length, return to MAIN.
'
if (TNA < TND) then
        TNOW=TNA
        if (TNOW >= LS) then return
        gosub ARRIVE
else
        TNOW=TND
        if (TNOW >= LS) then return
        gosub DEPART
end if
'
'   Call the STAT routine to update time based statistics at this event
'   time.
'
gosub STAT
'
'   Execute the loop to process the next event.
'
goto TOP
'
'******************** THE ARRIVAL EVENT ROUTINE ********************
'
ARRIVE:
'
'   Increment the total Number of Entities Entered the system.
'
NEE=NEE+1
'
'   Schedule the next arrival event
'
TNA=TNOW+MINIAT+(MAXIAT-MINIAT)*rnd(1)
'
'   If the server is busy increment the Length of Queue. If queue size
'   exceeds the limit, print error message and stop. Set the attribute
'   of the arriving entity equal to its arrival time. Return to the
'   Event routine.
'
if (BUSY = 1) then
        LQ=LQ+1
        if (LQ > QLIMIT) then print "Error - Queue Overflow!" : stop
        ATTRIB(LQ)=TNOW
        return
else
'
'   If the server is idle, make its status busy (place the entity just
'   arrived in the server station). Schedule the departure time for
'   the entity.
'
```

```
        BUSY=1
        TND=TNOW+MINST+(MAXST-MINST)*rnd(1)
end if
return
'
'******************* THE DEPARTURE EVENT ROUTINE ******************
'
DEPART:
'
'   Increment the total Number of Entities Served
'
NES=NES+1
'
'   If the queue is empty, make the server status idle. The next
'   departure may not be scheduled, so set Time of Next Departure equal
'   to a large number.
'
if (LQ = 0) then BUSY=0 : TND=99999 : return
'
'   If queue is not empty, compute the queue waiting time of the entity
'   that is about to start its service first entity in queue). Compute
'   the Total Queue time and decrement the length of the queue.
'
QT=TNOW-ATTRIB(1)
TQT=TQT+QT
LQ=LQ-1
'
'   Shift the attribute values by one position, as the queue moves
'   forward.
'
for I=1 to LQ
        ATTRIB(I)=ATTRIB(I+1)
next I
'
'   Schedule the next departure time, and return.
'
TND=TNOW+MINST+(MAXST-MINST)*rnd(1)
return
'
'***************** THE TIME BASED STATISTICS ROUTINE *****************
'
STAT:
'
'   Compute the Cumulative Length of Queue and Server Utilization by
'   updating the values they had at the last event time.
'
CLQ=CLQ+LQ*(TNOW-TLE)
CSUTIL=CSUTIL+BUSY*(TNOW-TLE)
return
'
'*********************** THE OUTPUT ROUTINE *************************
'
OUTPT:
```

```
'
'   Find the Average Waiting Time in Queue as an observation based
'   statistics.
'
AWTQ=TQT/NES
'
'   Compute the weighted average of time based statistics and print the
'   results.
'
ALQ=CLQ/TNOW
ASUTIL=CSUTIL/TNOW
print:print:print
print "SIMULATION OUTPUT:" :print
print "     Number of Entities Entered=";NEE
print "     Number of Entities Served=";NES
print "     Total Time Spent in Queue=";:print using "####.##";TQT
print "     Average Waiting Time in Queue=";:print using"###.##";AWTQ
print "     Average Length of Queue=";:print using"###.## ";ALQ
print "     Average Server Util.=";:print using " % ##.## ";ASUTIL*100
return
```

```
/*              SINGLE SERVER QUEUING SIMULATION PROGRAM (C VERSION)
                (Uniform Interarrival and Service Times)
*/

    #include <stdio.h>
    #include <stdlib.h>
    #include <math.h>

/* Symbol Definitions and Initializations:*/

    #define QLIMIT     100    /* Maximum Queue Capacity */
    #define BUSY         1
    #define IDLE         0

    int LQ =0,                /* Current Length of Queue */
        LS,                   /* Length of Simulation in time units */
        MAXIAT,               /* Maximum Inter-arrival Time */
        MAXST,                /* Maximum Service Time */
        MINIAT,               /* Minimum Inter-arrival Time */
        MINST,                /* Minimum Service Time */
        NEE=0,                /* Total Number of Entities Entered */
        NES=0,                /* Total Number of Entities Served */

        server_status=IDLE;   /* A binary (0,1) variable */

float ATTRIB[QLIMIT+1],  /* Attribute value (arrival time) of the Ith
                            entity in the queue */
        CLQ=0.0,              /* Cumulative Length of Queue */
        CSUTIL=0.0,           /* Cummulative Server Utilization */
        QT,                   /* Queue Time */
        TLE=0.0,              /* Time of Last Event */
        TNA,                  /* Time of Next Arrival */
        TND,                  /* Time of Next Departure */
        TNOW=0.0,             /* Current Simulation Time */
        TQT=0.0;              /* Total Queue Time */

     void INPT(), EVENTS(), ARRIVE(), DEPART(), OUTPT(), STAT();
     float gen_serv(), gen_arv();
     double rnd_no(), seed=950706376.0;

     main()
     {
       printf("\n *SINGLE SERVER QUEUING SYSTEM *\n");
       INPT();
       EVENTS();
       OUTPT(); return 0;
     }

     void INPT()
     {
       printf("\n INPUT PARAMETERS:\n");
       printf("\n Length of simulation in time units:");
```

```
      scanf("%d", &LS);
      printf("\n Minimum Inter-arrival Time:");
      scanf("%d", &MINIAT);
      printf("\n Maximum Inter-arrival Time:");
      scanf("%d", &MAXIAT);
      printf("\n Minimum Service Time:");
      scanf("%d", &MINST);
      printf("\n Maximum Service Time:");
      scanf("%d", &MAXST);
    }


    void EVENTS()
    {
/* Schedule the first arrival at time 0. Set the Time of Next Departure
   to a large number for now. */

   TNA=0; TND=99999.0;

/* If TNOW exceeds the simulation length, return to MAIN. */

   while (TNOW <= LS)
       {
/* Set the Time of Last Event equal to TNOW prior to advancing TNOW to
   Time of Next Event */
       TLE=TNOW;

/* Compare the two possible Next Event times and set TNOW equal to the
   smaller time.  Call the associated event subroutine. */

        if(TNA<TND) {
        TNOW=TNA;
        ARRIVE();
         }
         else {
        TNOW=TND;
        DEPART();
         }

/* Call the STAT routine to update time based statistics at this event
   time. */
         STAT();
        }
    }

    void ARRIVE()
    {
/* Increment the total Number of Entities Entered the system. */
      ++NEE;

/* Schedule the next arrival event. */
      TNA=TNOW+gen_arv();
```

```
/* If the server is busy increment the Length of Queue. If queue size
   exceeds the limit, print error message and stop. Set the attribute
   of the arriving entity equal to its arrival time. Return to the
   Events routine. */

         if (server_status == BUSY)
           {
             ++LQ;
             if (LQ > QLIMIT) { printf("\n Queue Overflow "); exit(2); }
             ATTRIB[LQ]=TNOW;
           }
         else
           {

/* If the server is idle, make its status busy (place the entity just
   arrived in the server station). Schedule the departure time for the
   entity. */

             server_status = BUSY;
             TND=TNOW+gen_serv();
           }
       }

       void DEPART()
       {
         int i;
/* Increment the total Number of Entities Served*/
         ++NES;

/* If the queue is empty, make the server status idle. The next
   departure may not be scheduled, so set Time of Next Departure equal
   to a large number. */
        if (LQ == 0) { server_status = IDLE; TND=99999.0; }

/* If queue is not empty, compute the queue waiting time of the entity
   that is about to start its service first entity in queue. Compute
   the Total Queue time and decrement the length of the queue. */
        else
          {
            QT = TNOW - ATTRIB[1];
            TQT += QT;
            --LQ;
/* Shift the attribute values by one position, as the queue moves
   forward. */

            for(i=1; i<=LQ; i++) ATTRIB[i]=ATTRIB[i+1];

/* Schedule the next departure time, and return. */
            TND=TNOW+gen_serv();
          }
      }

      void STAT()
      {
```

```
/* Compute the Cumulative Length of Queue and Server Utilization by
   updating the values they had at the last event time. */

      CLQ += LQ * (TNOW-TLE);
      CSUTIL += server_status * (TNOW-TLE);
    }

    void OUTPT()
    {
      printf("\n SIMULATION OUTPUT:\n");
      printf("\n Number of Entities Entered = %5d", NEE);
      printf("\n Number of Entities Served = %5d", NES);
      printf("\n Total Time Spent in Queue = %7.2f", TQT);
      printf("\n Average Waiting Time in Queue = %7.2f ", TQT/NES);
      printf("\n Average Length of Queue = %7.2f", CLQ/TNOW);
      printf("\n Average Serv. Util. = %7.2f \n", (100.0*CSUTIL)/TNOW);
     }
    float gen_serv()
    {
    return (MINST + (MAXST-MINST)*rnd_no());
    }
    float gen_arv()
    {
     return (MINIAT + (MAXIAT-MINIAT)*rnd_no());
    }

    double rnd_no()
    {
      int ir;
      double r;
/* The rand() function generates integer numbers between 0 and 32767 */
      ir=rand();
      r=ir/32767.0;
      return(r);
    }

/* If using UNIX C you may use the following function instead of the
   above:

        double rnd_no()
        {
        double a=16807.0, p=2147483647.0;
        seed=floor((a*seed)-floor(a*seed/p)*p);
        return(seed/p);
} */
/*
     A SAMPLE INPUT AND OUTPUT OF THE SINGLE SERVER QUEUING PROGRAM:


     INPUT PARAMETERS:

      Length of simulation in time units:1000
      Minimum Inter-arrival Time:10
      Maximum Inter-arrival Time:15
```

```
    Minimum Service Time:9
    Maximum Service Time:14


    SIMULATION OUTPUT:

    Number of Entities Entered = 81
    Number of Entities Served = 80
    Total Time Spent in Queue = 55.48
    Average (Expected) Waiting Time in Queue = 0.69
    Average (Expected) Length of Queue = 0.34
    Average Server Utilization = 47.21
*/
```

APPENDIX B

PROBABILITY DISTRIBUTIONS

Appendix B has been adapted with permission from *Introduction to Simulation Using SIMAN* by Pegden, Shannon, and Sadowski, McGraw-Hill, 1990.

Beta(α_1, α_2)

Density

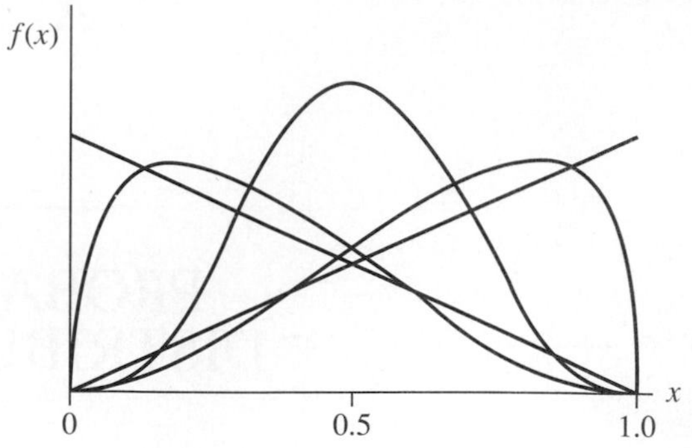

$$f(x) = \begin{cases} \dfrac{x^{\alpha_1 - 1}(1 - X)^{\alpha_2 - 1}}{B(\alpha_1, \alpha_2)} & \text{if } 0 < X < 1 \\ 0 & \text{otherwise} \end{cases}$$

where B is the beta function

Parameters Shape parameters Alpha$_1$ (α_1) and Alpha$_2$ (α_2) specified as nonnegative real numbers.

Range [0,1] (Can also be transformed to [a,b] as described below)

Mean $\dfrac{\alpha_1}{\alpha_1 + \alpha_2}$

Variance $\dfrac{\alpha_1 \alpha_2}{(\alpha_1 + \alpha_2)^2(\alpha_1 + \alpha_2 + 1)}$

Applications Because of its ability to take on a wide variety of shapes, this distribution is often used as a rough model in the absence of data. Because the range of the beta is from 0 to 1, the sample x is typically transformed to the scaled beta sample y with the range from a to b by using the equation $y = a + (b - a)x$. The beta is often used to represent random proportions, such as the proportion of defective items in a lot.

Erlang(β,k)

Density	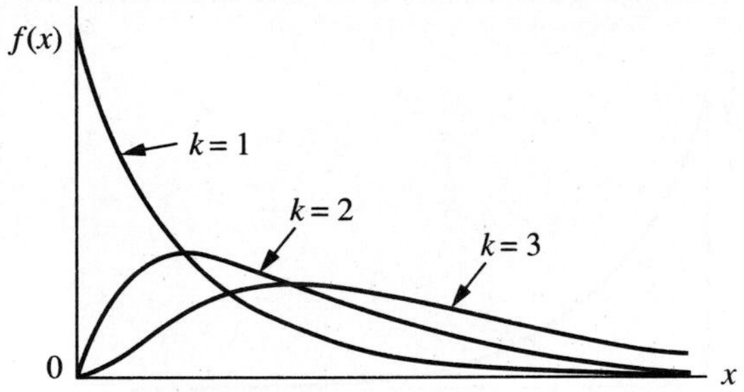 $f(x) = \dfrac{\beta^{-k} x^{k-1} e^{-x/\beta}}{(k-1)!}$
Parameters	If $X_1, X_2, \ldots, X_k$ are independent exponential samples, then the sum of these k samples has an Erlang-k distribution. The mean (β) of the exponential distribution and the number of exponential samples (k) are parameters of the distribution. The exponential mean is specified as a nonnegative real number, and k is specified as a positive integer.
Range	$[0, +\infty]$
Mean	$k\beta$
Variance	$k\beta^2$
Applications	Erlang distribution is used in situations in which an activity occurs in phases and each phase has an exponential distribution. For large k the Erlang approaches the normal distribution. The Erlang distribution is often used to represent the time required to complete a task. The Erlang distribution is a special case of the gamma distribution in which the shape parameter, α, is an integer.

Exponential(β)

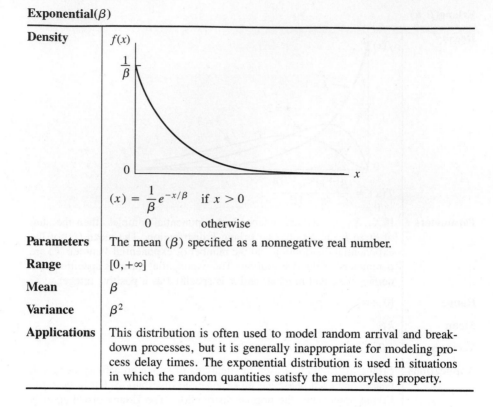

Density	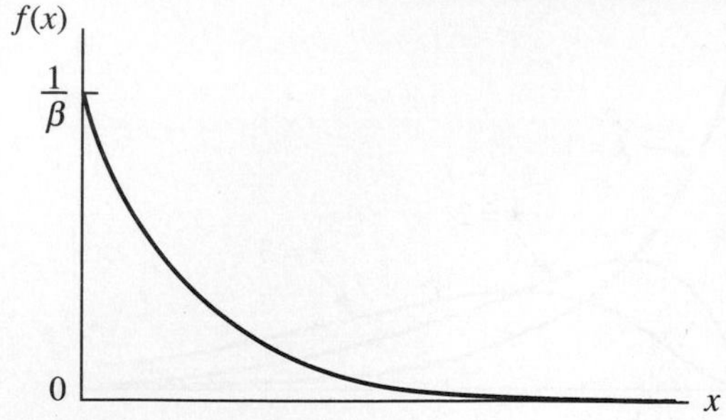 $(x) = \frac{1}{\beta}e^{-x/\beta}$ if $x > 0$ 0 otherwise
Parameters	The mean (β) specified as a nonnegative real number.
Range	$[0, +\infty]$
Mean	β
Variance	β^2
Applications	This distribution is often used to model random arrival and breakdown processes, but it is generally inappropriate for modeling process delay times. The exponential distribution is used in situations in which the random quantities satisfy the memoryless property.

Gamma(β, α)

Density	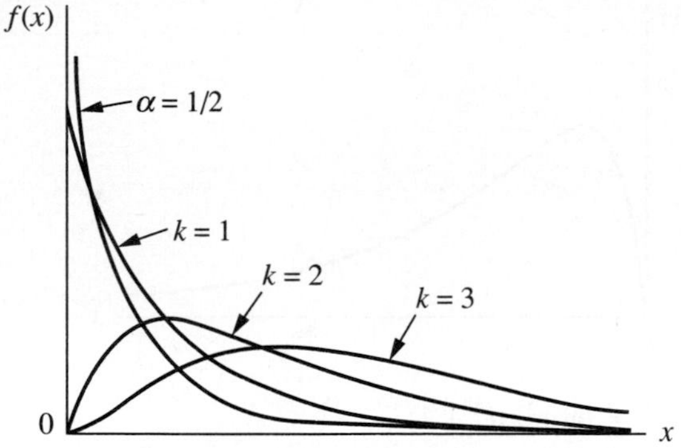 $f(x) = \begin{cases} \dfrac{\beta^{-\alpha} x^{\alpha-1} e^{-x/\beta}}{\Gamma(\alpha)} & \text{if } x > 0 \\ 0 & \text{otherwise} \end{cases}$ where Γ is the gamma function
Parameters	Shape parameter (α) and scale parameter (β) specified as nonnegative real values.
Range	$[0, +\infty]$
Mean	$\alpha\beta$
Variance	$\alpha\beta^2$
Applications	For integer shape parameters, the gamma is the same as the Erlang distribution. The gamma is often used to represent the time required to complete some task, e.g., a machining time or machine repair time.

Lognormal(μ, σ)

Density

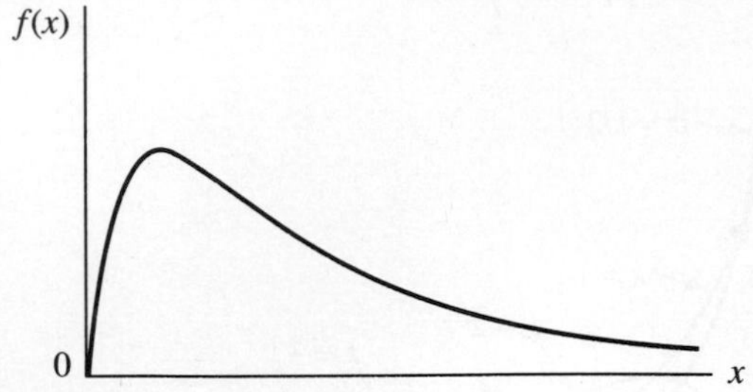

Let

$$\begin{aligned} \sigma_n^2 &= \ln(\sigma^2/\mu^2 + 1) \\ \mu_n &= \ln(\mu) - \sigma_n^2/2 \end{aligned}$$

Then

$$f(x) = \begin{cases} \dfrac{1}{\sigma x \sqrt{2\pi}} e^{-(\ln(x)-\mu)^2/2\sigma^2} & \text{if } x > 0 \\ 0 & \text{otherwise} \end{cases}$$

Parameters The mean (μ) and standard deviation (σ) specified as nonnegative real numbers.

Range $[0, +\infty]$

Mean μ

Variance σ^2

Applications The lognormal distribution is used in situations in which the quantity is the product of a large number of random quantities. It is also frequently used to represent task times that have a nonsymmetric distribution.

Normal(μ, σ)

Density	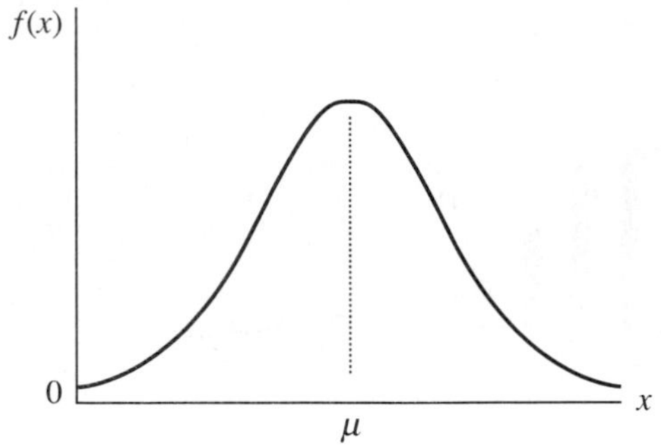 $f(x) = \frac{1}{\sigma\sqrt{2\pi}} e^{-(x-\mu)^2/2\sigma^2}$
Parameters	The mean (μ) is specified as a real number, and the standard deviation (σ) specified as a nonnegative real number.
Range	$[\infty, +\infty]$
Mean	μ
Variance	σ^2
Applications	The normal distribution is used in situations in which the central limit theorem applies — i.e., quantities that are sums of other quantities. It is also used empirically for many processes that are known to have a symmetric distribution and for which the mean and standard deviation can be estimated. Because the theoretical range is from $-\infty$ to $+\infty$, the distribution should only be used for processing times when the mean is at least three standard deviations above 0.

Poisson(λ)

Density	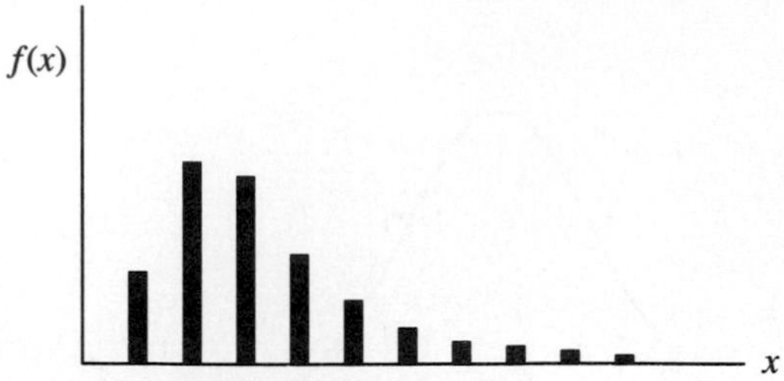 $p(x) = \frac{e^{-\lambda}\lambda^x}{x!}$ if $x \varepsilon \{0, 1, \ldots\}$ 0 otherwise
Parameters	The mean (λ) specified as a nonnegative real number.
Range	[0,1,2, . . . ,]
Mean	λ
Variance	λ
Applications	The Poisson distribution is a discrete distribution that is often used to model the number of random events occurring in an interval of time. If the time between events is exponentially distributed, then the number of events that occur in a fixed time interval has a Poisson distribution. The Poisson distribution is also used to model random variations in batch sizes.

Triangular(a,m,b)

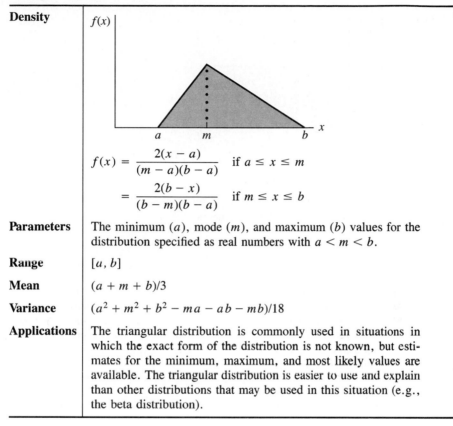

Density	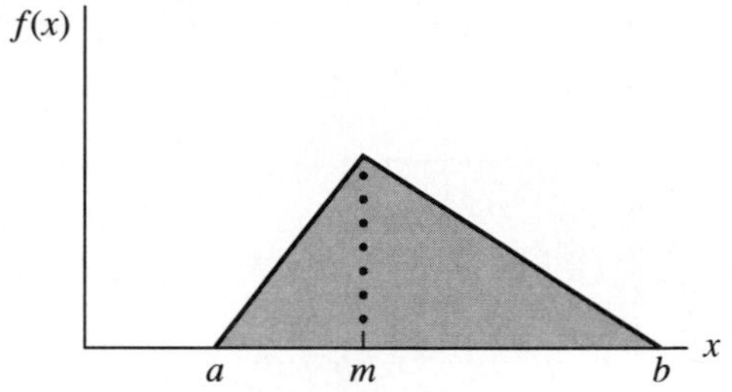 $f(x) = \dfrac{2(x-a)}{(m-a)(b-a)}$ if $a \le x \le m$ $= \dfrac{2(b-x)}{(b-m)(b-a)}$ if $m \le x \le b$
Parameters	The minimum (a), mode (m), and maximum (b) values for the distribution specified as real numbers with $a < m < b$.
Range	$[a, b]$
Mean	$(a + m + b)/3$
Variance	$(a^2 + m^2 + b^2 - ma - ab - mb)/18$
Applications	The triangular distribution is commonly used in situations in which the exact form of the distribution is not known, but estimates for the minimum, maximum, and most likely values are available. The triangular distribution is easier to use and explain than other distributions that may be used in this situation (e.g., the beta distribution).

Uniform(a,b)

Density

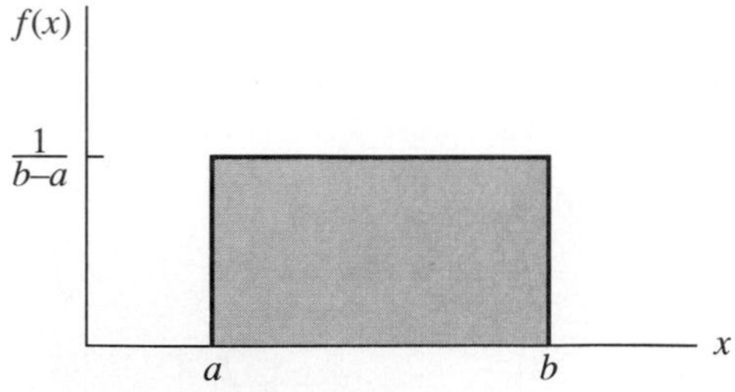

$$f(x) = \begin{cases} \dfrac{1}{b-a} & \text{if } a \leq x \leq b \\ 0 & \text{otherwise} \end{cases}$$

Parameters	The minimum (a), mode (m), and maximum (b) values for the distribution specified as real numbers with $a < b$.
Range	$[a, b]$
Mean	$(a + b)/2$
Variance	$(b - a)^2/12$
Applications	The uniform distribution is used when all values over a finite range are considered to be equally likely. It is sometimes used when no information other than the range is available. The uniform distribution has a larger variance than other distributions that are used when information is lacking (e.g., the triangular distribution). Because of its large variance, the uniform distribution generally produces "worst case" results.

Weibull(β, α)

Density	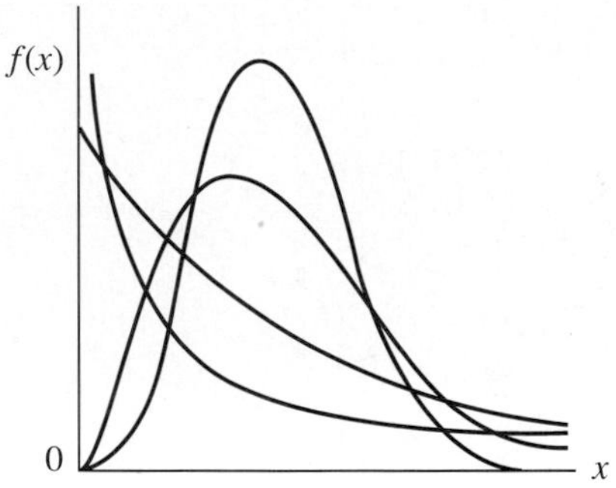 $f(x) = \alpha\beta^{-\alpha}x^{\alpha-1}e^{-(x/\beta)^{\alpha}}$ if $x > 0$ 0 otherwise
Parameters	Shape parameter (α) and scale parameter (β) specified as nonnegative real numbers.
Range	$[0, +\infty]$
Mean	$\frac{\beta}{\alpha}\Gamma\left(\frac{1}{\alpha}\right)$, where Γ is the gamma function.
Variance	$\frac{\beta^2}{\alpha}\left\{2\Gamma\left(\frac{2}{\alpha}\right) - \left(\frac{1}{\alpha}\right)\left[\Gamma\frac{1}{\alpha}\right]^2\right\}$
Applications	The Weibull distribution is widely used in reliability models to represent the lifetime of a device. If a system consists of a large number of parts that fail independently, and if the system fails when any single part fails, then the time between failures can be approximated by the Weibull distribution. This distribution is also used to represent nonnegative task times that are skewed to the left.

APPENDIX C

A HISTOGRAM PROGRAM

```
110 REM ********************** HISTOGRAM PROGRAM ****************************
120 REM
130 REM
140 REM THIS PROGRAM GENERATES A HISTOGRAM OF A DATA SET ON THE SCREEN AS DATA
150 REM ARE ENTERED ONE AT A TIME. FOLLOWING ARE THE HISTOGRAM PARAMETERS:
160 REM
170 REM          N: NUMBER OF CELLS
180 REM          W: CELL WIDTH
190 REM          L0: LOWER BOUND OF FIST CELL
200 REM  ENTER A NEGATIVE NUMBER TO STOP THE PROGRAM
210 CLS
220 N=15:W=5:L0=10
230 DPN=1
240 UN=L0+W*N
250 L=L0:U=L0+W
260 DIM FREQ(60)
270 LOCATE 1,2:PRINT" L U F
280 LOCATE 2,2:PRINT"-------------"
290 FOR I=1 to N
300 LOCATE 3+I,2:PRINT L;:PRINT"-";:PRINT U
310 LOCATE 3+I,16:PRINT"!"
320 L=L+W:U=U+W
330 NEXT I
340 LOCATE 1,50:PRINT "DATA POINT NO.";DPN;:INPUT X
350 IF (X<0) THEN STOP
360 DPN=DPN+1
370 C=INT(X/W)+1-INT(L0/W)
380 IF (X<L0) THEN C=1
390 IF (X>UN) THEN C=N
400 FREQ(C)=FREQ(C)+1
```

```
410 LOCATE 3+C,12:PRINT FREQ(C)
420 LOCATE 3+C,17+FREQ(C):PRINT "*"
430 LOCATE 1,50:PRINT"
440 GOTO 340
```

Sample histogram created by the program:

```
 L    U    F
--------------
 10 - 15   2  | **
 15 - 20   2  | **
 20 - 25   4  | ****
 25 - 30   6  | ******
 30 - 35   7  | *******
 35 - 40   15 | ***************
 40 - 45   17 | *****************
 45 - 50   12 | ************
 50 - 55   9  | *********
 55 - 60   8  | ********
 60 - 65   8  | ********
 65 - 70   4  | ****
 70 - 75   2  | **
 75 - 80   2  | **
 80 - 85   1  | *
```

APPENDIX D

STATISTICAL TABLES

This appendix contains tables of critical values for the chi-square and Kolmogorov–Smirnov goodness-of-fit tests, and values of the normal and student-*t* distributions.

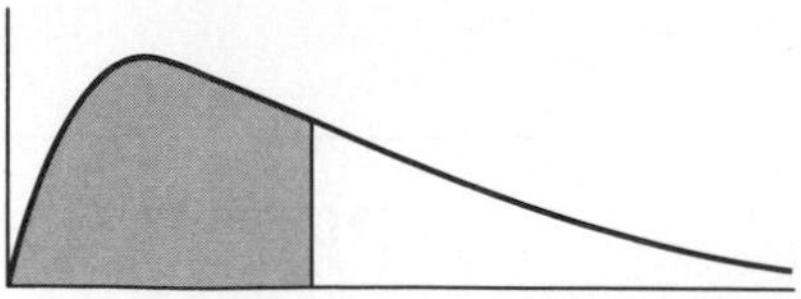

TABLE D.1
Percentile values (x^2) for the chi-square distribution with v degrees of freedom (shaded area = p)

v	$x^2_{.995}$	$x^2_{.99}$	$x^2_{.975}$	$x^2_{.95}$	$x^2_{.90}$
1	7.88	6.63	5.02	3.84	2.71
2	10.60	9.21	7.38	5.99	4.61
3	12.84	11.34	9.35	7.81	6.25
4	14.96	13.28	11.14	9.49	7.78
5	16.7	15.1	12.8	11.1	9.2
6	18.5	16.8	14.4	12.6	10.6
7	20.3	18.5	16.0	14.1	12.0
8	22.0	20.1	17.5	15.5	13.4
9	23.6	21.7	19.0	16.9	14.7
10	25.2	23.2	20.5	18.3	16.0
11	26.8	24.7	21.9	19.7	17.3
12	28.3	26.2	23.3	21.0	18.5
13	29.8	27.7	24.7	22.4	19.8
14	31.3	29.1	26.1	23.7	21.1
15	32.8	30.6	27.5	25.0	22.3
16	34.3	32.0	28.8	26.3	23.5
17	35.7	33.4	30.2	27.6	24.8
18	37.2	34.8	31.5	28.9	26.0
19	38.6	36.2	32.9	30.1	27.2
20	40.0	37.6	34.2	31.4	28.4
21	41.4	38.9	35.5	32.7	29.6
22	42.8	40.3	36.8	33.9	30.8
23	44.2	41.6	38.1	35.2	32.0
24	45.6	43.0	39.4	36.4	33.2
25	49.6	44.3	40.6	37.7	34.4
26	48.3	45.6	41.9	38.9	35.6
27	49.6	47.0	43.2	40.1	36.7
28	51.0	48.3	44.5	41.3	37.9
29	52.3	49.6	45.7	42.6	39.1
30	53.7	50.9	47.0	43.8	40.3
40	66.8	63.7	59.3	55.8	51.8
50	79.5	76.2	71.4	67.5	63.2
60	92.0	88.4	83.3	79.1	74.4
70	104.2	100.4	95.0	90.5	85.5
80	116.3	112.3	106.6	101.9	96.6
90	128.3	124.1	118.1	113.1	107.6
100	140.2	135.8	129.6	124.3	118.5

TABLE D.2
Kolmogorov–Smirnov critical values

Degrees of freedom, N	$D_{0\cdot 10}$	$D_{0\cdot 05}$	$D_{0\cdot 01}$
1	0.950	0.975	0.995
2	0.776	0.842	0.929
3	0.642	0.708	0.828
4	0.564	0.624	0.733
5	0.510	0.565	0.669
6	0.470	0.521	0.618
7	0.438	0.486	0.577
8	0.411	0.457	0.543
9	0.388	0.432	0.514
10	0.368	0.410	0.490
11	0.352	0.391	0.468
12	0.338	0.375	0.450
13	0.325	0.361	0.433
14	0.314	0.349	0.418
15	0.304	0.338	0.404
16	0.295	0.328	0.392
17	0.286	0.318	0.381
18	0.278	0.309	0.371
19	0.272	0.301	0.363
20	0.264	0.294	0.356
25	0.24	0.27	0.32
30	0.22	0.24	0.29
35	0.21	0.23	0.27
Over 35	$\frac{1.22}{\sqrt{N}}$	$\frac{1.36}{\sqrt{N}}$	$\frac{1.63}{\sqrt{N}}$

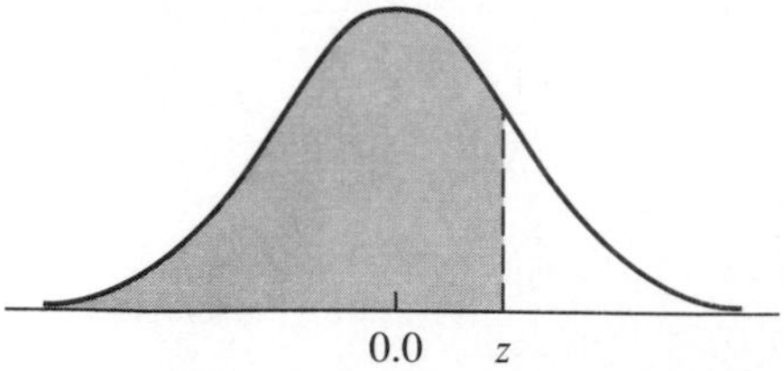

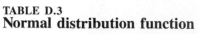

TABLE D.3
Normal distribution function

$$F(z) = \frac{1}{\sqrt{2\pi}} \int_{-\infty}^{z} e^{-(1/2)^2} dt$$

z	**0.00**	**0.01**	**0.02**	**0.03**	**0.04**	**0.05**	**0.06**	**0.07**	**0.08**	**0.09**
0.0	0.5000	0.5040	0.5080	0.5120	0.5160	0.5199	0.5239	0.5279	0.5319	0.5359
0.1	0.5398	0.5438	0.5478	0.5517	0.5557	0.5596	0.5636	0.5675	0.5714	0.5653
0.2	0.5793	0.5832	0.5871	0.5910	0.5948	0.5987	0.6026	0.6064	0.6103	0.6141
0.3	0.6179	0.6217	0.6255	0.6293	0.6331	0.6368	0.6406	0.6443	0.6480	0.6517
0.4	0.6554	0.6591	0.6628	0.6664	0.6700	0.6736	0.6772	0.6808	0.6844	0.6879
0.5	0.6915	0.6950	0.6985	0.7019	0.7054	0.7088	0.7123	0.7157	0.7190	0.7224
0.6	0.7257	0.7291	0.7324	0.7357	0.7389	0.7422	0.7454	0.7486	0.7517	0.7549
0.7	0.7580	0.7611	0.7642	0.7673	0.7704	0.7734	0.7764	0.7794	0.7823	0.7852
0.8	0.7881	0.7910	0.7939	0.7967	0.7995	0.8023	0.8051	0.8078	0.8106	0.8133
0.9	0.8159	0.8186	0.8212	0.8238	0.8264	0.8289	0.8315	0.8340	0.8365	0.8389
1.0	0.8413	0.8438	0.8461	0.8485	0.8508	0.8531	0.8554	0.8577	0.8599	0.8621
1.1	0.8643	0.8665	0.8686	0.8708	0.8729	0.8749	0.8770	0.8790	0.8810	0.8830
1.2	0.8849	0.8869	0.8888	0.8907	0.8925	0.8944	0.8962	0.8980	0.8997	0.9015
1.3	0.9032	0.9049	0.9066	0.9082	0.9099	0.9115	0.9131	0.9147	0.9162	0.9177
1.4	0.9192	0.9207	0.9222	0.9236	0.9251	0.9265	0.9279	0.9292	0.9306	0.9319
1.5	0.9332	0.9345	0.9357	0.9370	0.9382	0.9394	0.9406	0.9418	0.9429	0.9441
1.6	0.9452	0.9463	0.9474	0.9484	0.9495	0.9505	0.9515	0.9525	0.9535	0.9545
1.7	0.9554	0.9564	0.9573	0.9582	0.9591	0.9599	0.9608	0.9616	0.9625	0.9633
1.8	0.9641	0.9649	0.9656	0.9664	0.9671	0.9678	0.9686	0.9693	0.9699	0.9706
1.9	0.9713	0.9719	0.9726	0.9732	0.9738	0.9744	0.9750	0.9756	0.9761	0.9767
2.0	0.9772	0.9778	0.9783	0.9788	0.9793	0.9798	0.9803	0.9808	0.9812	0.9817
2.1	0.9821	0.9826	0.9830	0.9834	0.9838	0.9842	0.9846	0.9850	0.9854	0.9857
2.2	0.9861	0.9864	0.9868	0.9871	0.9875	0.9878	0.9881	0.9884	0.9887	0.9890
2.3	0.9893	0.9896	0.9898	0.9901	0.9904	0.9906	0.9909	0.9911	0.9913	0.9916
2.4	0.9918	0.9920	0.9922	0.9925	0.9927	0.9929	0.9931	0.9932	0.9934	0.9936
2.5	0.9938	0.9940	0.9941	0.9943	0.9945	0.9946	0.9948	0.9949	0.9951	0.9952
2.6	0.9953	0.9955	0.9956	0.9957	0.9959	0.9960	0.9961	0.9962	0.9963	0.9964
2.7	0.9965	0.9966	0.9967	0.9968	0.9969	0.9970	0.9971	0.9972	0.9973	0.9974
2.8	0.9974	0.9975	0.9976	0.9977	0.9977	0.9978	0.9979	0.9979	0.9980	0.9981
2.9	0.9981	0.9982	0.9982	0.9983	0.9984	0.9984	0.9985	0.9985	0.9986	0.9986
3.0	0.9987	0.9987	0.9987	0.9988	0.9988	0.9989	0.9989	0.9989	0.9990	0.9990
3.1	0.9990	0.9991	0.9991	0.9991	0.9992	0.9992	0.9992	0.9992	0.9993	0.9993
3.2	0.9993	0.9993	0.9994	0.9994	0.9994	0.9994	0.9994	0.9995	0.9995	0.9995
3.3	0.9995	0.9995	0.9995	0.9996	0.9996	0.9996	0.9996	0.9996	0.9996	0.9997
3.4	0.9997	0.9997	0.9997	0.9997	0.9997	0.9997	0.9997	0.9997	0.9997	0.9998

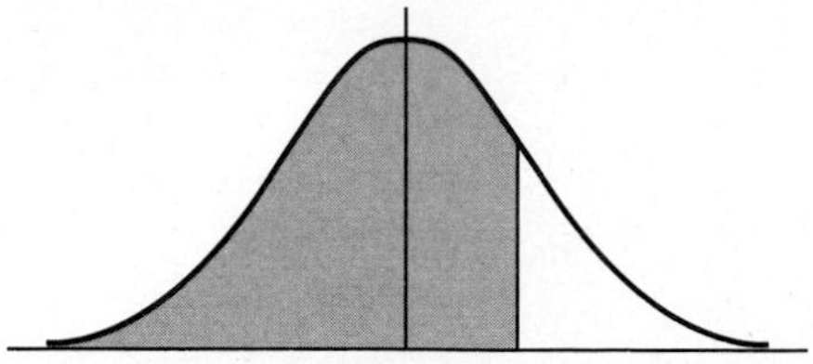

TABLE D.4
Values of student-t distribution with v degrees of freedom (shaded area = p)

v	$t_{.995}$	$t_{.99}$	$t_{.975}$	$t_{.95}$	$t_{.90}$
1	63.66	31.82	12.71	6.31	3.08
2	9.92	6.96	4.30	2.92	1.89
3	5.84	4.54	3.18	2.35	1.64
4	4.60	3.75	2.78	2.13	1.53
5	4.03	3.36	2.57	2.02	1.48
6	3.71	3.14	2.45	1.94	1.44
7	3.50	3.00	2.36	1.90	1.42
8	3.36	2.90	2.31	1.86	1.40
9	3.25	2.82	2.26	1.83	1.38
10	3.17	2.76	2.23	1.81	1.37
11	3.11	2.72	2.20	1.80	1.36
12	3.06	2.68	2.18	1.78	1.36
13	3.01	2.65	2.16	1.77	1.35
14	2.98	2.62	2.14	1.76	1.34
15	2.95	2.60	2.13	1.75	1.34
16	2.92	2.58	2.12	1.75	1.34
17	2.90	2.57	2.11	1.74	1.33
18	2.88	2.55	2.10	1.73	1.33
19	2.86	2.54	2.09	1.73	1.33
20	2.84	2.53	2.09	1.72	1.32
21	2.83	2.52	2.08	1.72	1.32
22	2.82	2.51	2.07	1.72	1.32
23	2.81	2.50	2.07	1.71	1.32
24	2.80	2.49	2.06	1.71	1.32
25	2.79	2.48	2.06	1.71	1.32
26	2.78	2.48	2.06	1.71	1.32
27	2.77	2.47	2.05	1.70	1.31
28	2.76	2.47	2.05	1.70	1.31
29	2.76	2.46	2.04	1.70	1.31
30	2.75	2.46	2.04	1.70	1.31
40	2.70	2.42	2.02	1.68	1.30
60	2.66	2.39	2.00	1.67	1.30
120	2.62	2.36	1.98	1.66	1.29
∞	2.58	2.33	1.96	1.645	1.28

Index

About the diskette:

The accompanying diskette contains two main files, EZ.EXE and MODEL.EXE, and several subdirectories, which contain the example files from Chapters 3, 5, 7, and 8 and the programs in Appendices A and C.

Before running the program, make a copy of the diskette using the DOS DISKCOPY command. EZSIM may be run using the floppy disk; however, loading the software into the memory will be considerably faster if a hard disk is used. To use your hard disk, make a directory called EZSIM. Change the active directory to EZSIM by entering the following command: CD EZSIM. Place the EZSIM diskette in the floppy drive (say, A) and issue the command XCOPY A:*.*/S. This command will copy the root and all subdirectory files of the diskette to your hard disk.

To use EZSIM with your mouse, enter EZM (you must have a Microsoft-compatible mouse driver already installed in your computer memory). If you do not intend to use a mouse, enter EZ to run the software.

To be able to run EZSIM from all subdirectories, it is best to issue a path command (e.g., PATH C:\EZSIM) for your EZSIM directory prior to running the program. You may modify your AUTOEXEC.BAT file to include the path command so that the command is automatically issued every time you turn on your computer. Another method of running EZSIM from different subdirectories is to run the program using the command \EZSIM\EZ. Alternatively, when the active directory is EZSIM and the program is executed, model filenames in other subdirectories may be given to the program by specifying the model subdirectory (e.g., \EZSIM\CH8\AGV).

The MODEL.EXE program is for printing the network and the model content description. This program is automatically called by EZSIM when the Print Model option is selected in the Output menu. A copy of the printed model information (in ASCII format) is stored on the disk with extension name MOD whenever the model is printed. MODEL.EXE may be run independently by issuing the following command: MODEL *filename*.EZ (be sure to include the extension name EZ). This command generates the file *filename*.MOD, which may be printed using the PRINT command of DOS, or it may be printed by a word processor in much the same way as the *filename*.OUT files. To print the network, your printer should be set to print the IBM graphics characters (e.g., for HP laser printers the PC-8 symbol set should be selected). Consult your printer manual for proper driver setup or DIP switch setup.

The student version of EZSIM limits the model networks to a maximum of 25 nodes. A maximum of 200 concurrent entities are allowed in the model during run time (there is no limit on the number of entities that can be generated). Zero-capacity queues are not allowed in the student version. To avoid memory conflicts it is best to run EZSIM outside of the Windows environment. For information about the professional version of EZSIM, please contact

Dr. B. Khoshnevis
Director, Manufacturing Engineering
University of Southern California
Los Angeles, CA 90089-0193